Coordination Chemistry

Papers Presented in Honor of Professor John C. Bailar, Jr.

John C. Bailar Jr.

Coordination
Chemistry

Papers Presented in Honor of Professor John C. Bailar, Jr.

Proceedings of the John C. Bailar, Jr. Symposium on Coordination Chemistry, Honoring Professor John C. Bailar, Jr. on the Occasion of His Sixty-fifth Birthday, June 22-25, 1969, University of Illinois, Urbana, Illinois, U.S.A.

Edited by
Stanley Kirschner
Professor of Chemistry
Wayne State University, Detroit, Michigan

℗ Springer Science+Business Media, LLC • 1969

Library of Congress Catalog Card Number 77-81522

ISBN 978-1-4899-6256-0   ISBN 978-1-4899-6555-4 (eBook)
DOI 10.1007/978-1-4899-6555-4
© **1969** Springer Science+Business Media New York
Originally published by Plenum Press in 1969.
Softcover reprint of the hardcover 1st edition 1969

DEDICATION

This book is dedicated by his students with much affection and gratitude for his guidance, understanding, and inspiration to

PROFESSOR JOHN C. BAILAR, JR.

and it is also dedicated to

MRS. JOHN C. BAILAR, JR. (FLORENCE)

in affectionate remembrance of her warmth, kindness, and generosity to these same students (and their families) who, at times, must have appeared as a "hungry horde."

PREFACE

Rarely in one's lifetime does one have the opportunity to come into close contact with a great man. It is likely that many scientists at one time or another in their careers meet a great scientist. Frequently, those in education have an association with a great educator, and often many administrators work with a great colleague. But it is truly rare for most men to know a great man.

Those who have come into contact with Professor John C. Bailar, Jr.—especially those who have had the opportunity to work with him directly—realize that not only is he a great scientist, educator, and administrator—he is also a great man.

It is not easy to define those qualities essential to greatness in a man—it seems to be simpler to recognize a great man than to describe the essence of greatness. For John C. Bailar, Jr., the spirit of his greatness probably lies in his personal realtionships with people. He recognizes the fundamental aspects of a problem—be it scientific, educational, administrative, or personal—and he invariably suggests sound, human routes to its solution. Humanity, probably more than any other quality, has endeared Professor Bailar to his colleagues all over the world, and perhaps this is the reason that so many people have sought and continue to seek Professor Bailar's advice and counsel.

Included in this volume are letters from the International Union of Pure and Applied Chemistry and the American Chemical Society expressing their appreciation for his work on behalf of chemistry. Elsewhere in this work a biography of Professor John C. Bailar, Jr. is given which includes many of his activities, and some of the honors he has received during his illustrious career.

Professor Bailar's Ph.D. students could think of no better way to mark his sixty-fifth birthday (and his forty-first year at the University of Illinois) than to hold a symposium in his honor. This volume includes papers presented at the Symposium by twenty-two of the scientists who obtained their doctoral degree training under his direction. A complete list of the 84 scientists who received their Ph.D. training under Professor Bailar's aegis from 1934 to 1969 is included along with a list of Professor Bailar's scientific publications.

To trace the scientific genealogy of Professor John C. Bailar, Jr. backward in time to the middle of the eighteenth century, and to show some of the roots of his greatness, the following "academic family tree" is given:

John C. Bailar, Jr.
(1904–   )
|
Moses Gomberg
(1866–1897)
|
Victor Meyer
(1848–1897)
|
Robert Bunsen
(1811–1889)
|
Friedrich Stromeyer
(1776–1835)
|
Louis N. Vauquelin
(1763–1829)
|
Antoine F. de Fourcroy
(1755–1809)
|
Jean B. M. Bucquet
(1746–1780)
|
Antoine L. Lavoisier
(1743–1794)

The friends of Professor John C. Bailar, Jr. wish him a long life of continued productivity.

Detroit, Michigan   Stanley Kirschner
March, 1969

JOHN C. BAILAR, JR.

John C. Bailar, Jr. was born on May 27, 1904 in Golden, Colorado. His father was both a chemist and educator. John, Jr. received his early education in Colorado, and earned the B.A. degree (*magna cum laude*, 1924) and the M.A. degree (1925) from the University of Colorado. He received the Ph.D. degree from the University of Michigan (organic chemistry, 1928), and he also received honorary Sc.D. degrees from the University of Colorado (1959) and the University of Buffalo (1959). He came to the University of Illinois as an Instructor in Chemistry in 1928 and rose through the ranks to full Professor in 1943. He was Secretary of the Chemistry Department there from 1937 to 1951 and was Head of the Division of Inorganic Chemistry from 1941 to 1966. He has served on many committees in the Department and at the University of Illinois, and has acted as Dean of both the College of Liberal Arts and Sciences and of the Graduate College during his career. He has lectured in undergraduate courses at freshman and higher levels, and has lectured in graduate courses and directed the Ph.D. research of 84 pre-doctoral and 18 post-doctoral research students from all over the world. He has been active as an advisor to the Zeta Chapter of Alpha Chi Sigma at the University for many years, and was honored by that professional chemistry fraternity with the John R. Kuebler Award in 1962 for his contributions to the profession of chemistry and to the Alpha Chi Sigma Fraternity.

He has served the American Chemical Society extensively, holding most of the offices of the Illinois Section, including Chairman (1936); he has been active with the Divisions of Physical and Inorganic Chemistry, Chemical Education, and Inorganic Chemistry, and has served as Chairman of each of these Divisions. He has also been active with the Committees on National Meetings, Pre-Doctoral Fellowships, Nominations and Elections, Divisional Officers Group, Committee on National Meetings on Divisional Activities, and Visiting Associate for the Committee on Professional Training. In addition, he has served as President of the American Chemical Society itself. Further, he has been a member of the Electrochemical Society and he has also served on many committees of the National Research Council, including the Committees on the Codification of Chemical Compounds, Inorganic Nomenclature, Pre-Doctoral and Post-Doctoral Fellowships, Design of Chemical Laboratories, Scientific Personnel, and Inorganic Chemistry.

Still further, Professor Bailar has rendered much service to the Editorial Boards of the *Journal of Chemical Education,* the *Journal of the American Chemical Society, Chemical Reviews, Journal of Inorganic and Nuclear Chemistry, Inorganic Syntheses, Inorganic Chemistry, Journal of Applied Chemistry* and the *Revue de Chimie Minerale.* He has also been very active in the affairs of the International Union of Pure and Applied Chemistry, and he has been a delegate from the United States to several IUPAC Conferences, the Chairman of the Ad Hoc Finance Committee, a member of the Committee to Revise the Statutes and By-Laws, and he is currently serving as Treasurer of IUPAC. He was also the Chairman of the VIth International Conference on Coordination Chemistry in 1961, which was co-sponsored by IUPAC.

Professor Bailar served his country during the Second World War as an official investigator of the National Defense Research Advisory Committee, Basic Physical Science, Panel on Chemistry, Research and Development Board. He is currently a member of the Board of Directors of Monmouth College. Professor Bailar has delivered hundreds of lectures in the United States and abroad on chemistry and chemical education, and has been awarded the Welch Lectureship, the Foster Lectureship at the University of Buffalo (twice), the Merck Lectureship at Bucknell University, the Clark E. Friend Lectureship at the University of West Virginia, and the American Cyanamid Lectureship at the University of Connecticut. He has been an author, editor, or co-author of four books and 145 scientific papers as well as many book reviews, lectures, etc.

In addition to receiving the John R. Kuebler Award of Alpha Chi Sigma, Professor Bailar has also received the American Chemical Society Award in Chemical Education sponsored by the Scientific Apparatus Makers Association (1964), the Priestley Medal of the American Chemical Society (1964), the Francis P. Dwyer Medal of the Chemical Society of New South Wales (1965), the Alfred Werner Gold Medal of the Swiss Chemical Society (1966), and the Synthetic Chemical Manufacturers Association Award in the Teaching of Chemistry (1968), and has also received Honorary Membership in Phi Lambda Upsilon—the national honorary chemistry fraternity (1959).

John C. Bailar, Jr. married Florence Catherwood on August 8, 1931, and they have two sons, John C. Bailar III, M.D., who is Director of the Demography Division of the United States Public Health Service, Bethesda, Maryland, and Benjamin F. Bailar, who is Vice-President of the American Can Company, New York, New York. Professor Bailar is also a member of the Phi Beta Kappa Fraternity and of the Society of the Sigma Xi, and he serves as an Elder in the McKinley Presbyterian Church.

PH.D. STUDENTS OF
PROFESSOR JOHN C. BAILAR, JR.

The names of the students who did their Ph.D. research with Professor John C. Bailar, Jr. are given below (the year in which each student was awarded the Ph.D. degree is given in parentheses):

J. H. Balthis, Jr. (1934)
T. Parsons, Jr. (1934)
E. H. Huffman (1937)
C. A. Stiegman (1937)
J. P. McReynolds (1938)
E. B. Middleton (1938)
D. F. Peppard (1939)
C. L. Rollinson (1939)
R. N. Keller (1940)
L. B. Clapp (1941)
T. D. O'Brien (1941)
Betty R. Tarr (1941)
R. C. Brasted (1942)
J. B. Work (1942)
F. Basolo (1943)
H. B. Jonassen (1946)
Margaret Kramer (1946)
J. A. Mattern (1946)
A. L. Oppegard (1946)
R. W. Parry (1946)
J. V. Quagliano (1946)
M. M. Woyski (1946)
C. F. Callis (1948)
R. D. Johnson (1948)
P. G. Arvan (1949)
B. P. Block (1949)
B. E. Douglas (1949)
Sr. Mary M. Hagan (1949)

T. H. Dexter (1950)
G. L. Eichhorn (1950)
L. F. Heneghan (1950)
A. L. McClelland (1950)
W. B. Schaap (1950)
R. L. Dalton (1951)
J. P. Kuebler, Jr. (1951)
Jennie C. Liu (1951)
O. F. Williams (1951)
A. D. Gott (1952)
P. A. Horrigan (1953)
E. H. Lyons, Jr. (1953)
N. C. Nielsen (1953)
D. H. Busch (1954)
W. E. Cooley (1954)
S. Kirschner (1954)
R. L. Rebertus (1954)
F. McCollough, Jr. (1955)
H. F. Holtzclaw, Jr. (1955)
R. L. Rau (1955)
M. M. Chamberlain (1956)
W. C. Drinkard (1956)
C. F. Liu (1957)
B. A. Ferrone (1957)
J. Selbin (1957)
C. E. Wymore (1957)
G. L. Johnson (1958)
M. L. Judd (1958)

R. D. Archer (1959)
J. P. Dismukes (1959)
J. A. McLean (1959)
Mary J. Crespi (1960)
L. F. Dempsey (1960)
S. K. Madan (1960)
R. E. Sievers (1960)
E. J. Friihauf (1961)
C. Fujikawa (1961)
J. S. Oh (1961)
T. A. Donovan (1962)
R. M. Klein (1963)
L. V. Interrante (1963)
E. L. Safford (1963)
G. J. Tennenhouse (1963)
L. B. Boucher (1964)
R. C. Burrows (1964)
J. E. George (1964)
R. W. Ohmke (1964)
L. Jane Park (1965)
H. E. LeMay (1966)
H. A. Tayim (1967)
R. E. Wagner (1967)
Bobbye K. Baylis (1968)
D. F. DeCraene (1969)
Ieva O. Hartwell (1969)
J. T. Murphy (1969)
E. Herzenberg (1969)

PUBLICATIONS OF
PROFESSOR JOHN C. BAILAR, JR.

A. PUBLICATIONS IN COORDINATION CHEMISTRY (1934–1968)

1. J. C. Bailar, Jr., and Robert W. Auten, "The Stereochemistry of Complex Inorganic Compounds. I. The Walden Inversion as Exhibited by Diethylenediaminocobaltic Compounds," *J. Am. Chem. Soc.* **56**: 774 (1934).
2. J. C. Bailar, Jr., "The Preparation of Anhydrous Ethylenediamine," *J. Am. Chem. Soc.* **56**: 955 (1934).
3. J. H. Balthis, Jr. and J. C. Bailar, Jr., "Some Chromous and Chromic Ammines," *J. Am. Chem. Soc.* **58**: 1474–6 (1936).
4. J. C. Bailar, Jr., "The Stereochemistry of Complex Inorganic Compounds," *Chem. Rev.* **19**: 67 (1936).
5. J. C. Bailar, Jr., Frank G. Jonelis and E. H. Huffman, "The Stereochemistry of Complex Inorganic Compounds. II. The Reaction of Carbonates with Dichloro-bis-ethylenediamine Cobaltic Chloride," *J. Am. Chem. Soc.* **58**: 2224 (1936).
6. J. C. Bailar, Jr., J. H. Haslam, and Eldon M. Jones, "The Stereochemistry of Complex Inorganic Compounds. III. The Reaction of Ammonia with levo-Dichloro-bis-ethylenediamine Cobaltic Chloride," *J. Am. Chem. Soc.* **58**: 2226 (1936).
7. J. C. Bailar, Jr., E. H. Huffman and A. R. Wreath, "Configuration Changes in the Reactions of Complex Inorganic Compounds," *Science* (New Series) **85** (1937).
8. J. C. Bailar, Jr., "Coordination Tendency of the Metallic Ions," *Chem. Rev.* **23**: 65 (1938).
9. J. P. McReynolds and J. C. Bailar, Jr., "The Catalytic Reaction Between Sodium Nitrite and Dichloro-bis-ethylenediamine Cobaltic Chloride," *J. Am. Chem. Soc.* **60**: 2817 (1938).
10. J. C. Bailar, Jr., C. A. Stiegman, J. H. Balthis, and E. H. Huffman, "The Stereochemistry of Complex Inorganic Compounds. IV. The Introduction of Racemic Organic Molecules into Some Optically Active Complex Ions of Cobalt and Chromium," *J. Am. Chem. Soc.* **61**: 2402 (1939).
11. J. C. Bailar, Jr., and J. P. McReynolds, "The Stereochemistry of Complex Inorganic Compounds. V. The Reaction of Carbonates with Dichlorodipropylenediamine Cobaltic Chloride. A New Method of Determining Relative Configurations," *J. Am. Chem. Soc.* **61**: 3199 (1939).
12. J. C. Bailar, Jr., and D. F. Peppard, "The Stereochemistry of Complex Inorganic Compounds. VI. A Study of the Stereoisomers of Dichloro-diammino-ethylenediamine Cobaltic Ion," *J. Am. Chem. Soc.* **62**: 105 (1940).
13. J. C. Bailar, Jr., and D. F. Peppard, "The Stereochemistry of Complex Inorganic Compounds. VII. The Mechanism of the Walden Inversion in Some Reactions Leading to the Formation of the Carbonato-diethylenediamine Cobaltic Ion," *J. Am. Chem. Soc.* **62**: 820 (1940).

14. J. C. Bailar, Jr., M. J. Copley and L. S. Foster, "The Stabilization of Valences by Coordination," *Chem. Rev.* **30**: 227 (1942).
15. Carl L. Rollinson and J. C. Bailar, Jr., "Studies in the Chromammines. II. Preparation of Luteo Salts," *J. Am. Chem. Soc.* **65**: 250–254 (1943).
16. Carl L. Rollinson and J. C. Bailar, Jr., "Studies in the Chromammines. III. Preparation of Triethylenediamine Luteo Salts," *J. Am. Chem. Soc.* **66**: 641–644 (1944).
17. J. C. Bailar, Jr., "The Oxidation States of Silver," *J. Chem. Educ.* **21**: 523 (1944).
18. J. C. Bailar, Jr., and Leallyn B. Clapp, "The Preparation and Properties of Inorganic Coordination Compounds. I. The Action of Some Organic Amines Upon Dichloro-diethylenediamine Cobaltic Chloride," *J. Am. Chem. Soc.* **67**: 171 (1945).
19. J. C. Bailar, Jr., and J. B. Work, "The Role of Catalysis in the Preparation and Reactions of Some Cobaltic and Chromic Ammines," *J. Am. Chem. Soc.* **67**: 176 (1945).
20. Thomas D. O'Brien and J. C. Bailar, Jr., "Studies in the Chromammines. IV. Thermal Decomposition of Luteo Salts," *J. Am. Chem. Soc.* **67**: 1856 (1945).
21. J. C. Bailar, Jr., and J. B. Work, "Some Coordination Compounds of Cobalt Containing Trimethylenediamine and Neopentanediamine," *J. Am. Chem. Soc.* **68**: 232 (1946).
22. Margaret Davis Kramer, Sherlock Swann, Jr., and J. C. Bailar, Jr., "The Electrodeposition of Cobalt and Nickel from Coordination Compounds," *Trans. Electrochem. Soc.* Preprint 90-27 (1946).
23. R. W. Parry, Sherlock Swann, Jr., and J. C. Bailar, Jr., "Coordination Compounds in the Electrodeposition of Chromium," *Electrochem. Soc.* Preprint 92-27 (1947).
24. Thomas D. O'Brien, James P. McReynolds, and J. C. Bailar, Jr., "The Stereochemistry of Complex Inorganic Compounds. VIII. Configurations of Some Compounds as Revealed by Rotatory Dispersion Curves," *J. Am. Chem. Soc.* **70**: 749 (1948).
25. Hans B. Jonassen, J. C. Bailar, Jr., and E. H. Huffman, "The Stereochemistry of Complex Inorganic Compounds. IX. The Diastereoisomers of *dextro*-tartratro-bis-ethylenediamine Cobaltic Ion," *J. Am. Chem. Soc.* **70**: 756 (1948).
26. H. A Laitinen, J. C. Bailar, Jr., Henry F. Holtzclaw, Jr., and J. V. Quagliano, "Polarographic Investigations of Hexamminecobaltic Chloride in Various Supporting Electrolytes," *J. Am. Chem. Soc.* **70**: 2999 (1948).
27. J. C. Bailar, Jr., "Some Special Features of the Stereochemistry of the Metal Ammines," *Record Chem. Progr.* **10**: 17 (1949).
28. H. A. Laitinen, E. I. Onstott, J. C. Bailar, Jr., and Sherlock Swann, "Polarography of Copper Complexes. I. Ethylenediamine, Propylenediamine, Diethylenetriamine, and Glycine Complexes," *J. Am. Chem. Soc.* **71**: 1550 (1949).
29. Fred Basolo, J. C. Bailar, Jr., and Betty Rapp Tarr, "The Stereochemistry of Complex Inorganic Compounds. X. The Stereoisomers of Dichloro-bis-ethylenediamine-platinum-(IV) Chloride," *J. Am. Chem. Soc.* **72**: 2433 (1950).
30. Bodie E. Douglas, H. A. Laitinen and J. C. Bailar, Jr., "Polarography of Some Cadmium Complexes," *J. Am. Chem. Soc.* **72**: 2484 (1950).
31. B. P. Block and J. C. Bailar, Jr., "The Reaction of Gold(III) Chloride with Some Bidentate Coordinating Groups," *J. Am. Chem. Soc.* **73**: 4722 (1951).
32. B. P. Block, J. C. Bailar, Jr., and D. W. Pearce, "The Reaction of Silver Acetate with 8-Quinolinol," *J. Am. Chem. Soc.* **73**: 4971 (1951).
33. Jeanne C. I. Liu and J. C. Bailar, Jr., "The Stereochemistry of Complex Inorganic Compounds. XI. The Resolution of Bis-(8-quinolino-5-sulfonic Acid) Zinc(II)," *J. Am. Chem. Soc.* **73**: 5432 (1951).
34. Sister Mary Martinette, BVM, and J. C. Bailar, Jr., "The Stereochemistry of Complex Inorganic Compounds. XII. The Diastereoisomers of Carbonato-bis-levo-Propylene-diamine Cobaltic Ion," *J. Am. Chem. Soc.* **74**: 1054 (1952).

35. Hans Jonassen, J. C. Bailar, Jr., and Allan Gott, "The Stereoisomerism of Complex Inorganic Compounds. XIII. A Partial Resolution of Racemic Tartaric Acid by Means of
Different Stabilities of Isomers of Complex Ions," *J. Am. Chem. Soc.* **74**: 3131 (1952).

36. Clayton F. Callis, Niels Nielsen and J. C. Bailar, Jr., "Some Metal Derivatives of Azo and
Azomethine Dyes," *J. Am. Chem. Soc.* **74**: 3461 (1952).

37. John R. Kuebler, Jr., and J. C. Bailar, Jr., "The Stereoisomerism of Complex Inorganic
Compounds. XIV. Studies Upon the Stereochemistry of Saturated Tervalent Nitrogen
Compounds," *J. Am. Chem. Soc.* **74**: 3535 (1952).

38. Allan D. Gott and J. C. Bailar, Jr., "The Stereochemistry of Complex Inorganic Compounds. XV. A Partial Resolution of Racemic Mixtures of Organic Acids by Means of
Preferential Coordination," *J. Am. Chem. Soc.* **74**: 4820 (1952).

39. J. C. Bailar, Jr., and Clayton F. Callis, "The Structures and Properties of Some Metal
Derivatives of Azo and Azomethine Dyes," *J. Am. Chem. Soc.* **74**: 6018 (1952).

40. Leo F. Heneghan and J. C. Bailar, Jr., "The Stereoisomers of Complex Inorganic Compounds. XVI. The Stereoisomers of Dichloro-bis-(ethylenediamine)-platinum(IV) Salts,"
J. Am. Chem. Soc. **75**: 1840 (1953).

41. Gunther L. Eichhorn and J. C. Bailar, Jr., "Metal Ion Catalysis in the Hydrolysis of Schiff
Bases," *J. Am. Chem. Soc.* **75**: 3051 (1953).

42. Robert L. Rebertus, H. A. Laitinen, and J. C. Bailar, Jr., "A Polarographic Study of the
Hydrazine Complexes of Zinc," *J. Am. Chem. Soc.* **75**: 3051 (1953).

43. Daryle H. Busch and J. C. Bailar, Jr., "The Stereochemistry of Complex Inorganic Compounds. XVII. The Stereochemistry of Hexadentate Ethylenediaminetetraacetic Acid
Complexes," *J. Am. Chem. Soc.* **75**: 4574 (1953).

44. Basudeb Das Sarma and J. C. Bailar, Jr., "The Stereochemistry of Metal Chelates of a
Polydentate Ligand," *J. Am. Chem. Soc.* **76**: 4051 (1954).

45. Ernest H. Lyons, Jr., J. C. Bailar, Jr., and H. A. Laitinen, "Electronic Configuration in
Electrodeposition from Aqeuous Solution. III. Metal Deposition from Certain Complex
Ions," *J. Electrochem. Soc.* **101**: 410 (1954).

46. Daryle H. Busch and J. C. Bailar, Jr., "The Optical Stability of Beryllium Complexes,"
J. Am. Chem. Soc. **76**: 5352 (1954).

47. Ward B. Schaap, H. A. Laitinen and J. C. Bailar, Jr., "Polarography of Iron Oxalates,
Malonates, and Succinates," *J. Am. Chem. Soc.* **76**: 5868 (1954).

48. Basudeb Das Sarma and John C. Bailar, Jr., "The Stereochemistry of Metal Chelates with
Polydentate Ligands. Part I," *J. Am. Chem. Soc.* **77**: 5476 (1955).

49. Basudeb Das Sarma and J. C. Bailar, Jr., "The Stereochemistry of Complex Inorganic
Compounds. XVIII. A New Method for the Preparation of Inorganic Complexes in Their
Optically Active Forms," *J. Am. Chem. Soc.* **77**: 5480 (1955).

50. Daryle H. Busch and John C. Bailar, Jr., "Stereochemistry of Complex Inorganic Compounds. XX. The Tetradentate and Bidentate Complexes of EDTA," *J. Am. Chem. Soc.* **78**:
716 (1956).

51. Fred McCollough and J. C. Bailar, Jr., "The Stereochemistry of Complex Inorganic Compounds. XIX. The Resolution of Bis-Ethylenediamine (2,2′, diaminobiphenyl) Cobalt(III)
Chloride," *J. Am. Chem. Soc.* **78**: 714 (1956).

52. Basudeb Das Sarma and J. C. Bailar, Jr., "Partial Resolution of Diamines, Amino Acids,
and Dicarboxylic Acids Through Coordination with Optically Active Ligands," *J. Am.
Chem. Soc.* **78**: 895 (1956).

53. Edited by J. C. Bailar, Jr., *The Chemistry of the Coordination Compounds*, Reinhold Publishing Corp., New York 1956.

54. Daryle H. Busch and J. C. Bailar, Jr., "The Iron(II)-Methine Chromophore," *J. Am. Chem.
Soc.* **78**: 1137 (1956).

55. J. C. Bailar, Jr., "The Numbers and Structures of Isomers of Hexacovalent Complexes," *J. Chem. Educ.* **34**: 334 (1957).

56. Joel Selbin and J. C. Bailar, Jr., "The Kinetics of Aquation of *cis*-Dichloro-bis-(ethylene-diamine)-chromium(III) Ion," *J. Am. Chem. Soc.* **79**: 4285 (1957).

57. J. C. Bailar, Jr., William C. Drinkard, Jr., and Malcolm L. Judd, "Polymerization Through Coordination," WADC Technical Report 57-391, ASTIA Document No. AD 131100 (1957).

58. Stanley Kirschner, Yung-Kang Wei, and J. C. Bailar, Jr., "The Stereochemistry of Complex Inorganic Compounds. XXI. The Resolution of Racemic Substances Through Optically Active Inorganic Complexes," *J. Am. Chem. Soc.* **79**: 5877 (1957).

59. Carl S. Marvel, Ludwig F. Audrieth and J. C. Bailar, Jr., "High Polymeric Materials," WADC Technical Report No. 58-51, ASTIA Document No. 151177.

60. J. C. Bailar, Jr., "Some Problems in the Stereochemistry of Coordination Compounds," Proceedings of the International Symposium on the Chemistry of Coordination Compounds, La Ricerca Scientifica, **28**: 165 (1958). Also in *J. Inorg. Nucl. Chem.* **8**: 165 (1958).

61. J. C. Bailar, Jr., K. V. Martin, Malcolm L. Judd and John McLean, "Polymerization Through Coordination," WADC Technical Report 57-391, Part II, ASTIA Document No. 155799.

62. C. S. Marvel, L. F. Audrieth, and J. C. Bailar, Jr., "High Polymeric Materials," WADC Technical Report 58-51, Part II, ASTIA Document No. 213598.

63. E. J. Corey and J. C. Bailar, Jr., "The Stereochemistry of Complex Inorganic Compounds. XXII. Stereospecific Effects in Complex Ions," *J. Am. Chem. Soc.* **81**: 2620 (1959).

64. William E. Colley, Chui Fan Liu and J. C. Bailar, Jr., "The Stereochemistry of Complex Inorganic Compounds. XXIII. Double Optical Isomerism and Optical-Geometric Isomerism in Cobalt(III) Complexes," *J. Am. Chem. Soc.* **81**: 4189 (1959).

65. Oren F. Williams and J. C. Bailar, Jr., "The Stereochemistry of Complex Inorganic Compounds. XXIV. Cobalt Stilbenediamine Complexes," *J. Am. Chem. Soc.* **81**: 4464 (1959).

66. William C. Drinkard and J. C. Bailar, Jr., "Copper Phthalocyanine Polymers," *J. Am. Chem. Soc.* **81**: 4795 (1959).

67. Mark M. Chamberlain and J. C. Bailar, Jr., "The Infrared Spectra of Some Thiocyanato-cobalt Ammines," *J. Am. Chem. Soc.* **81**: 6412 (1959).

68. Joel Selbin and J. C. Bailar, Jr., "The Stereochemistry of Complex Inorganic Compounds. XXV. A *trans* Complex of Triethylenetetramine," *J. Am. Chem. Soc.* **82**: 1524 (1960).

69. C. S. Marvel, L. F. Audrieth, T. Moeller and J. C. Bailar, Jr., "High Polymeric Materials," WADC Technical Report 58-51, Part III (1960).

70. William C. Drinkard, H. Frederick Bauer, and J. C. Bailar, Jr., "Reactivity of Organic Hydroxy Groups β to the Coordination Site in Cobalt(III) Complexes," *J. Am. Chem. Soc.* **82**: 2992 (1960).

71. C. Elmer Wymore and J. C. Bailar, Jr., "Uncommon Coordinating Agents I. P,P,P′,P′-Tetraethylethylenediphosphine," *J. Inorg. Nuclear Chem.* **14**: 42 (1960).

72. Robert L. Rau and J. C. Bailar, Jr., "Bridged Complexes and the Deposition of Tin-Nickel Alloys," *J. Electrochem. Soc.* **107**: [9] 745 (1960).

73. Ronald D. Archer and J. C. Bailar, Jr., "Stereochemistry of Inorganic Complexes. XXVI. The Ammonation of Two Optically Active Cobalt(III) Complexes," *J. Am. Chem. Soc.* **83**: 812 (1961).

74. J. P. Dismukes, L. H. Jones and J. C. Bailar, Jr., "The Measurement of Metal-Ligand Bond Vibrations in Acetylacetonate Complexes," *J. Phys. Chem.* **65**: 792 (1961).

75. Harold A. Goodwin and J. C. Bailar, Jr., "Coordination Compounds Prepared from Polymeric Schiff's Bases," *J. Am. Chem. Soc.* **83**: 2467 (1961).

76. J. C. Bailar, Jr., "Coordination Polymers," A Chapter in "Inorganic Polymers, An International Symposium," pages 51–66, The Chemical Society, London, 1961.

77. C. S. Marvel, T. Moeller, and John C. Bailar, Jr., WADD Technical Report 61-12 to Wright Air Development Division (April, 1961).

78. Robert E. Sievers and J. C. Bailar, Jr., "Some Metal Chelates of Ethylenediaminetetraacetic Acid, Diethylenetriaminepentaacetic Acid and Triethylenetetraaminehexaacetic Acid," Inorg. Chem. **1** : 174 (1962).

79. Badar-Ud-Din and J. C. Bailar, Jr., "Observations on the Oxidation and Reduction of Platinum(II) Nitro Complexes," *J. Inorg. Nuclear Chem.* **22** : 241 (1961).

80. Edward J. Friihauf and J. C. Bailar, Jr., "The Heat Stabilities of Some Polymeric Metal Complexes," *J. Inorg. Nuclear Chem.* **24** : 1205 (1962).

81. J. S. Oh and J. C. Bailar, Jr., "Some Coordination Polymers Prepared from Bis-β-Diketones," *J. Inorg. Nuclear Chem.* **24** : 1225 (1962).

82. T. Moeller and J. C. Bailar, Jr., "High Polymeric Materials," Technical Documentary Report No. ASD-TDR-62-209, Part II, February 1963.

83. E. J. Olszewski, L. J. Boucher, R. W. Oehmke, J. C. Bailar, Jr., and Dean F. Martin, "Amine Catalyzed Condensation of β-Diketones Using a Metal-Chelate Template," *Inorg. Chem.* **2** : 661 (1963).

84. Richard M. Klein and J. C. Bailar, Jr., "Preparation and Reduction of Bis-(3-nitro-2,4-pentanediono)-beryllium," *Inorg. Chem.* **2** : 1187 (1963).

85. Richard M. Klein and J. C. Bailar, Jr., "Reactions of Coordination Compounds. Polymers from 3-Substituted Bis-(β-diketone)-beryllium Complexes," *Inorg. Chem.* **2** : 1190 (1963).

86. B. Das Sarma, K. R. Ray, Robert E. Sievers and J. C. Bailar, Jr., "The Stereochemistry of Metal Chelates with Multidentate Ligands II," *J. Am. Chem. Soc.* **86** : 14 (1964).

87. T. A. Donovan and J. C. Bailar, Jr., "An Investigation of the Coordination Chemistry of Glyoxal-bis-(Guanylhydrazone) and Methylglyoxal-bis-(Guanylhydrazone)," *J. Inorg. Nuclear Chem.* **26** : 1283 (1964).

88. G. L. Johnson, J. C. Bailar, Jr., and R. H. Herber, "Kinetics of the Isotopic Halogen Exchange Between trans-Dichloro-bis-ethylenediamine-platinum(IV) Ion and Aqueous Chloride," *J. Inorg. Nuclear Chem.* **26** : 1061 (1964).

89. L. J. Boucher and J. C. Bailar, Jr., "The Preparation and Properties of Some Complexes of the Type β-Diketonatobis-(ethylenediamine)-cobalt(III) Iodide and Bis-(β-diketonato)-tetrakis-(ethylenediamine)dicobalt(III) Iodide," *Inorg. Chem.* **3** : 589 (1964).

90. Chiu Fan Liu, Nora C. Liu and J. C. Bailar, Jr., "The Stereochemistry of Complex Inorganic Compounds. XXVII. Asymmetric Syntheses of Tris-Bipyridine Complexes of Ruthenium-(II) and Osmium(III)," *Inorg. Chem.* **3** : 1197 (1964).

91. Chiu Fan Liu, Nora C. Liu and J. C. Bailar, Jr., "A Specific Synthesis for Bis-Bipyridine Ruthenium Compounds," *Inorg. Chem.* **3** : 1197 (1964).

92. J. C. Bailar, Jr., "Coordination Polymers," A Chapter in Preparative Inorganic Reactions (p. 1–27), Interscience Publishers, 1964.

93. L. J. Boucher, Eishin Kyuno and J. C. Bailar, Jr., "Stereochemistry of Complex Inorganic Compounds XXVIII. The Walden Inversion in the Base Hydrolysis of Optically Active Cobalt(III) Complexes," *J. Am. Chem. Soc.* **86** : 3656 (1964).

94. Leonard V. Interrante and J. C. Bailar, Jr., "Preparation and Thermal Stability of Azo and Axomethine Coordination Polyesters," *Inorg. Chem.* **3** : 1339 (1964).

95. J. C. Bailar, Jr., "The Walden Inversion in the Reactions of Cobalt Complexes," *Chemicke zvesti,* **19** [3] : 153 (1965).

96. L. J. Boucher and J. C. Bailar, Jr., "The Preparation and Properties of Sodium Dinitro-Bis-Acetylaceonato-Cobaltate(III) and Some Complexes of the Type Nitro-Ammine-Bis-Acetylacetonato-Cobalt(III)," *J. Inorg. Nucl. Chem.* **27** : 1093 (1965).

97. Stanley K. Madan, Wm. M. Reiff, and J. C. Bailar, Jr., "Substitution Reactions of *cis*-Dichloro-(β,β',β"-triamintriethylamine)cobalt(III) Complex Ion. Kinetics of Aquation of *cis*-Dichloro-(triaminotriethylamine)cobalt(III) Ion," *Inorg. Chem.* **4**: 1366 (1965).

98. R. W. Oehmke and J. C. Bailar, Jr., "Some Coordination Compounds of the 2,2-Diamino-methyl-1,3-Diamino-Propane Moiety," *J. Inorg. Nucl. Chem.* **27** [10]: 2199 (1965).

99. R. W. Oehmke and J. C. Bailar, Jr., "Some Coordination Compounds of 2-(Salicylidene-aminomethyl-)pyridine," *J. Inorg. Nucl. Chem.* **27** [10]: 2209 (1965).

100. Eishin Kyuno, L. J. Boucher and J. C. Bailar, Jr., "Stereochemistry of Complex Inorganic Compounds. XXIX. The Base Hydrolysis of the Optically Active α-Dichlorotriethylene-tetramine Cobalt(III) Cation," *J. Am. Chem. Soc.* **87**: 4458 (1965).

101. J. C. Bailar, Jr., and Hiroshi Itatani, "Hydridochlorobis-(triphenylphosphine)platinum(II) and Some Related Compounds," *Inorg. Chem.* **4**: 1618 (1965).

102. Eishin Kyuno and J. C. Bailar, Jr., "The Stereochemistry of Complex Inorganic Compounds. XXX. The Base Hydrolysis of Some Optically Active β-Dihalotriethylenetetra-minecobalt(III) Cations," *J. Am. Chem. Soc.* **88**: 1120 (1966).

103. Eishin Kyuno and J. C. Bailar, Jr., "The Stereochemistry of Complex Inorganic Compounds. XXXI. Optical Inversions in the Reactions of Some Optically Active α-Dihalo-triethylenetetraminecobalt(III) Cations with Ammonia and Ethylenediamine," *J. Am. Chem. Soc.* **88**: 1125 (1966).

104. Teiji Habu and J. C. Bailar, Jr., "The Stereochemistry of Complex Inorganic Compounds. XXXII. The Stereochemistry of the Ethylenediamine-2,2'-diaminobiphenylplatinum(II) Ion, and the bis-Diaminobiphenyl-platinum(II) Ion," *J. Am. Chem. Soc.* **88**: 1128 (1966).

105. Gerald J. Tennenhouse and J. C. Bailar, Jr., "Coordination Compounds of Triethylene-diamine," *J. Inorg. Nucl. Chem.* **28**: 682 (1966).

106. J. C. Bailar, Jr., and H. Itatani, "Catalytic Hydrogenation of Soybean Oil Methyl Ester and Some Related Compounds," Proceedings of the Symposium on Coordination Chemistry, Tihany, Hungary, 67–81 (1964) published in 1965.

107. J. C. Bailar, Jr., "Optical Inversions in the Reactions of Cobalt Complexes," *Rev. Pure Appl. Chem.* **16**: 91 (1966).

108. J. C. Bailar, Jr., and Hiroshi Itatani, "Catalytic Hydrogenation of Soybean Oil Methyl Ester and Some Related Compounds," *J. Am. Oil Chem. Soc.* **43**: 337 (1966).

109. R. C. Burrows and J. C. Bailar, Jr., "Iron(III) and Cobalt(III) Complexes of Some N-Salicylideneamino Acids," *J. Am. Chem. Soc.* **88**: 4150 (1966).

110. Eishin Kyuno and J. C. Bailar, Jr., "The Stereochemistry of Complex Inorganic Compounds. XXXIII. Reactions of Optically Active α-Dichlorotriethylenetetramine Cobalt(III) Cation with Optically Active Propylenediamine," *J. Am. Chem. Soc.* **88** [23]: 5447 (1966).

111. J. C. Bailar, Jr., "Modern Inorganic Synthesis," *Pure Appl. Chem.* **10** [4]: 495 (1965).

112. J. C. Bailar, Jr., Hiroshi Itatani, Mary Jean Sirotek Crespi and John Geldard, "Some Recent Developments in Coordination Chemistry," Advances in Chemistry Series, American Chemical Society, No. 62, p. 103 (1967).

113. J. C. Bailar, Jr., "Some Developments in the Stereochemistry of Complexes Since the Time of Werner," *Helv. Chim. Acta*, 82–92 (1967).

114. Hiroshi Itatani and J. C. Bailar, Jr., "Homogeneous Catalysis in the Reactions of Olefinic Substances. V. Hydrogenation of Soybean Oil Methyl Ester with Triphenylphosphine and Triphenylarsine Palladium Catalysts," *J. Am. Oil Chem. Soc.* **44**: 147 (1967).

115. J. C. Bailar, Jr., and Hiroshi Itatani, "Homogeneous Catalysis in the Reactions of Olefinic Substances. VI. Selective Hydrogenation of Methyl Linoleate and Isomerization of Methyl Oleate by Homogeneous Catalysis with Platinum Complexes Containing Triphenyl-phosphine, Arsine or Stibine," *J. Am. Chem. Soc.* **89**: 1592 (1967).

116. Hiroshi Itatani and J. C. Bailar, Jr., "Homogeneous Catalysis in the Reactions of Olefinic

Substances. VII. Hydrogenation and Isomerization of Methyl Linoleate with Bistri-phenylphosphine Nickel Halides," *J. Am. Chem. Soc.* **89**: 1600 (1967).

117. J. C. Bailar, Jr., and L. F. Dempsey, "The Stereochemistry of Complex Inorganic Compounds. XXXIV. The Stereochemistry of the β,β',β''-Triaminoethylamine-2,2'-diamino-biphenylcobalt(III) Ion," *Rev. Chim. Minerale* **3**: 1029 (1966).

118. R. C. Burrows and J. C. Bailar, Jr., "Solubility Behavior of Some N-Salicylideneamino Acid Complexes," *J. Inorg. Nucl. Chem.* **29**: 709 (1967).

119. E. N. Frankel, E. A. Emken, Hiroshi Itatani and J. C. Bailar, Jr., "Homogeneous Hydrogenation of Methyl Linolenate Catalyzed by Platinum-Tin Complexes," *J. Org. Chem.* **32**: 1447 (1967).

120. Hassan A. Tayim and J. C. Bailar, Jr., "Homogeneous Catalysis in the Reactions of Olefinic Substances. VIII. Isomerization of 1,5-Cyclooctadiene with Dichlorobis-(triphenyl-phosphine)platinum(II)," *J. Am. Chem. Soc.* **89**: 3420 (1967).

121. Hassan A. Tayim and J. C. Bailar, Jr., "Homogeneous Catalysis in the Reactions of Olefinic Substances. IX. Homogeneous Catalysis of the Specific Hydrogenation of Polyolefins by Some Platinum and Palladium Complexes," *J. Am. Chem. Soc.* **89**: 4330 (1967).

122. H. Eugene LeMay and J. C. Bailer, Jr., "The Solid Phase *trans*-to-cis Isomerization of [Co(NH$_3$)$_4$Cl$_2$]IO$_3$·2H$_2$O," *J. Am. Chem. Soc.* **89**: 5577 (1967).

123. B. Das Sarma, Gerald J. Tennenhouse, and J. C. Bailar, Jr., "Complexes of N-Hydroxy-ethylethylenediamine," *J. Am. Chem. Soc.* **90**: 1362 (1968).

124. J. C. Bailar, Jr., and H. E. LeMay, Jr., "The Solid Phase Racemization of *l*-cis-[Cren$_2$Cl$_2$]-Cl·H$_2$O," *J. Am. Chem. Soc.* **90**: 1729 (1968).

125. J. C. Bailar, Jr., Hiroshi Itatani, and Hassan Tayim, "Homogeneous Catalysis in the Reactions of Olefinic Substances. X. The Isomerization and Hydrogenation of Polyolefinic Substances," *J. Japan. Chem.* **22**: 41 (1968).

126. R. W. Adams, G. E. Batley and J. C. Bailar, Jr., "Homogeneous Catalytic Hydrogenation and Isomerization of Olefins with Dichlorobis(triphenylphosphine)platinum(II)-Tin Chloride Catalyst," *Inorg. Nucl. Chem. Letters* **4**: 455 (1968).

127. R. W. Adams, G. E. Batley and J. C. Bailar, Jr., "Homogeneous Catalysis in the Reactions of Olefinic Substances. XI. Homogeneous Catalytic Hydrogenation of Short Chain Olefins with Dichlorobis(triphenylphosphine)platinum(II)-Tin(II) Chloride Catalyst," *J. Am. Chem. Soc.* **90**: 6051 (1968).

128. G. E. Batley and J. C. Bailar, Jr., "Palladium(II)-Tin(II) Complexes," *Inorg. Nucl. Chem. Letters* **4**: 577 (1968).

129. Robert E. Wagner and J. C. Bailar, Jr., "Some Complexes of S-Substituted Cysteines and Their Reactions with Heavy Metal Ions," *J. Am. Chem. Soc.* (In Press.)

130. B. Das Sarma and J. C. Bailar, Jr., "Chemistry of Metal Complexes with Polydentate Ligands; Complexes of N-Hydroxyethylethylenediamine," *J. Am. Chem. Soc.* (In Press.)

131. J. E. House, Jr., and J. C. Bailar, Jr., "Kinetics of the Linkage Isomerization in Iron(II) Hexacyanochromate," *Inorg. Chem.* (In Press.)

132. J. E. House, Jr., and J. C. Bailar, Jr., "The Solid State Deamination of Tris-(ethylenediamine) and Tris-(propylenediamine)chromium(III) Thiocyanate," *J. Am. Chem. Soc.* (In Press.)

133. Walter J. Kasowski and J. C. Bailar, Jr., "Steric Effects in Coordination of 1,1,1-Tris-(dimethylaminomethyl)ethane and of 1,1,1-Tris-(monomethylaminomethyl)ethane. Part I. Zinc Complexes," *J. Am. Chem. Soc.* (In Press.)

B. OTHER PUBLICATIONS (1925–1965)

1. H. B. VanValkenburgh and John C. Bailar, Jr., "Nitrogen Tetrasulfide and Nitrogen Tetraselenide," *J. Am. Chem. Soc.* **47** : 2134 (1925).
2. M. Gomberg and John C. Bailar, Jr., "Halogen Substituted Aromatic Pinacols and the Formation of Ketyl Radicals, R_2COMgI," *J. Am. Chem. Soc.* **51** : 2229 (1929).
3. John C. Bailar, Jr., "An Experiment to Illustrate the Law of Multiple Proportions," *J. Chem. Educ.* **6** : 1759 (1929).
4. John C. Bailar, Jr., "The Effect of Substituents Upon the Rearrangement of Benzopinacol," *J. Am. Chem. Soc.* **52** : 3596 (1930).
5. John C. Bailar, Jr., "The Study of Isomerism in Courses in General Chemistry," *J. Chem. Educ.* **8** : 310 (1931).
6. John C. Bailar, Jr., "Isomerism in Inorganic Chemistry," *Trans. Ill. State Acad. Sci.* **23** : 307 (1931).
7. John C. Bailar, Jr., "Some Studies in the Pinacol Series," *Trans. Ill. State Acad. Sci.* **23** : 310 (1931).
8. John C. Bailar, Jr., "Comparison of Solubilities of Calcium and Strontium p-Bromobenzoates in Acetone-Water Mixtures," *Ind. Eng. Chem. (Anal. Ed.)* **3** : 362 (1931).
9. John C. Bailar, Jr., "Variation in the Prices of Metals in the Last Twenty Years," *J. Chem. Educ.* **10** : 99 (1933).
10. John C. Bailar, Jr., *Laboratory Assignments for Chemistry 2*, (Textbook), Edwards Bros., Ann Arbor, Mich., 1933.
11. John C. Bailar, Jr., "The Preparation of Anhydrous Ethylenediamine," *J. Am. Chem. Soc.* **56** : 955 (1934).
12. T. Parsons, Jr., and John C. Bailar, Jr., "The Preparation of Methyl Substituted Azobenzenes and the Rearrangement of Methyl Substituted Azoxybenzenes," *J. Am. Chem. Soc.* **58** : 268 (1936).
13. John C. Bailar, Jr., Allan J. Barney, and R. F. Miller, "The Action of Alkalies on Mixtures of Aromatic Aldehydes," *J. Am. Chem. Soc.* **58** : 2110 (1936).
14. John C. Bailar, Jr., "Chemistry for Professional Value," *The Science Teacher* **4** : No. 2, 5 (1937).
15. John C. Bailar, Jr., "Introduction to the Symposium (On Complex Inorganic Compounds), *Chem. Rev.* **21** : 1 (1937).
16. John C. Bailar, Jr., "Improving Instruction in Chemistry," *The Science Teacher* **8** : 8 (1941).
17. B. S. Hopkins and John C. Bailar, Jr., *Essentials of Chemistry*, (Textbook), D. C. Heath and Co., Boston, Mass., 1946.
18. John C. Bailar, Jr., "The General Chemistry Laboratory," *J. Chem. Educ.* **24** : 327 (1947).
19. John C. Bailar, Jr., "Chemical Education at the University of Illinois," *J. Chem. Educ.* **24** : 550 (1947).
20. John C. Bailar, Jr., Karl F. Heumann, and Edwin J. Seiferle, "The Use of Punched Card Techniques in the Coding of Inorganic Compounds," *J. Chem. Educ.* **25** : 142 (1948).
21. John C. Bailar, Jr., and D. R. Martin, *Laboratory Exercises for Chemistry 2*, Stipes Publ. Co., Champaign, Ill., 1949.
22. John C. Bailar, Jr., "On Choosing a Job," *Chem. Eng. News* **28** : 2959 (1950).
23. B. S. Hopkins and John C. Bailar, Jr., *General Chemistry for Colleges*, D. C. Heath and Co., 4th Ed., 1951; 5th Ed., 1956.
24. John C. Bailar, Jr., "The General Chemistry Laboratory," Ch. 4 of Part 2 in Coleman *Laboratory Design*, 95 (1951).
25. Milton Tamres and John C. Bailar, Jr., "The Course in General Chemistry," *J. Chem. Educ.* **29** (1952).

26. John C. Bailar, Jr. (Editor), *Inorganic Syntheses*, Vol. IV (1953).
27. John C. Bailar, Jr., *The Encyclopedia of Chemistry*, Reinhold Publishing Corp., New York 208 (1957). Article on Chelation.
28. John C. Bailar, Jr., *The Encyclopedia of Chemistry*, Reinhold Publ. Corp., New York 853 (1957). Article on Sequestering Agents.
29. John C. Bailar, Jr., "On Choosing a Job," *Chem. Eng. News* **35**: No. 14, Part II, 4 (1957). This is a condensation of the earlier article on the same subject. It appeared in *Chem. Eng. News* **28**: 2959 (1950).
30. John C. Bailar, Jr., "Fifty Years of Inorganic Chemistry," *Ind. Eng. Chem.* **50**: 40A (1958).
31. John C. Bailar, Jr., "The Chemical Profession Without the ACS (Guest Editorial), *Chem. Eng. News* **37** [1]: 7 (1959).
32. John C. Bailar, Jr., "The Changing Role of the Science Teacher," (An editorial), *J. Chem. Educ.* **36**: 155 (1959).
33. John C. Bailar, Jr., "The Utilization of Our National Resources," *Chem. Eng. News* **37** No. 39: 116 (1959).
34. John C. Bailar, Jr., *Chemistry and Chemical Engineering in the United States*, Chemische Rundschau, No. 20 (1960).
35. John C. Bailar, Jr., "Chemical Education—Then and Now," (Address given at St. Louis as the Award Address for the ACS Award in Chemical Education), *J. Chem. Educ.* **38**: 434 (1961).
36. S. S. Hsu, P. N. Yocum, T. C. C. Cheng, K. B. Oldham, C. E. Meyers, K. Gingerich, C. H. Travaglini, John C. Bailar, Jr., H. A. Laitinen, and S. Swann, Jr., "Preparation and Properties of Phosphides and Silicides and Some Electrochemical Studies," U.S. Dept. Comm., Office Tech. Serv., P. B. Rept. 147,079, 5 pp. (1961).
37. John C. Bailar, Jr. Preface to "Reactions of Coordinated Ligands and Homogeneous Catalysis," *Adv. in Chem. Series* **37**: ACS (1963), p. VII.
38. John C. Bailar, Jr., "The Expanding Universe of Chemistry," (The Priestley Medal Address), *Chem. Eng. News* **42**: 110 (1964).
39. John C. Bailar, Jr., "The Expanding Universe of Chemistry," (Abridged Version), *Chemistry* **37**: 18 (1964).
40. John C. Bailar, Jr., Therald Moeller, and Jacob Kleinberg, *University Chemistry*, D. C. Heath and Co., Boston, Mass., 1965.
41. John C. Bailar, Jr., "Evaluation of Research from the Viewpoint of the University Professor," *Research Management*, **8**: 133 (1965).
42. John C. Bailar, Jr., Summer Symposium on Mechanisms of Inorganic Reactions. University of Kansas, 1964. Introduction to the Symposium. Adv. in Chem. Series, No. 49, 1 (1965).

INTERNATIONAL UNION OF PURE AND APPLIED CHEMISTRY

c/o F. Hoffmann-La Roche & Co. Ltd.,
4002 Basle, Switzerland
Telegrams: IUPAC Basle – Telephone 32 38 20

—

RUDOLF MORF, Secretary General

New Address: Post Office Box No. 165
8058 Zürich-Airport   Tel. 051 83 99 65
Switzerland

517/RM/ek

Zürich, December 13, 1968

Wayne State University
Detoit, Michigan 48202
U.S.A.

Gentlemen:

We have learned that a Symposium will be held in honor of the
sixty-fifth birthday of Professor John C. Bailar, Jr. at Urbana,
Illinois in June, 1969. On behalf of the International Union of
pure and Applied Chemistry, we wish to express its appreciation
for all of the many contributions which Professor Bailar has made
to international chemistry and understanding during his productive
career.

He has served the International Union of Pure and Applied
Chemistry as its Treasurer. In addition, he was Chairman of the
Executive Committee which organized the Sixth International
Conference on Coordination Chemistry in the United States in 1961,
a Conference which was co-sponsored by the International Union of
Pure and Applied Chemistry. He has also been very active in the
organization of the International Conferences on Coordination
Chemistry throughout their existence. In addition, Professor
Bailar has contributed generously of his time in visiting many
countries in order to assist with the development of their educational programs, and to deliver lectures on chemistry and chemical
education to the scientists and students of those countries.

Perhaps his most important contribution has been the effort
he has devoted to increasing understanding among scientists from
many countries who held many views about various aspects of science.

For these endeavors and for his efforts in strengthening the
International Union of Pure and Applied Chemistry, we wish to
express our appreciation and our hope that he will continue to be
active in this area for many years.

Sincerely yours,

Prof. V.N. Kondratiev   Dr. Rudolf Morf
President   Secretary General

AMERICAN CHEMICAL SOCIETY
1155 SIXTEENTH STREET, N. W.
WASHINGTON, D. C. 20036

December 6, 1968

Professor Stanley Kirschner
Department of Chemistry
Wayne State University
Detroit, Michigan 48202

Dear Professor Kirschner:

It gives us great pleasure to note that a symposium
is to be held at Urbana, Illinois, in June, 1969, honoring
Professor John C. Bailar, Jr., on the occasion of his sixty-
fifth birthday. We are particularly mindful of the many ACS
offices held by Professor Bailar over the years. Before being
elected in 1959 to the highest office of the Society, the
Presidency, he was chairman of the following: University of
Illinois Section, Division of Chemical Education, Division of
Physical and Inorganic Chemistry, Division of Inorganic Chemistry, Council Committee on National Meetings and Divisional
Activities and of its Subcommittee on Constitution and Bylaws,
and the Council Policy Committee. He has been a local section
and division councilor and an associate editor of the JOURNAL
OF THE AMERICAN CHEMICAL SOCIETY and of CHEMICAL REVIEWS. He
also has given liberally of his time as a lecturer, and has
served our profession effectively through his active membership on committees of the International Union of Pure and
Applied Chemistry and the National Research Council. He has
been active in the Visiting Scientists Program of the Division of Chemical Education.

Professor Bailar's achievements as both professional
chemist and educator are widely recognized. In 1961 he received the ACS Award in Chemical Education and, in 1964, the
Priestley Medal, the highest honor in American chemistry. He
also has received the John R. Kuebler Award of Alpha Chi Sigma,
of which he is a member. He holds honorary doctorates from
the Universities of Colorado and Michigan and is a member of
Phi Beta Kappa, Phi Lambda Upsilon, and Sigma Xi.

On behalf of the American Chemical Society, we welcome this opportunity to express to Professor Bailar our sincere appreciation for his years of leadership as chemist, educator, and ACS member, and to congratulate him on attaining this milestone in a long and most distinguished career.

Sincerely yours,

B R Stanerson | Robert W. Cairns

B. R. Stanerson
Executive Secretary

Robert W. Cairns
President

CONTENTS

THE STEREOCHEMISTRY OF FIVE- AND SIX-MEMBERED CHELATE RINGS— STEREOCHEMICAL ANALYSIS OF A MACROCYCLIC COMPLEX CONTAINING SIX ASYMMETRIC CENTERS

L. G. Warner and D. H. Busch

The Ohio State University
Columbus, Ohio

INTRODUCTION

The two sets of isomers, Ni(1,4-CTH)$^{2+}$ (5,7,7,12,12,14-hexamethyl-1,4,8,11-tetraazacyclotetradecanenickel(II), Structure (I) and Ni(1,7-CTH)$^{2+}$ (5,5,7,12,12,14-hexamethyl-1,4,8,11-tetraazacyclotetradecanenickel(II), Structure (II), are distinguishable primarily on the basis of methyl group positions.

```
   H3C   CH2   CH3              H3C   CH2   CH3
H3C—C       C—H             H3C—C       C—H
   H |       | H               H |       | H
 H2C—N       N—CH2          H2C—N       N—CH2
        Ni2+                        Ni2+
 H2C—N       N—CH2          H2C—N       N—CH2
   H |       | H               H |       | H
H3C—C       C—H              H—C       C—CH3
    CH3  CH2  CH3               CH3  CH2  CH3
          I                          II
```

Within each set are nonintercovertible diastereoisomers which result from the presence of two asymmetric carbons: Ni(*meso-* or *rac*-1,7-CTH)$^{2+}$ and Ni(*meso-* or *rac*-1,4-CTH)$^{2+}$. On the basis of symmetry and steric arguments Curtis[1-3] assumed that in the Ni(*meso*-1,7-CTH)$^{2+}$ and Ni(*rac*-1,4-CTH)$^{2+}$ isomers the macrocyclic ligand coordinates only in a planar fashion, while in Ni(*rac*-1,7-CTH)$^{2+}$ and Ni(*meso*-1,4-CTH)$^{2+}$ it can coordinate in both planar and folded forms. This has been confirmed *via* x-ray crystallographic analysis of Ni(II) complexes containing C-*meso*-1,7-CTH and C-*rac*-1,7-CTH.[1,3,4] Curtis[1,4,5] has reported that the diastereoisomeric sets Ni(*rac*-1,7-CTH)$^{2+}$ and Ni(*meso*-1,4-CTH)$^{2+}$ each contain an α and β-isomer.

Very recently, detailed studies[5] on the stereoisomerism of the $Ni(CTH)^{2+}$ system have confirmed the existence of some eight isomers and the results of physical measurements assist in the assignment of structures to these isomers. A number of significant stereochemical relationships can be deduced from the relative stabilities of the various possible isomeric structures (40 in all) since the complicated array of possible structures include many stereochemical features, e.g., ring conformations, absolute and relative configurations of asymmetric centers, and axial and equatorial distinctions between bulky groups.

STEREOCHEMICAL RELATIONSHIPS

There are five basic forms for the *trans* or planar metal complexes which contain 1,4,8,11-tetraazacyclotetradecane, cyclam, or the hexamethyl analog, CTH. These forms arise due to the combinations of the configurations of the four asymmetric coordinated secondary amine nitrogens present in these systems. Each secondary amine proton is approximately in an axial position and either resides above or below the plane containing the nickel and four nitrogen atoms. Three of these forms are nonenantiomeric (I, II, and III, Fig. 1) and two are enantiomeric (IV and V, Fig. 1). The numerals 3 and 2 in Fig. 1 represent the three and two carbon chains, respectively, which span adjacent coordinated secondary amines. The + and − notation represents secondary amine protons above and below the NiN_4 plane, respectively.

Assuming chelate ring conformations, which are analogous to those predicted to be the most strain free for $trans\text{-}Co(cyclam)X_2^{+n}$ [6] there are 8 *meso* and 32 *racemic* diastereoisomers for the $Ni(CTH)^{2+}$ system. These arise from the asymmetry of the four nitrogen donor atoms and from the asymmetry of two carbon atoms present in the macrocyclic ligand. The asymmetric carbons occur at positions 7 and 14 in 1,7-CTH (Structure II), and at positions 5 and 14 in 1,4-CTH (Structure I). Consideration of absolute and relative configurations is useful in the identification of various

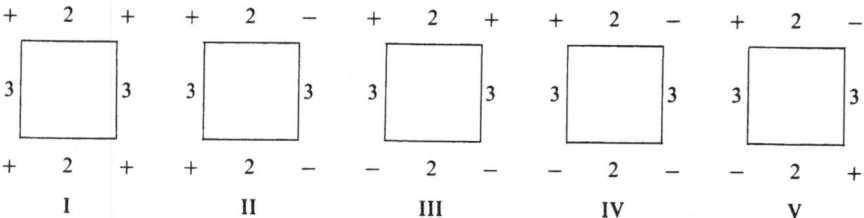

Fig. 1. Basic Sets of Asymmetric Nitrogen Configurations in $M(cyclam)^{n+}$ and $M(CTH)^{n+}$ Complexes.

asymmetric relationships for the two pairs of asymmetric coordinated secondary amine nitrogens. The two nitrogens adjacent to the geminal carbons (Figs. 2–5) are designated as nitrogen set N_1 and the two nitrogen atoms adjacent to the two asymmetric carbons are designated as nitrogen set N_2. If both N_1 nitrogens are of the same R or S configuration[7] then this set is N_1-*racemic*. If one nitrogen is R and one is S, then the set is N_1-*meso*. The same applies to the N_2 set.

The occurrence of a center of symmetry i, reflection plane σ, or improper rotation axis S_n in the point group describing a given $Ni(CTH)^{2+}$ isomer identifies the structure as a *meso*-diastereoisomer. The absence of these elements of symmetry affords the possibility of racemic forms. The *meso*-diastereoisomers of the $Ni(CTH)^{2+}$ system belong either to the C_i or the C_s point group, which have, in addition to identity, E, only a center of inversion or a plane of reflection, respectively. The racemic forms belong either to the C_2 point group, whose only element of symmetry besides identity is a two fold rotation axis, or to the C_1 point group which only contains the identity element, E.[10]

Possible Diastereoisomers of $Ni(1,7-CTH)^{2+}$

There are 10 possible diastereoisomers for the $Ni(meso$-$1,7$-$CTH)^{2+}$ isomer; four are meso and six are racemic—idealized structures m-1.7-1 through 10, Fig. 2. Structures m-1,7-1 and 2, and m-1,7-3 and 4 are of the basic forms II and III (Fig. 1), respectively, and are *meso*-diastereoisomers of C_i symmetry. Structures m-1,7-1 and 3 have the methyl groups attached to the asymmetric carbons in equatorial positions, while in isomers m-1,7-2 and 4 these methyl groups are in axial positions.

The basic form I (Fig. 1) yields a single racemate of C_1 symmetry in the case of the C-*meso*-1,7-CTH ligand, m-1,7-5. The basic forms IV and V (Fig. 1) are in themselves enantiomeric in contrast to the previous three forms; therefore, they yield only racemic diastereoisomers. Basic form IV has a unique, coordinated secondary amine. Considering only the six-membered chelate ring containing this unique nitrogen, it is readily seen that the asymmetric carbon can have either the R or S configuration and be either across the six-membered chelate ring from, or adjacent to, the unique nitrogen. This gives rise to four possible structures within this form (IV), m-1,7-6 through 9. In structures m-1,7-6 and 7 the asymmetric carbon is across the chelate ring from the unique nitrogen. Within each of these structures, one methyl group attached to the asymmetric carbon is axial and one is equatorial. In structures m-1,7-8 and 9 the asymmetric carbons are adjacent to the unique nitrogen. In the former structure the methyl groups attached to the asymmetric carbons are axial and in the latter they are equatorial. Each of these four isomers has C_1 symmetry. The structure

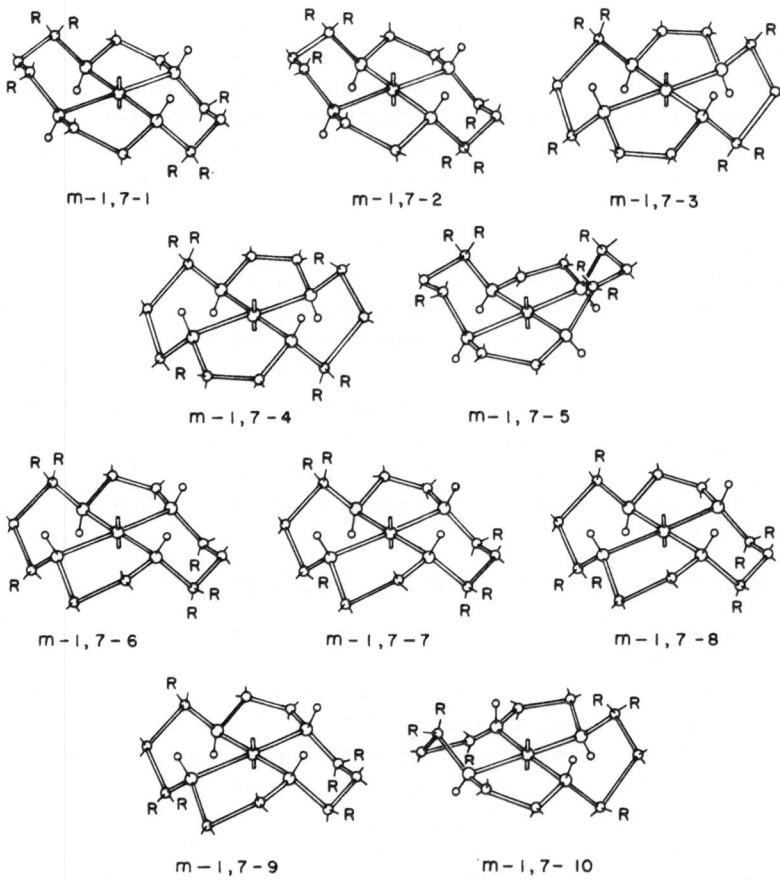

Fig. 2. Idealized Structures of the Stereoisomers of Ni(*meso*-1,7-CTH)$^{2+}$.

m-1,7-10 (Fig. 2) is of the basic form V (Fig. 1). Within this structure, which has C_1 symmetry, one of the methyl groups attached to an asymmetric carbon is axial and one is equatorial.

There are also 10 possible diastereoisomers of the planar Ni(*rac*-1,7-CTH)$^{2+}$ system (Fig. 3). All are necessarily racemates. Structures *r*-1,7-1 and 2 are of the basic form I and are of C_2 symmetry. The methyl groups attached to the asymmetric carbons are both equatorial in *r*-1,7-1 and both axial in *r*-1,7-2. Structures *r*-1,7-3 and 4 are of the basic forms II and III, respectively, and have C_1 symmetry. In each of these structures one methyl group attached to an asymmetric carbon is axial and one is equatorial.

All of the structures *r*-1,7-5 through 8 have the basic form IV and are of C_1 symmetry. When the asymmetric carbon is across the six-membered

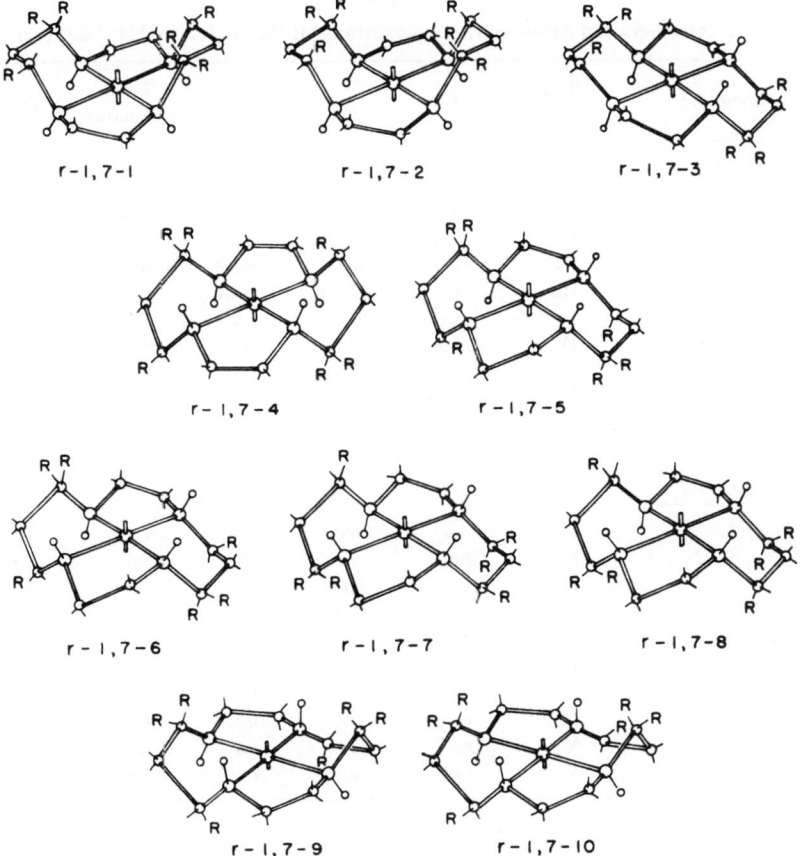

Fig. 3. Idealized Structures of the Stereoisomers of Ni(rac-1,7-CTH)$^{2+}$.

chelate ring from the unique nitrogen, the methyl groups attached to the asymmetric carbons are both axial in r-1,7-5 and both equatorial in r-1,7-6. In structures r-1,7-7 and 8, the asymmetric carbon is adjacent to the unique nitrogen and one methyl group attached to the asymmetric carbons is axial and one equatorial.

The structures r-1,7-9 and 10 are of the basic form V and have C$_2$ symmetry. The methyl groups attached the asymmetric carbons are both axial in r-1,7-9 and both equatorial in r-1,7-10.

Table 1 summarizes the point groups and asymmetric designations for all the diastereoisomeric structures possible for the Ni(1,7-CTH)$^{2+}$ system. There are several important observations that should be made with respect to these and related structures. The conformations of the two, six-membered

Table 1
Diastereoisomers and Asymmetric Designations for the Ni(1,7-CTH)$^{2+}$ System

Structure	Basic form[a]	C	N_1	N_2	Total designation	Symmetry
Ni(*meso*-1,7-CTH)$^{2+}$						
m-1,7-1[b]	II	*meso*	*meso*	*meso*	*meso*	C_i
2	II	*meso*	*meso*	*meso*	*meso*	C_i
3	III	*meso*	*meso*	*meso*	*meso*	C_i
4	III	*meso*	*meso*	*meso*	*meso*	C_i
5	I	*meso*	*rac*	*rac*	*rac*	C_1
6	IV	*meso*	*meso*	*rac*	*rac*	C_1
7	IV	*meso*	*meso*	*rac*	*rac*	C_1
8	IV	*meso*	*rac*	*meso*	*rac*	C_1
9	IV	*meso*	*rac*	*meso*	*rac*	C_1
10	V	*meso*	*rac*	*rac*	*rac*	C_1
Ni(*rac*-1,7-CTH)$^{2+}$						
r-1,7-1[c]	I	*rac*	*rac*	*rac*	*rac*	C_2
2	I	*rac*	*rac*	*rac*	*rac*	C_2
3	II	*rac*	*meso*	*meso*	*rac*	C_1
4	III	*rac*	*meso*	*meso*	*rac*	C_1
5	IV	*rac*	*meso*	*rac*	*rac*	C_1
6	IV	*rac*	*meso*	*rac*	*rac*	C_1
7	IV	*rac*	*rac*	*meso*	*rac*	C_1
8	IV	*rac*	*rac*	*meso*	*rac*	C_1
9	V	*rac*	*rac*	*rac*	*rac*	C_2
10	V	*rac*	*rac*	*rac*	*rac*	C_2

[a]Fig. 1.
[b]Fig. 2.
[c]Fig. 3.

chelate rings in a given structure are identical for each of the basic forms except for form IV. One six-membered chelate ring of this latter form is in a chair conformation (Structure III), and the other is in a twist conformation (Structure IV). Both six-membered chelate rings in structures of basic forms III and V have the twist conformation, while both six-membered chelate

Structure III   Structure IV   Structure V rings in structures of the basic forms I and II have chair conformations. A boat conformation (Structure V) is also possible as an alternate to the chair conformation, however, the chair form is preferred[8] and has been found to occur in $Ni(cyclam)Cl_2$.[11]

Every racemic structure having C_2 symmetry and every meso structure of C_i symmetry has both of those methyl groups attached to the asymmetric carbons either axial or equatorial. Both methyl groups attached to the asymmetric carbons in structures having C_1 symmetry can be axial or equatorial, or one may be axial and one equatorial.

Possible Diastereoisomers of Ni(1,4-CTH)$^{+2}$

There are 10 racemic structures possible for the $Ni(rac\text{-}1,4\text{-}CTH)^{2+}$ system and these are shown in Fig. 4. Structures r-1,4-1 and 4 have the basic

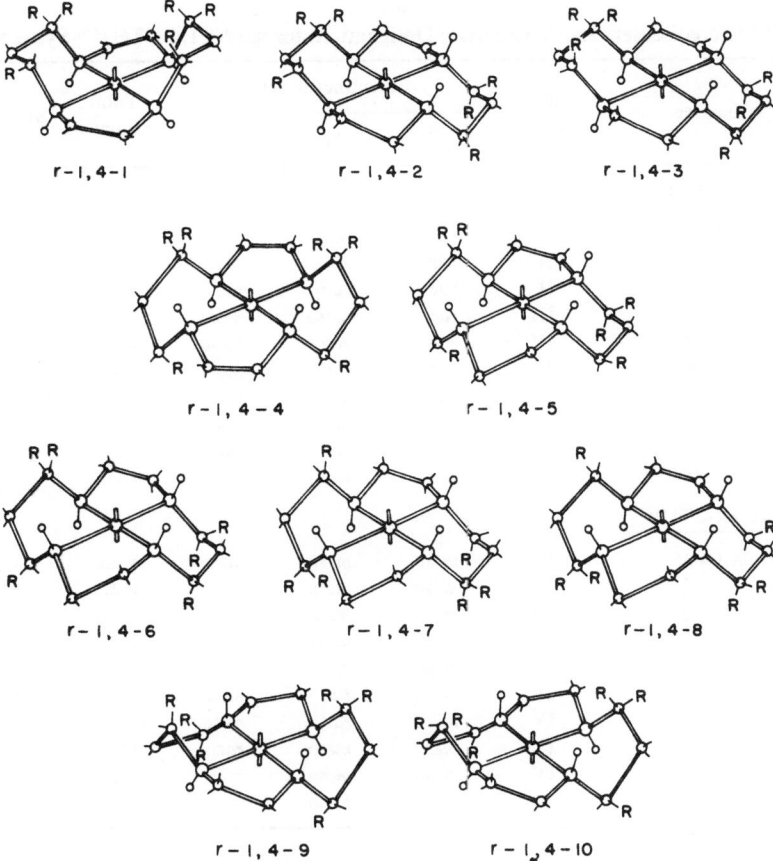

r-1,4-1   r-1,4-2   r-1,4-3

r-1,4-4   r-1,4-5

r-1,4-6   r-1,4-7   r-1,4-8

r-1,4-9   r-1,4-10

Fig. 4. Idealized Structures of the Stereoisomers of $Ni(rac\text{-}1,4\text{-}CTH)^{2+}$.

forms I and III (Fig. 1), respectively, and are of C_1 symmetry, while structures r-1,4-2 and 3 are of the basic form II and have C_2 symmetry. Structures r-1,4-5 through 8 are of the basic form IV and have C_1 symmetry, while structures r-1,4-9 and 10 are of the basic form V and have C_2 symmetry.

Within the Ni($meso$-1,4-CTH)$^{2+}$ system there are possible four meso, m-1,4-1 through 4 (Fig. 5) and six racemic, m-1,4-5 through 10 (Fig. 5), structures. The meso structures are of the basic forms I and III and have C_s symmetry. The racemic structures have the basic forms II, IV, and V and are of C_1 symmetry.

Table 2 summarizes all of the diastereoisomers, their symmetries, and asymmetric designations for the Ni(1,4-CTH)$^{2+}$ system. All of the $meso$-diastereoisomers have C_s symmetry and have the methyl groups attached to the asymmetric carbons either both axially or both equatorially oriented.

Table 2

Diastereoisomers and Asymmetric Designations for the Ni(1,4-CTH)$^{2+}$ System

Isomer	Basic form[a]	Asymmetric sets			Total designation	Symmetry
		C	N_1	N_2		
Ni(rac-1,4-CTH)$^{2+}$						
r-1,4-1[b]	I	rac	meso	meso	rac	C_1
2	II	rac	rac	rac	rac	C_2
3	II	rac	rac	rac	rac	C_2
4	III	rac	meso	meso	rac	C_1
5	IV	rac	rac	meso	rac	C_1
6	IV	rac	rac	meso	rac	C_1
7	IV	rac	meso	rac	rac	C_1
8	IV	rac	meso	rac	rac	C_1
9	V	rac	rac	rac	rac	C_2
10	V	rac	rac	rac	rac	C_2
Ni($meso$-1,4-CTH)$^{2+}$						
m-1,4-1[c]	I	meso	meso	meso	meso	C_s
2	I	meso	meso	meso	meso	C_s
3	III	meso	meso	meso	meso	C_s
4	III	meso	meso	meso	meso	C_s
5	II	meso	rac	rac	meso	C_1
6	IV	meso	rac	meso	rac	C_1
7	IV	meso	rac	meso	rac	C_1
8	IV	meso	meso	rac	rac	C_1
9	IV	meso	meso	rac	rac	C_1
10	V	meso	meso	rac	rac	C_1

[a]Fig. 1.
[b]Fig. 4.
[c]Fig. 5.

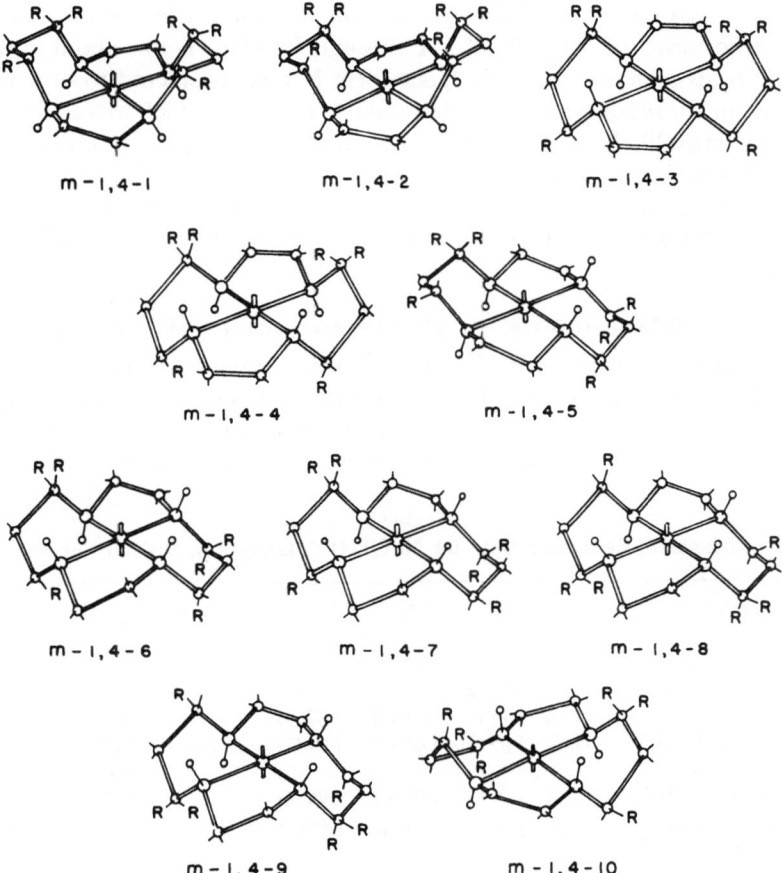

Fig. 5. Idealized Structures of the Stereoisomers of Ni(*meso*-1,4-CTH)$^{2+}$.

All of the other observations made above for the Ni(1,7-CTH)$^{2+}$ system with regard to the axial and equatorial nature of the methyl groups attached to the asymmetric carbons as a function of symmetry apply also to the Ni(1,4-CTH)$^{2+}$ system.

Examination of Drieding stereo-models of these diastereoisomers suggests that, in all those cases where methyl groups attached to the asymmetric carbons are axial and on the same side of the NiN$_4$ plane as the axial geminal methyl group associated with the same chelate ring, significant steric interactions occur between these axial groups. This eliminates the following structures from consideration when structural assignments are made for the known isomers of Ni(CTH)$^{2+}$: *m*-1,7-2,5,6,8 (Fig. 2); *r*-1,7-2,3,5,8, (Fig. 3); *r*-1,4-1,3,6,7 (Fig. 4); and *m*-1,4-2,5,6,9 (Fig. 5). It has been estimated that the amount of strain and the nonbonding interactions in the four basic forms of M(cyclam)$^{n+}$, I through IV (Fig. 1) are about equal[6] for the *trans* planar complexes. Basic form V was considered to be highly strained in a *trans* or planar complex and it is this form which was concluded to be the most likely to fold and give *cis*, strain-free complexes. Indeed, only structures having a *trans* pair of secondary amines of such configuration that their protons are on the same side of the coordination plane are expected to be capable of folding. In addition to basic form V, this includes only basic form IV.

Structural Assignments for Diastereoisomers of Ni(CTH)$^{2+}$

Structural assignments for the diastereoisomers isolated for the Ni(CTH)$^{2+}$ system rest heavily upon proton magnetic resonance spectra of the diperchlorate derivatives.[5] The methyl proton resonance assignments are listed in Table 3. The pmr spectra were obtained in D_6-DMSO (*vs.*

Table 3
Proton Magnetic Resonance Data for Ni(CTH)$^{2+}$ Isomers in DMSO Solution[a,b]

Isomer	Me-I	Me-II	Me-III[a]	Me-III[b]	Δ(II-III)	^{J}Me-I[c]
Ni(*meso*-1.7-CTH)$^{2+}$	1.08 (3)	1.15 (3)	1.15 (6)	1.72 (6)	—	0.57	4.3
β-Ni(*rac*-1,7-CTH)$^{2+}$	0.97 (3)	1.07 (3)	1.11 (6)	2.12 (6)	—	1.01	6.0
Ni(*rac*-1,4-CTH)$^{2+}$	1.01 (3)	1.11 (3)	1.11 (6)	1.87 (6)	—	0.76	6.1
β-Ni(*meso*-1,4-CTH)$^{2+}$	1.00 (3)	1.09 (3)	1.06 (6)	1.94 (3)	2.24 (3)	0.88	5.4
						1.18	
γ-Ni(*meso*-1,4-CTH)$^{2+}$	0.99 (3)	1.07 (3)	1.07 (6)	1.96 (6)	—	0.89	3.7
δ-Ni(*meso*-1,4-CTH)$^{2+}$	1.00 (3)	1.11 (3)	1.07 (6)	1.87 (3)	1.97 (3)	0.80	4.0
						0.90	

[a]All chemical shifts in ppm downfield from internal TMS.
[b]Estimated relative intensities in parenthesis.
[c]Units in cps.

Internal TMS). Solutions of α-Ni(*rac*-1,7-CTH)(ClO$_4$)$_2$ and Ni(*meso*-1,4-CTH)(ClO$_4$)$_2$ are paramagnetic; therefore, their spectra were not obtained. The most prominent features of these pmr spectra are the three to five sharp, intense bands which appear between 0.95 and 2.20 ppm. These are due to the six methyl groups within each isomer. The orientations of the geminal and asymmetric carbon atoms in the six-membered chelate rings in the twist (IV) or in the chair conformation (III) produces axial and equatorial distinctions between the geminal methyl groups. The methyl groups attached to the asymmetric carbons are either axial or equatorial also. A change of a six-membered chelate ring from the chair to the boat conformation results in the loss of the axial and equatorial distinctions, and no methyl groups lie over the NiN_4 plane (V). The resonances due to the methyl groups attached to the asymmetric carbons Me-I, are doublets with coupling constants of 3.7 to 6.1 cps, in the high-field portion (~ 1.1 ppm) of the pmr spectrum of each isomer.[8] In the cases of $Ni(meso\text{-}1,7\text{-}CTH)^{2+}$ and $Ni(rac\text{-}1,4\text{-}CTH)^{+2}$ the low-field component of this doublet overlaps completely with the resonance due to the equatorial geminal methyl protons, Me-II.

The remaining methyl resonances are due to the axial and equatorial geminal methyl groups Me-III and Me-II, respectively. The large chemical shift differences, Δ(II-III) (Table 3), indicate that a relatively large non-bonding effect is occurring. This effect is identical to that observed for the geminal methyl groups of the $Ni(CT)^{2+}$ system[9] and is explained on the same basis. Protons which lie over the plane of square planar d^8 systems experience a low field shift with respect to protons which lie in the plane or out away from the metal ion center.[10] Therefore, the equatorial and axial geminal methyl groups are assigned to the high- and low-field methyl resonances, respectively.

One consequence of the pmr assignments shown in Table 3 is that *both methyl groups attached to asymmetric carbons* (Me-I) *must be equatorial in all the* $Ni(CTH)^{2+}$ *isomers studied*, because no doublets ($J = 2$ to 9 cps) occur in the low-field portion of the pmr spectra. A second general conclusion is also readily drawn. In those cases where only one kind of axial geminal methyl resonance (Me-III) occurs, the two, six-membered chelate rings within a given structure must have the same conformation, and be of the basic forms I, II, III, or V (Fig. 1). When two Me-III resonances appear, the two, six-membered chelate rings within that isomer must be of different conformations; therefore, such an isomer must have basic form IV. The magnitude of the deshielding of protons over the plane of square planar d^8 complexes is a function of the radial and angular orientation of the protons with respect to the axial site of the metal ion center. Therefore, as the six-membered chelate ring conformations change, the pmr chemical shift of the axial geminal methyl groups change because their orientation over the NiN_4 plane changes.

Six-membered chelate rings in the boat conformation, as mentioned above, eliminate the axial-equatorial distinctions. As a consequence, no methyl group would lie over the NiN_4 plane and all methyl proton resonances would appear in the high-field portion of the pmr spectra. This is clearly not the case for any of the $Ni(CTH)^{2+}$ isomers studied and reported previously[5] or herein.

The following summarizes the conclusions that have been made on the basis of the methyl proton resonance assignments (Table 3) and the observations made with respect to methyl group orientations and chelate ring conformations as a function of the symmetry (Table 1 and 2) for each of the

40 possible diastereoisomeric structures of the $Ni(CTH)^{2+}$ system (Figs. 2 to 5): (a) Those isomers which exhibit only one axial methyl proton resonance, Me-III, must be *meso*-diastereoisomers of C_i or C_s symmetry or *racemic*-diastereoisomers of C_2 symmetry; (b) The isomers which exhibit two, low field Me-III singlet resonances in their pmr spectra must be *racemic*-diastereoisomers of C_1 symmetry, and be of the basic form IV; (c) All isomers contain exclusively, six-membered chelate rings of the twist and chair conformations; and (d) The methyl groups attached to the asymmetric carbons are always equatorial.

Isomers of $Ni(1,7\text{-CTH})^{2+}$

The sources, designations and interconversions of the three known isomers[5] of $Ni(1,7\text{-CTH})^{2+}$ are summarized in Fig. 6. The structures of these species can be inferred on the basis of chemical and physical properties.

The pmr spectrum requires that $Ni(meso\text{-}1,7\text{-CTH})^{2+}$ be a true *meso* isomer of C_i symmetry [see conclusion (a) and Table 1]. This accords with the preliminary results of an x-ray study on the single isomer of the copper(II) complex with *meso*-1,7-CTH, for that compound is reported to have an inversion center.[1] Of the true meso forms (four in number, Table 1), m-1,7-2 and 4 can be eliminated because the Me-I groups would be axial and this is inconsistent with the pmr [conclusion (d) above].

Structures m-1,7-1 and 3 are consistent with the pmr spectrum. Structure m-1, 7-3 is of the basic form III and have two twist, six-membered chelate rings and two, eclipsed five-membered chelate rings. Eclipsed five-membered chelate rings are not common in metal chelate systems.[11] The complex $Ni(cyclam)Cl_2$ is directly analogous to $Ni(CTH)^{2+}$ and its structure is identical with structures m-1,7-1 or 2 minus (the methyl groups). Therefore, structure m-1,7-1 is assigned to the Ni $(meso\text{-}1,7\text{-CTH})^{2+}$ isomer.

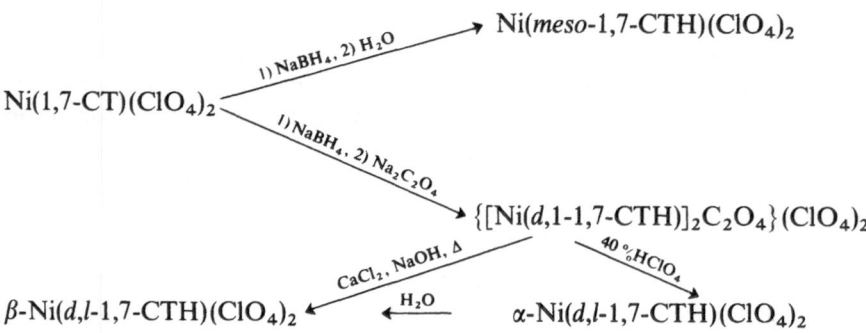

Fig. 6. Formation and Interconversion of the Isomers of $Ni(1,7\text{-CTH})^{2+}$.

The α-Ni(rac-1,7-CTH)(ClO$_4$)$_2$ isomer is paramagnetic in DMSO and is only slightly soluble in other solvents; its pmr spectrum cannot be obtained. The structure of this isomer is inferred from the known x-ray structure of [Ni(rac-1,7-CTH)(CH$_3$CO$_2$)]-(ClO$_4$). The macrocyclic ligand is coordinated in its cis-folded form; acetate functions as a bidentate ligand. This structure is very similar to that of the basic form V (Fig. 1) which was predicted by Bosnich et $al.$ to be the most likely to form cis-Co(cyclam)X$_2^{n+}$ complexes.[6] It is presumed that the ligand form in the oxalate bridged dimer {[Ni(rac-1,7-CTH)]$_2$(C$_2$O$_4$)}(ClO$_4$)$_2$, which is the parent to α-Ni(rac-1,7-CTH) (ClO$_4$)$_2$, is identical to that of this acetato-perchlorate derivative. Thus, α-Ni(rac-1,7-CTH)(ClO$_4$)$_2$ must also contain the same ligand form. This form, when coordinated in a planar fashion, has one pair of trans-secondary amine protons directed up and the other pair pointed down, because it is formed under conditions which require retention of the configurations of the asymmetric coordinated secondary amine nitrogens, namely, a strongly acidic medium. The α-Ni(rac-1,7-CTH)$^{2+}$ isomer has structure r-1,7-10, Fig. 3. A Drieding stereomodel of this structure strongly suggests that significant strain exists throughout the complex. Only when the ligand folds along the line which includes the nitrogens adjacent to the geminal carbons, set N$_1$, and the metal ion is this strain reduced. This would explain why this particular isomer is most readily formed under those conditions where the ligand must fold to accommodate a bidentate ligand. It also explains why in solutions under conditions which allow secondary amine proton dissociation (neutral or basic aqueous media) this isomer readily isomerizes to give an equilibrium mixture of α and β-Ni(rac-1,7-CTH)$^{2+}$, where the β isomer represents more than 90 % of the total.

On the basis of observation (1), the pmr spectrum of the β-Ni(rac-1,7-CTH)(ClO$_4$)$_2$ is consistent only with structures r-1,7-1 or 10 (Fig. 3), for these are racemic and have C$_2$ symmetry (Table 1). Since structure r-1,7-10 has already been assigned to the α-isomer, structure r-1,7-1 is assigned to this β-isomer.

The five-membered chelate rings within this structure are in the presumably less stable eclipsed conformation. In spite of this fact, structure r-1,7-1 is the thermodynamically most stable for planar Ni(rac-1,7-CTH)$^{2+}$. This suggests that the relative stabilities of the six-membered chelate rings are more important than those of the five-membered chelate rings, and that the chair conformation of six-membered chelate rings is more stable than the twist or boat conformations.

Isomerization from the α- to β-Ni(rac-1,7-CTH)$^{2+}$ isomer involves the dissociation of the secondary amine protons from and configurational inversion of the two nitrogens of set N$_1$ in a, presumably, step-wise process. Structure r-1,7-6 (Fig. 3) is a necessary intermediate. There is no evidence that this intermediate ever builds up to significant concentrations in solution.

Structures of the Ni(1,4-CTH)$^{2+}$ isomers

The sources and interconversion of the isomers of Ni(1,4-CTH)$^{2+}$ are summarized in Fig. 7.

Original identification of the ligand diastereoisomers of the 1,4-CTH system was made on the basis that one isomer forms only planar nickel(II) complexes, Ni(rac-1,4-CTH)$^{2+}$, while the other forms both planar complexes and cis six-coordinated derivatives containing bidentate ligands and the folded form of the macrocyclic ligand.[3] Evidence, supporting these assignments, has been found in the partial resolution of Ni(rac-1,4-CTH)(BF$_4$)$_2$ into enantiomers by a chromatographic technique which utilizes 10^{-2}N HBF$_4$ solutions and starch as the optically active adsorbent. The specific rotations at 492 mμ observed for the first and last fractions after one pass through the column were $+10°$ and $-6°$, respectively. Resolution was increased by recrystallization of partially resolved samples from acidic media (racemate least soluble) and by passage of partially resolved samples through the starch column a second time.

Figure 8 shows the absorption, ORD, and CD spectra for the most highly resolved sample of Ni(rac-1,4-CTH)$^{2+}$: $[\alpha]_{510}^{25°} = +34°$, $[\alpha]_{426}^{25°} = -26$, and $\Delta\varepsilon_{464} = +0.14$. This partially resolved sample was made basic, pH about 10, with sodium hydroxide and allowed to stand for 12 hr. After acidification with HBF$_4$ to a pH of about 2, the absorbance, ORD and CD spectra were again obtained. The ORD and CD spectra were nearly identical with those originally obtained, $[\alpha]_{510}^{25°} = +31$, $[\alpha]_{426}^{25°} = -28°$, and $\Delta\varepsilon_{464} = +0.14$. Thus, no mutarotation or racemization is induced by a strongly basic medium. This indicates that the asymmetry which gives rise to these enantiomers of this isomer is principally due to the asymmetric carbons, and not due to the four asymmetric coordinated secondary nitrogens.

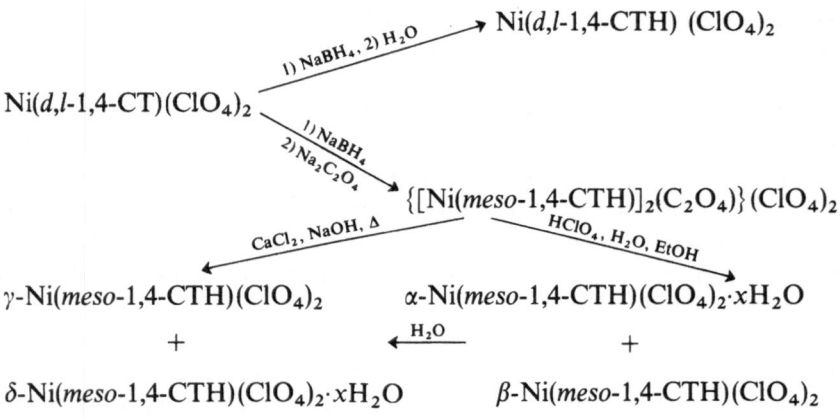

Fig. 7. Formation and Interconversion of the Isomers of Ni(1,4-CTH)$^{2+}$.

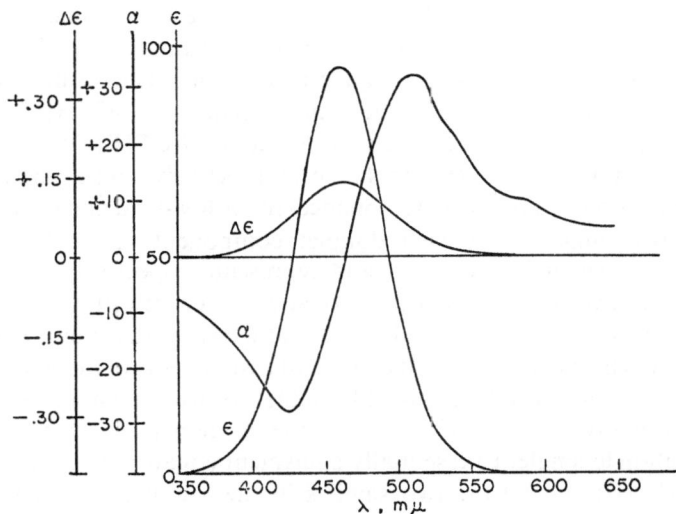

Fig. 8. The absorption (ε), optical rotatory dispersion (α), and circular dichroism ($\Delta\varepsilon$) spectra of partially resolved Ni(rac-1,4-CTH)(BF$_4$)$_2$.

Therefore, the macrocyclic ligand in this complex must be the rac-1,4-CTH diastereoisomer. Only two structures, r-1,4-2 and 10 (Fig. 4), are consistent with the pmr spectrum (Table 3) and the summary observations (a) through (d) (see above). On the basis of its similarity with the known structure of Ni(cyclam)Cl$_2$[12] and the assigned structure of Ni($meso$-1,7-CTH)$^{2+}$, (m-1,7-1 Fig. 2) the structure r-1,4-2 is assigned to Ni(rac-1,4-CTH)$^{2+}$. The assignment of this structure rather than the structure r-1,4-10 is strongly supported by the presumed greater stability of the chair conformation for the six-membered chelate ring and the usual preference for gauche or staggered rather than eclipsed, five-membered chelate rings. The symmetrical nature of the axial coordination sites in this isomer is indicated by the occurrence of only one ν(CN of NCS) in the infrared spectrum of a chloroform solution of the dithiocyanato derivative.

The α-Ni($meso$-1,4-CTH)(ClO$_4$)$_2$ isomer only forms paramagnetic five-coordinated derivatives in the solid state and in most solutions, therefore, no pmr spectra have been obtained.[3] The only experimental evidence which directly relates to the structure of this isomer is found in the fact that it is formed in the acid decomposition of an oxalate-bridged dimer. This dimer contains the ligand, $meso$-1,4-CTH, coordinated in a cis-folded manner and, by reasoning analogous to that used in the assignment of structure of the α-Ni(rac-1,7-CTH)$^{2+}$ isomer, it is concluded that α-Ni($meso$-1,4-CTH)$^{2+}$ is of the basic form V and has the structure m-1,4-10 (Fig. 5). As in the case of α-Ni(rac-1,7-CTH)$^{2+}$, α-Ni($meso$-1,4-CTH)$^{2+}$ readily isomerizes under conditions which allow amine proton exchange and amine nitrogen configurational inversion (basic or neutral aqueous solutions). The resulting equilibrium mixture contains primarily the γ and δ-Ni($meso$-1,4-CTH)$^{2+}$ isomers.

The pmr spectra of the β- and δ-Ni($meso$-1,4-CTH)(ClO$_4$)$_2$ isomers both show a doublet, $J = 5.0$ and 4.0 cps, respectively, and a singlet in the high-field portion and two sharp resonances in the low-field portion (Table 3). The relative intensities of each doublet component and of the singlet resonance at high field are $1:1:2$ and represent, respectively, 3, 3, and 6 protons. The two, sharp low-field resonances, each of estimated intensity equaling 3 protons, are separated by 18 and 6 cps for the β and δ isomers, respectively. The most logical and consistent assignments of these resonances for each isomer are: (a) the high-field doublet to two, essentially equivalent equatorial-methyl groups attached to the asymmetric carbons (Me-I); (b) the high-field singlet to essentially equivalent equatorial geminal methyl groups (Me-II); and (c) the two singlets in the low field portion; one to each of two axial geminal methyl groups attached to six-membered chelate rings of different conformations (Table 3). Only one of the 10 structures possible for Ni($meso$-1,4-CTH)$^{2+}$ is consistent with these assignments: Structure m-1,4-7 (Fig. 5), which is a racemate of C_1 symmetry and the basic form IV, and in which both the twist and chair conformations occur for the six-membered chelate rings. It is barely possible that both contain ligand configuration m-1,4-7, but that the ligand is folded in one structure.

Only two of the 10 possible structures (Fig. 5) are consistent with the pmr spectrum of γ-Ni($meso$-1,4-CTH)(ClO$_4$)$_2$: m-1,4-1 and 3; both are $meso$-diastereoisomers of C_s symmetry and of the basic forms I and III, respectively. On the basis of its similarity to the assigned structure of β-Ni(rac-1,7-CTH)$^{2+}$ (r-1,7-1, Fig. 4) and the presumed greater stability of the chair over the twist conformation of six-membered chelate rings. Structure m-1,4-1 is assigned to the γ-Ni($meso$-1,4-CTH)$^{2+}$ isomer.

ACKNOWLEDGMENT

This research was made possible by a grant from the National Science Foundation. L.G.W. expresses his gratitude for a fellowship sponsored by the Chrysler Corp.

REFERENCES

1. N. F. Curtis, *Coord. Chem. Revs.* 3: 3 (1968).
2. N. F. Curtis, *J. Chem. Soc.* 2644 (1964).
3. N. F. Curtis, *ibid.*, (C), 1979 (1967).

4. M. Bailey, private communication, 1968.
5. L. G. Warner and D. H. Busch, (in press); L. G. Warner, Ph.D. Thesis, The Ohio State University, 1968.
6. B. Bosnick, C. K. Poon, and M. L. Tobe, *Inorg. Chem.* **4**: 1102 (1965).
7. E. L. Eliel, *Stereochemistry of Carbon Compounds*, McGraw-Hill Book Co., N.Y., 1962, p. 92.
8. L. M. Jackman, *Applications of Nuclear Magnetic Resonance Spectroscopy in Organic Chemistry*, Pergamon Press, N.Y., 1959, p. 124, 129.
9. L. G. Warner, N. J. Rose, and D. H. Busch, *J. Am. Chem. Soc.* **90**:6938 (1968).
10. D. A. Buckingham and P. J. Stevens, *J. Chem. Soc.* 4583 (1964).
11. E. J. Corey and J. C. Bailar, Jr., *J. Am. Chem. Soc.* **81**: 2620 (1959).
12. B. Bosnick, R. Mason, P. Pauling, G. B. Robenison, and M. L. Tobe, *Chem. Comm.* 97 (1965).

THE OCTAHEDRAL COBALT(III) BAILAR INVERSIONS

R. D. Archer

University of Massachusetts
Amherst, Massachusetts

INTRODUCTION

Although the Bailar inversion (an octahedral complex substitution D* → L* conversion, where D* represents a right-handed screw relative to the C_2 symmetry axis of an octahedral complex ion of the type *cis*-$[Co(en)_2Cl_2]^+$, in which en = ethylenediamine, as shown in Fig. 1) was first reported by John C. Bailar, Jr. and R. W. Auten in 1934,[1] the exact mechanism of this type of reaction is still unknown. Originally, the reaction was considered to be due to the reaction of Ag_2CO_3 with the $[Co(en)_2Cl_2]^+$ ion

$$\text{D*-}[Co(en)_2Cl_2]^+ \xrightarrow{Ag_2CO_3} [Co(en)_2(CO_3)]^+$$
$$\text{L*} > \text{D*}$$

(1)

with the relative configurations determined by the reverse reactions with HCl and a consideration of other reactions which appeared to give retention.[1,2] Mathieu[3] verified the chemical results with optical rotatory dispersion and circular dichroism studies. Later, Bailar and McReynolds[4]

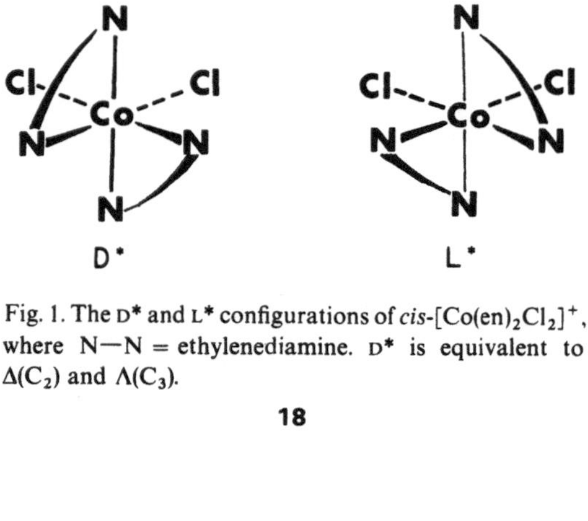

Fig. 1. The D* and L* configurations of *cis*-$[Co(en)_2Cl_2]^+$, where N—N = ethylenediamine. D* is equivalent to $\Delta(C_2)$ and $\Lambda(C_3)$.

18

showed that the silver ion is unnecessary, i.e.,

$$\text{D*-}[Co(en)_2Cl_2]^+ \xrightarrow[0°]{K_2CO_3,\ aqueous\ paste} [Co(en)_2CO_3]^+$$

$$L^* > D^* \tag{2}$$

Even so, more than two decades later when Dwyer, Sargeson, and Reid[5] showed that hydroxide rather than carbonate ion actually causes the inversion, the Australian group obtained inversion only with added silver ion.

$$\text{D*-}[Co(en)_2Cl_2]^+ \xrightarrow{Ag^+, OH^-} \xrightarrow{H^+} \xrightarrow{HCO_3^-} [Co(en)_2CO_3]^+$$

$$L^* > D^* \tag{3}$$

but

$$\text{D*-}[Co(en)_2Cl_2]^+ \xrightarrow{OH^-} \xrightarrow{H^+} \xrightarrow{HCO_3^-} [Co(en)_2CO_3]^+$$

$$D^* > L^* \tag{4}$$

Quenching the silver-catalyzed base hydrolysis reaction with acid and isolating the reaction products as dinitro complexes also shows inversion and considerable isomerism to the *trans* species

$$\text{D*-}[Co(en)_2Cl_2]^+ \xrightarrow{Ag^+, OH^-} \xrightarrow{H^+} \xrightarrow{NO_2^-} [Co(en)_2(NO_2)_2]^+$$

$$trans > L^* > D^* \tag{5}$$

The same investigation[5] noted that reactions between the hydroxo complexes with acid and either bicarbonate or nitrite go with a predominant retention of configuration. Furthermore, the reaction

$$\text{D*-}[Co(en)_2(NO_2)(OH)]^+ \xrightarrow{Ag^+, OH^-} \xrightarrow{H^+} \xrightarrow{NO_2^-} [Co(en)_2(NO_2)_2]^+$$

$$\tag{6}$$

also goes with retention of configuration, so the basic Bailar inversion reaction appeared to be

$$\text{D*-}[Co(en)_2Cl_2]^+ \xrightarrow{Ag^+, OH^-} L\text{*-}[Co(en)_2Cl(OH)]^+ \tag{7}$$

The study was also consistent with the results of Chan and Tobe,[6] who had shown that neither step of the two-step base hydrolysis reaction

$$\text{D*-}[Co(en)_2Cl_2]^+ \xrightarrow{OH^-} [Co(en)_2Cl(OH)]^+$$

$$trans > D^* > L^* \text{ in dilute solution} \tag{8}$$

and

$$\text{D*-}[Co(en)_2Cl(OH)]^+ \xrightarrow{OH^-} [Co(en)_2(OH)_2]^+$$

$$D^* > L^* \text{ in dilute solution} \tag{9}$$

involves a predominance of D* to L* inversion. However, the first step does involve considerable *cis* to *trans* isomerization.

Subsequently, Boucher, Kyuno, and Bailar[7] showed once again that the silver ion is unnecessary if the hydroxide and complex ion concentrations are large. The second hydrolysis step (equation 9) gives retention regardless of hydroxide, silver, or complex ion concentrations.[7] These results are consistent with the inversion occurring in the first step, i.e.,

$$\text{D*-}[Co(en)_2Cl_2]^+ \xrightarrow{\text{OH}^-} [Co(en)_2Cl(OH)]^+$$

$$\text{L* > D* in concentrated solution} \tag{10}$$

Increased inversion with an added silver ion was also observed, but the amounts of *trans* products were not determined. Detailed interpretation of the above inversion reactions is difficult because ion-exchange experiments have indicated[8] that incomplete reaction occurs under most of the silver-free reaction conditions. In other words, the increasing negative rotations observed with increasing base concentrations are due to two factors, less starting material remaining at the time of acidification and more inverted product. Therefore, a study was embarked upon to determine the stereochemistry and reaction parameters associated with this inversion reaction, making use of both ion-exchange techniques and computer analysis of the reaction products. But before discussing the base hydrolysis results with the D*-$[Co(en)_2Cl_2]^+$ ion, a look at some related inversions is appropriate.

Most of the inversion reactions related to the above Bailar inversion have also been studied either by Bailar, his students, and his students' students, or by the capable Australian chemists following in the footsteps of the late Frank Dwyer. Bailar has reviewed the contributions made at the University of Illinois at a Dwyer memorial lecture.[9]

Closely related inversions were observed by Bailar and Peppard[10] for the bromochloro and dibromo ions of the $[Co(en)_2XY]^+$ series. The rotations observed for the bromochloro reaction were similar to those reported earlier for the dichloro complex, and the one rotation reported for the dibromo complex ion reaction was appreciably more negative than any observed for the other starting materials. Insufficient details are available to draw definite conclusions for these reactions or for the corresponding inversion observed for optically active $[Co([-]pn)_2Cl_2]^+$,[4] where [-]pn = 1,2-diaminopropane, specifically the isomer which rotates 589 nm light to the left. As discussed later in this paper, definitive results for this reaction should be most informative.

The reaction between D*-$[Co(en)_2Cl_2]^+$ and ammonia also shows a Bailar inversion.[11] Correlation of the appropriate optical rotatory dispersion curves and the temperature dependence of the optical rotatory dispersion curves of the resultant products indicate that the inversion predominates at low temperature, but at room temperature and above retention exceeds inversion.[12] Furthermore, the second step gives retention at all temperatures;[12,13] hence, inversion apparently occurs in the first reaction step. Isomerization information through spectroscopic investigations[12,14] has been interpreted to mean that

$$D^*-[Co(en)_2Cl_2]^+ \xrightarrow[\text{low temp.}]{NH_3} [Co(en)_2Cl(NH_3)]^{2+} \xrightarrow{NH_3}$$

$$\text{trans} > L^* > D^* \tag{11}$$

$$[Co(en)_2(NH_3)_2]^{3+}$$

$$\text{trans} > L^* > D^*$$

Attempts to study the rates of the low-temperature ammonation reaction have been plagued by catalysis problems.[14]

No other Bailar inversions have been observed for any other complexes in which the coordination sphere is occupied by only monodentate and bidentate ligands, or for complexes with other metal ions. On the other hand, a sizable number of new $D^* \rightarrow L^*$ inversions have been observed which appear to be related to the steric restraints of polydentate ligands.

The base hydrolysis stereochemistry for D^*-α-cis- and L^*-β-cis-$[Co(trien)Cl_2]^+$ and the corresponding chlorohydroxo ions, where trien = triethylenetetramine, have been investigated by Kyuno and Boucher in collaboration with Bailar.[15,16] Whereas, the β isomers undergo base hydrolysis with complete retention of configuration, the α isomers show inversion during base hydrolysis. Furthermore, the products of the α isomer reactions were shown to be exclusively D^*-α-cis and L^*-β-cis isomers. The rotations of the pure D^*-α-$[Co(trien)(OH)_2]^+$ ion ($[\alpha]_D = +1950°$), the pure L^*-β-$[Co(trien)(OH)_2]^+$ ion ($[\alpha]_D = -700°$), the reaction

$$D^*-\alpha-[Co(trien)Cl(OH)]^+ \xrightarrow{OH^-} D^*-\alpha- \text{ and } L^*-\beta-[Co(trien)(OH)_2]^+$$
$$[\alpha]_D = +50° \tag{12}$$

and the reaction

$$D^*-\alpha-[Co(trien)Cl_2]^+ \xrightarrow{OH^-} D^*-\alpha- \text{ and } L^*-\beta-[Co(trien)(OH)_2]^+$$
$$[\alpha]_D = -400° \tag{13}$$

together indicate about 70% inversion for the second chloride removal (equation 12) and about 90% for the two steps together (equation 13). The 90% inversion observed for the overall two steps means that about 10% of the original dichloro species reacts via the path: D^*-α-dichloro to D^*-α-chlorohydroxo to D^*-α-dihydroxo, because no trans products were found and complete retention was found for the base hydrolysis of the L*-β-chloro-hydroxo ion. From the stereochemistry of the second step (equation 12) it follows that about 20% goes from the D*-α-chlorohydroxo intermediate to the L*-β-dihydroxo isomer as the 10% noted above follows the path from the D*-α-chlorohydroxo species to the D*-α-dihydroxo complex. The other 70% of inverted L*-β-dihydroxo product from the dichloro hydrolysis appears to have reacted via the route: D*-α-dichloro to L*-β-chlorohydroxo to L*-β-dihydroxo. The total stereochemistry for both α-cis isomers is shown in Fig. 2.

The cobalt(III) trien base hydrolysis inversion needs further elucidation with complete optical rotatory dispersion or circular dichroism curves, in order to determine whether the secondary nitrogen about which the rotation takes place during the α to β inversion is also inverted. The synthetic method of preparing β-cis isomers from α-cis isomers[17]

$$\alpha\text{-}[Co(trien)Cl_2]^+ \xrightarrow{Li_2CO_3} \beta\text{-}[Co(trien)(CO_3)]^+ \qquad (14)$$

gives products with the same nitrogen conformation as the α-starting materials (RR and SS), but at elevated temperatures the other β-conformers (SR and RS) are converted to the observed products.[18] Therefore, at present no definite conclusions can be drawn regarding the ring conformations of the direct base hydrolysis reaction.

Optical inversions also have been obtained for the reaction of D*-α-[Co(trien)Cl_2]Cl with liquid ammonia[19,20] and with ethylenediamine.[20] No inversion was observed for the reaction between the L*-β-[Co(trien)Cl_2]Cl

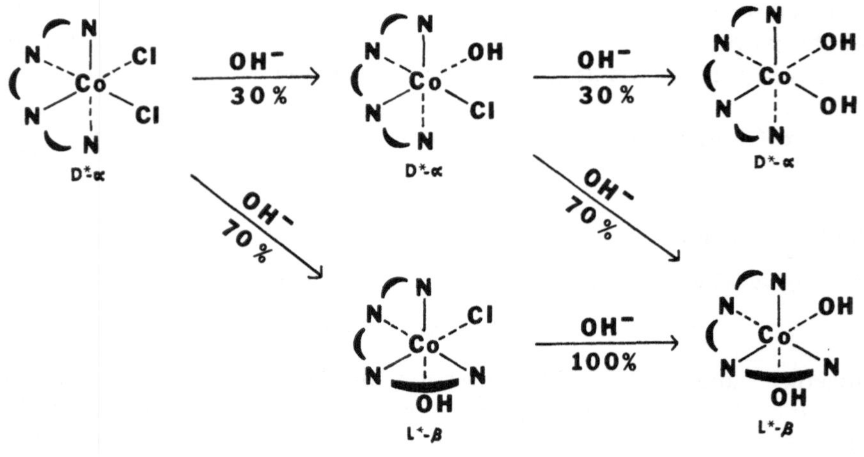

Fig. 2. The base hydrolysis stereochemistry of the D*-α-[Co(trien)Cl_2]^+ ion, where N—N—N—N = triethylenetetramine.

complex and ethylenediamine nor for the reactions between 1,10-phenan-throline or gaseous ammonia and the D*-α-dichloro ion.[20] The dibromo cations give results similar to the dichloro ion.[20] Whether α-[Co(trien)Cl-$(NH_3)]^{2+}$ gives retention or inversion is unknown—so the reaction route of the ammonia inversion is also unknown.

Closely related to the trien inversion are the interesting inversions observed by Asperger and Liu[21] during the addition of amino acids to the D*-cis-α-[Co(L,L-α,α'-dimethyltrien)Cl$_2$]$^+$ ion. Aquation to the correspond-ing aquochloro ion followed by addition of the amino acid gives a pre-dominance of inverted L*-β-product. The addition of the amino acid to the dichloro ion without prior aquation gives a retention product. Unfortunately, the intermediate steps in the inversion reaction are not known.

Surprisingly, no α-cis to β-cis conversions occur for $[Co(NSSN)CO_3]^+$ species in the presence of base (where NSSN = $NH_2CH_2CH_2SCH_2CH_2$-$SCH_2CH_2NH_2$ or $NH_2CH_2CH_2SCH(CH_3)CH_2SCH_2CH_2NH_2$).[22] Inver-sions similar to the trien series could occur since the β-cis isomer for any simple one ring movement of an α-cis isomer has the inverted configuration. (See Figs. 3 and 4, Ref. 15, or the general substitution reaction scheme[23] for more details.)

A reaction which involves the change of two terminal ends of a sexa-dentate ligand[24] has been discussed in terms of an inversion by MacDermott and Sargeson.[25,26] Undoubtedly, this inversion is a multistep reaction, but as in the cases of the trien and related multidentate inversions, the backbone and related steric factors contribute to the product stereochemistry.

The stereochemistry of the reaction of ethylenediamine with [Co-(PDTA)]$^-$ and [Co(EDTA)]$^-$, where PDTA and EDTA are propylene-diaminetetraacetate and ethylenediaminetetraacetate, respectively, illus-trates the problem of deducing what is meant by an inversion of configura-tion in multistep processes. Busch, Swaminatham, and Cooke [27] originally interpreted the reaction with [Co(PDTA)]$^-$ as an inversion reaction and that with [Co(EDTA)]$^-$ as primarily (2/3) retention. Since the same absolute configurations for these complexes had been subsequently considered of opposite handedness by MacDermott and Sargeson[25,26] and by Douglas and coworkers,[28,29] Basolo and Pearson[30] have discussed the same reactions as retention and 2/3 inversion, respectively.

RESULTS AND DISCUSSIONS

In order to more fully elucidate the Bailar inversion mechanism before considering other systems which exhibit sterochemical changes during substitution, Elaine Dittmar, one of my students, used a combination of visible spectrophotometry, polarimetry, and ion-exchange chromatography

to analyze the acid-quenched reaction solutions of D*-[Co(en)$_2$Cl$_2$]$^+$ and
D*-[Co(en)$_2$Cl(OH)]$^+$ at 0° at various concentrations. Specifically, cold
aqueous solutions of the complexes were mixed with cold aqueous sodium
hydroxide, and after a suitable time period the reaction solutions were
quenched with perchloric acid. Spectrophotometric measurements of the
solutions were analyzed with a least squares evaluation of 30 absorption
points per spectrum. Ion-exchange separations allowed an evaluation of the
optical rotations of the three *cis* components. Inasmuch as incomplete
separation often occurs with the ion-exchange method, and since *trans*
species also exist in the solutions, spectral analysis of the eluents was also
necessary. The spectral measurements were limited to the visible region
since the tails of the charge transfer bands, which are susceptible to ion-pair
effects, give small varying background absorptions below 400 nm. Above
600 nm two complexes absorb very little light; therefore, the precision of the
measurements is decreased if this region is included.

The base hydrolysis of D*-[Co(en)$_2$Cl(OH)]$^+$ produces mainly *cis*
product with essentially complete (>90%) retention of configuration. The
results of the D*-[Co(en)$_2$Cl$_2$]$^+$ base hydrolysis reaction indicate that the
D*-[Co(en)$_2$Cl(OH)]$^+$ ion predominates over the corresponding L* optical
isomer at all concentrations, while the reverse is true for the dihydroxo
enantiomers—under conditions which give dihydroxo species to a greater
extent than predicted by rate data in the literature.[30] Some L*-[Co(en$_2$-
(OH)$_2$]$^+$ and *trans*-[Co(en)$_2$(OH)$_2$]$^+$ appear to be formed above that
predicted by the individual reaction sequences deduced from kinetics
measurements in dilute solution. This effect had been noted earlier by
Pearson, Meeker, and Basolo,[31] but was later discounted by Chan and
Tobe, [32] who did not observe this phenomenon when hydroxide ion is slowly
added at 0°. The results (Table 1) agree with both observations.

The stereochemistry for the reaction sequence thus is

$$\text{D*-[Co(en)}_2\text{Cl}_2]^+ \xrightarrow{\text{OH}^-} \text{[Co(en)}_2\text{Cl(OH)]}^+$$
$$trans > \text{D*} > \text{L*}$$
$$\Updownarrow$$
$$\xrightarrow{\text{2OH}^-} \text{[Co(en)}_2\text{(OH)}_2]^+$$
$$\begin{cases} \text{L*} > \text{D*} \\ trans \end{cases}$$

The results with *trans*-[Co(en)$_2$Cl(OH)]$^+$ were suggestive of a rapid equilibrium between the *trans*-chlorohydroxo and the *trans*-dihydroxo ions,
but solutions of *trans*-[Co(en)$_2$Cl(OH)]$^+$ do not exchange chloride and
solutions of the dihydroxo ion do not react rapidly with chloride ion.[33]
The erroneous conclusion was based on the slightly different spectrum which

Table 1
Products Formed in the Base Hydrolysis of D*-cis-[Co(en)₂Cl₂]⁺

Approximate Molarity		Percent					Approximate reaction time, sec
D*-ClCl[a]	OH⁻	Total trans	D*-ClOH	L*-ClOH	D*-OHOH	L*-OHOH	
0.1[b]	0.03	77[c]	22	1	0	0	50[b]
0.03[b]	0.03	84[c]	12	1	1	3	45[b]
0.1[b]	0.1	81[c]	10	2	2	5	20[b]
0.1	0.2	72[c]	4	1	10	13	35
0.1	0.3	66[c]	3	2	12	16	25
0.3	0.3	63[c]	1	1	16	20	50

[a]Column headings indicate the two monodentate ligands coordinated in the bis(ethylenediamine) cobalt(III) species.
[b]Not all of the dichloro ion is consumed at the lower hydroxide concentrations.
[c]Breakdown of trans products is questionable, but total appears consistent.

was obtained each time base and then acid was added to the trans-chlorohydroxo ion compared to that obtained by just adding acid to an aqueous solution of the same species. The spectral differences between the two trans ions are not very great, so small errors in absorptions are also magnified in the data treatment.

The involvement of conjugate base species in such base hydrolysis reactions must be considered in arriving at a mechanistic path for the Bailar inversion. And even though a D* and L* inversion can be accommodated by an unsymmetrical trigonal bipyramidal conjugate base intermediate,[34] the concentration and silver ion dependence of the Bailar inversion require other explanations. Four such possibilities can be imagined for the concentration dependent reaction.

1. A trans attack inversion in which two groups move during the reaction sequence without the formation of a normal chlorohydroxo complex (see Fig. 3).
2. The formation of an intermediate with one monodentate ethylenediamine could also produce an inverted product through subsequent chloride replacement, before the ethylenediamine is rechelated.
3. Inversion could also be explained by a front side ion-pair mechanism in which the second step occurs prior to the diffusion of the first chloride from the second coordination sphere.
4. A trigonal twist could also explain the inversion, but is considered less likely.

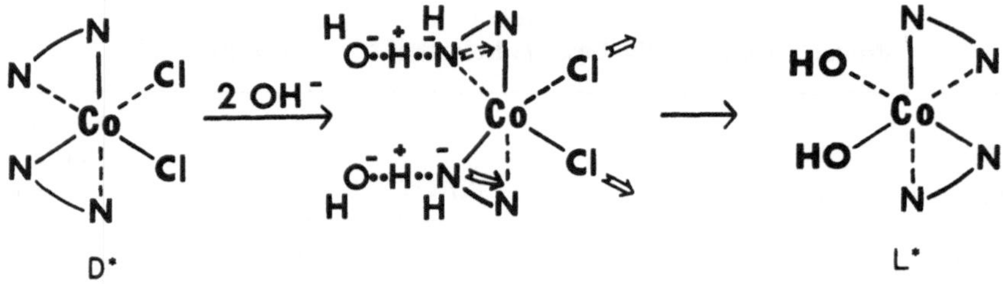

Fig. 3. One of the possible inversion mechanisms for the cis-[Co(en)$_2$Cl$_2$]$^+$ ion, where N—N = ethylenediamine. The hydrogens of the conjugate base amines are shown in the center figure. For more details, see Ref. 8.

These possible mechanistic paths have been evaluated elsewhere in some detail.[8] Elucidation of the stereochemistry of a Bailar inversion with an unsymmetrical diamine would distinguish between the *trans* attack mechanism and the other possibilities.

The stereochemical changes associated with the conjugate base reactions of cobalt(III) complexes are usually discussed[30,35,36] in terms of the lone pair of electrons on the amido group which can interact with the metal ion and its electrons.[37] The dontation can also be considered in terms of reduction[38] or in terms of an intermediate with unpaired electrons.[23] Although the latter was formulated as plausible on the basis of the energy levels of the cobalt(III) complexes which exhibit stereochemical changes during substitution reactions, the probability has increased with the isolation of paramagnetic conjugate base cobalt(III) complexes.[39] The magnetic susceptibilities are suggestive of triplet ground states, and some pseudo-octahedral iron(II) complexes have been prepared which have triplet ground states.[40] Furthermore, the bond strength parameters reported for the iron(II) triplet species are similar to those for similar iron complexes with singlet ground states. Thus, while the triplet species should have more mobility than the singlet ones—enough to allow the requisite stereochemical changes—thermodynamic equilibria would not be obtained. The actuality of such an intermediate in the conjugate base reactions awaits the appropriate ESR studies.

Two undiscovered but probable Bailar inversion areas for future research include nonaqueous cobalt(III) anation reactions (where ion multiplets appear to give large stereochemical changes[41]), and spin-paired iron(II) systems [where outer sphere interactions may produce an electronic effect analogous to the extra electron pair of the amido group of the conjugate base cobalt(III) reactions]. The author's efforts to produce similar effects with cobalt(III) species containing several oxygen donors have produced some interesting chemistry,[42–46] but no new Bailar inversions to date.

ACKNOWLEDGMENT

The author's deepest gratitude goes to Prof. Bailar, both for initiating this interesting research area and for his inspiration and guidance during my early scientific career. The help of my undergraduate and graduate students, particularly Dr. Elaine Dittmar, and the monetary support of the National Science Foundation is sincerely appreciated.

REFERENCES

1. J. C. Bailar, Jr. and R. W. Auten, *J. Am. Chem. Soc.* **56**: 774 (1934).
2. J. C. Bailar, Jr., F. G. Jonelis, and E. H. Huffman, *ibid.* **58**: 2224 (1936).
3. J.-P. Mathieu, *Bull. Soc. Chim. France* **3**: [5] 476 (1936).
4. J. C. Bailar, Jr. and J. P. McReynolds, *J. Am. Chem. Soc.* **61**: 3199 (1939).
5. F. P. Dwyer, A. M. Sargeson, and I. K. Reid, *ibid.* **85**: 1215 (1963).
6. S. C. Chan and M. L. Tobe, *J. Chem. Soc.* 4531 (1962).
7. L. J. Boucher, E. Kyuno, and J. C. Bailar, Jr., *J. Am. Chem. Soc.* **86**: 3656 (1964).
8. E. A. Dittmar and R. D. Archer, *ibid.* **90**: 1468 (1968).
9. J. C. Bailar, Jr., *Rev. Pure Appl. Chem.* **16**: 91 (1966).
10. J. C. Bailar, Jr. and J. C. Peppard, *J. Am. Chem. Soc.* **83**: 820 (1940).
11. J. C. Bailar, Jr., J. H. Haslam, and E. M. Jones, *ibid.* **58**: 2226 (1936).
12. R. D. Archer and J. C. Bailar, Jr., *ibid.* **83**: 812 (1961).
13. E. L. Greenwood, B.S. Thesis, University of Illinois, 1936.
14. R. D. Archer, Eighth Intl. Conf. on Coordination Chemistry, Vienna, 1964, pp. 111-113.
15. E. Kyuno, L. J. Boucher, and J. C. Bailar, Jr., *J. Am. Chem. Soc.* **87**: 4458 (1965).
16. E. Kyuno and J. C. Bailar, Jr., *ibid.* **88**: 1120 (1966).
17. A. M. Sargeson and G. H. Searle, *Inorg. Chem.* **6**: 787 (1967); *cf.*, G. H. Searle, Ph.D. Dissertation, Australian National University, Canberra, 1963.
18. D. A. Buckingham, P. A. Marzilli, and A. M. Sargeson, *Inorg. Chem.* **6**: 1032 (1967).
19. A. M. Sargeson, private communication, 1964.
20. E. Kyuno and J. C. Bailar, Jr., *J. Am. Chem. Soc.* **88**: 1125 (1966).
21. R. G. Asperger and C. F. Liu, *ibid.* **89**: 1533 (1967).
22. J. H. Worrell and D. H. Busch, private communications.
23. R. D. Archer, *Advances in Chemistry Series, No. 62*, American Chemical Society, Washington, D.C., 1967, pp. 452–468.
24. F. P. Dwyer, N. S. Gill, E. C. Gyarfas, and F. Lions, *J. Am. Chem. Soc.* **74**: 4188 (1952).
25. T. E. MacDermott and A. M. Sargeson, *Australian J. Chem.* **16**: 334 (1963).
26. A. M. Sargeson, in F. P. Dwyer and D. P. Mellor, *Chelating Agents and Metal Chelates*, Academic Press, New York, 1964, pp. 230–232.
27. D. H. Busch, K. Swaminathan, and D. W. Cooke, *Inorg. Chem.* **1**: 260 (1962).
28. B. E. Douglas, R. A. Haines, and J. G. Brushmiller, *ibid.* **2**: 1194 (1963).
29. J. I. Legg and B. E. Douglas, *J. Am. Chem. Soc.* **88**: 2697 (1966).
30. F. Basolo and R. G. Pearson, *Mechanisms of Inorganic Reactions*, John Wiley and Sons, Inc., New York, 1967, Chapters 3, 4.
31. R. G. Pearson, R. E. Meeker, and F. Basolo, *J. Am. Chem. Soc.* **78**: 2673 (1956).
32. S. C. Chan and M. L. Tobe, *J. Chem. Soc.* 4531 (1962).
33. M. E. Farago, B. Page, and M. L. Tobe, *Inorg. Chem.* **8**: 388 (1969).
34. R. G. Pearson and F. Basolo, *Inorg. Chem.* **4**: 1522 (1965).

35. C. H. Langford and H. B. Gray, *Ligand Substitution Processes*, W. A. Benjamin, Inc., New York, 1966, Chapter 3.
36. M. L. Tobe (ed. J. H. Ridd) *Studies on Chemical Structure and Reactivity*, Methuen and Co., Ltd., London, 1966, Chapter 11 and references cited therein.
37. R. G. Pearson and F. Basolo, *J. Am. Chem. Soc.* **78**: 4878 (1956).
38. R. D. Gillard, *J. Chem. Soc.*, Sect. A: 917 (1967).
39. G. W. Watt and J. F. Knifton, *Inorg. Chem.* **7**: 1159 (1968).
40. E. Konig and K. Madeja, *ibid.* **7**: 1848 (1968).
41. W. R. Fitzgerald and D. W. Watts, *J. Am. Chem. Soc.* **90**: 1734 (1968).
42. R. D. Archer and B. P. Cotsoradis, *Inorg. Chem.* **4**: 1584 (1965).
43. B. P. Cotsoradis and R. D. Archer, *ibid.* **6**: 800 (1967).
44. R. D. Archer and B. D. Catsikis, *J. Am. Chem. Soc.* **88**: 4520 (1966).
45. O. Hyodo and R. D. Archer, The Mechanism of Tricarbonatocobaltate(III) Ammonation, Northeast Regional Meeting, American Chemical Society, Boston, October, 1968 (to be published).
46. R. J. York, W. D. Bonds, Jr., B. P. Cotsoradis, and R. D. Archer, Tris(β-diketonato)cobalt(III) and Bis(acetylacetonato)diaminecobalt(III) Trichelate Species, 156th National American Chemical Society Meeting, Atlantic City, 1968 (in press, *Inorg. Chem.* Vol. 8, 1969).

SOME CONFORMATIONAL EFFECTS OF CHELATE RINGS

Bodie E. Douglas

University of Pittsburgh
Pittsburgh, Pennsylvania

INTRODUCTION

The basis for stereospecific or stereoselective processes is now understood from the classic paper by Corey and Bailar.[1] Their approach and predictions have been well substantiated; the paper is one of the most often cited references in the field of coordination chemistry. It provides the basis for interpreting pmr studies and for detailed stereochemical studies of metal chelates, including conformational effects of chelate rings.

CONFORMATIONAL CONTRIBUTIONS TO CIRCULAR DICHROISM SPECTRA

The chirality of complex ions such as $[Co(en)_3]^{3+}$ or cis-$[Co(en)_2Cl_2]^+$ can be attributed to the spiral arrangement of the chelate rings. Configurations of the enantiomers of these complex ions and those of multidentate ligands have been related by the ring pairing method of Legg and Douglas[2] and the octant rule of Hawkins and Larsen.[3] Shimura[4] showed that an optically active ligand in the complex ion $[Co(NH_3)_4(S\text{-leucinato})]^{2+}$ causes Cotton effects in the region of the cobalt d–d absorption bands. He referred to this as a "vicinal" effect. This effect had been observed by Pfeiffer,[5] who introduced the term "vicinal" effect for Cotton effects caused by optically active ligands. The compounds Pfeiffer studied were not well characterized and did not clearly lack a spiral arrangement of chelate rings.

The optical isomers* Λ-$(+)$-$[Co(en)_2S\text{-pala}]^{2+}$ and Δ-$(-)$-$[Co(en)_2S\text{-pala}]^{2+}$ are not enantiomers, nor are the CD spectra for these internal diastereomers mirror images. In fact, they even differ in the number of CD peaks (Fig. 1). The contributions from the Λ or Δ arrangement of the chelate

---

*Absolute configurations of the complex ions are designated Δ or Λ in accordance with a forthcoming I.U.P.A.C. tentative rule corresponding to the left spiral of the chelate rings about the C_3 axis in Λ-$(+)$-$[Co(en)_3]Cl_3$.

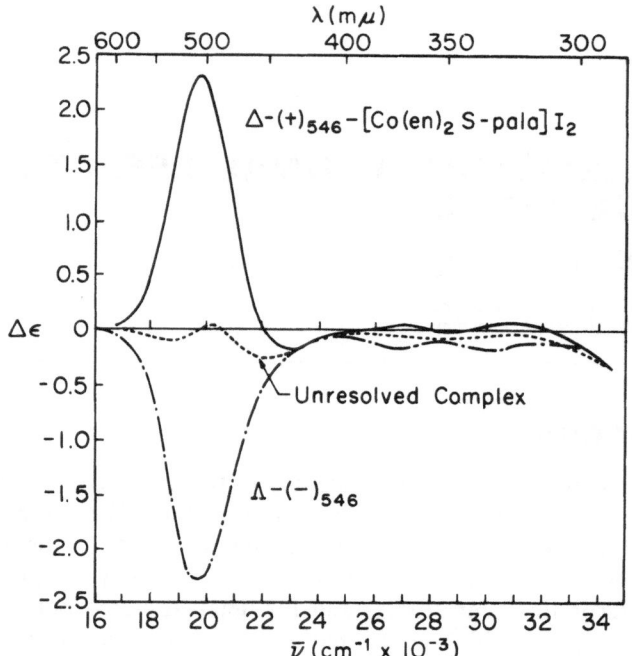

Fig. 1. Circular dichroism curves for the isomers of [Co(en)$_2$S-phenylalaninato]I$_2$.

rings, termed the configurational contribution, and from the presence of the optically active ligand, termed the "vicinal" contribution, have been shown to be separable and reasonably additive.[6] Examination of these separate contributions accounts for the marked differences between CD curves for the Δ and Λ isomers. Further evidence for the additivity of such contributions has appeared in numerous papers, from which only similar cases are cited.[7-11]

The contribution of an optically active ligand in a complex ion to the CD spectrum is expected to be predominately a conformational effect if it forms a chelate ring. There might be smaller effects resulting from changes in the conformations of other chelate rings caused by the optically active ligand in a preferred conformation. A vicinal effect, though of smaller magnitude, has been reported for unidentate ligands.[12-14] This, presumably, is the result of having a dissymmetric center near the chromophore, a true vicinal effect.

Dingle and Ballhausen[15] found that the trigonal splitting in 2[Co(en)$_3$Cl$_3$]·NaCl·6H$_2$O is very small (perhaps 10 cm^{-1}) from polarized crystal spectra down to 4.2°K. They concluded that the two CD peaks of opposite sign in the T$_{1g}$ band region for [Co(en)$_3$]$^{3+}$ in solution cannot result from trigonal splitting, as generally assumed. They attributed the two peaks to the presence of two conformational isomers, as suggested by Woldbye.[16] In the case of $[Co(pn)_3]^{3+}$, however, one or two CD peaks are observed in the first absorption band region, depending upon the conformational isomers present.[7,8] Thus, two CD peaks are observed for Λ-(+)-$[Co(+)-pn_3]^{3+}$, where only one conformational isomer is present because of the optically active propylenediamine.

Interpretations of CD spectra of cobalt(III) complexes in solution have been based upon Mason's assignment[17] of the two peaks of opposite sign in the T_{1g} band region for $[Co(en)_3]^{3+}$ as E_a (dominant) and A_2, using D_3 symmetry. They have suggested[18] that only one CD peak is observed in this region for some combinations of chelate ring conformations in pn complexes because of a reduction in the frequency interval between the E_a and A_2 transitions, causing complete cancellation of the weaker A_2 peak.

The CD curves of $[Co(-)-pn(NH_3)_4]Cl_3$,[8,19] $trans$-$[Co(-)$-$pn_2(NH_3)_2]Cl_3$,[18] and the vicinal or conformational effects of $(-)pn$[8,18] in tris(diamine) complexes are strikingly similar. Such similarities through the series do not lead one to expect significant changes in the frequency intervals for tris-(diamine) complexes as chelate ring conformations change.

The contradiction between the apparently great trigonal splitting from solution CD spectra and the negligible splitting from polarized crystal spectra has not been resolved. The extent of correspondence of CD data for solutions and crystals has not been established, however. Electrolytes have a great effect upon CD spectra of complex ions in solution,[20,21] but the influence of the ionic lattice is unknown.

Yasui et al.[13] found that the CD curve in the first band region for a series of "typical" optically active amino acid complex ions of the type $[Co(NH_3)_4(aa)]^{2+}$ (aa = amino acid anion) were of the same form as reported for the "vicinal" effect of optically active amino acids in complex ions of the type $[Co(en)_2(aa)]^{2+}$ (Fig. 1). Cyclic amino acids and some others gave different types of CD curves.[13] The CD curves for "vicinal" effects are also of the same type for some isomers of $[Co\ trien(aa)]^{2+}$ complex ions, $vide\ infra$. The combination of such an effect with the configurational contribution can account for the appearance of one or two CD peaks in the first band region for complex ions of the $[Co(en)_2aa]^{2+}$ type[6,10] without assuming a change in frequency interval for the two transitions.

GEOMETRICAL ISOMERISM IN TRIETHYLENETETRAAMINE COMPLEXES

Sargeson and Searle[22] demonstrated that triethylenetetraamine can wrap around a cobalt(III) ion in three different ways. The trans- and two

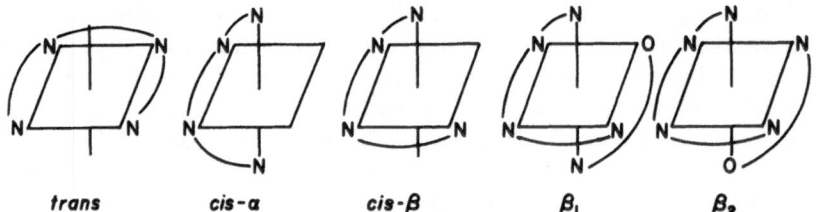

Fig. 2. Geometrical isomers of some complexes of triethylenetetraamine.

cis- (α and β) isomers are shown in Fig. 2. This was an important step in the study of complexes of multidentate ligands. Earlier, one might have expected either that one isomer would be strongly favored over others, or that it would not be possible to separate and identify the two very similar *cis*-isomers. While C.-Y. Lin[23] was studying a series of amino acid complex ions of the type $[Co(trien)aa]^{2+}$ as an outgrowth of earlier work on $[Co(en)_2aa]^{2+}$ complex ions,[6] Marzilli and Buckingham[24] reported two β-isomers of the complex ion $[Co(trien)gly]^{2+}$ (β_1 and β_2, see Fig. 2) and one isomer (β_2) of $[Co(trien)sar]^{2+}$ (gly = glycinate ion and sar = sarcosinate ion). These isomers were distinguished and identified on the basis of the visible, infrared, and pmr spectra, and the stereospecificity of sarcosine.[25]

CONFORMATIONAL ISOMERS

Marzilli and Buckingham recognized the possibility of conformational isomers for complexes of trien in a β-isomer (Fig. 3). Since they observed no evidence for such isomerism, their complexes were presumed to have the SS conformation of trien* because of the greater stability expected from a conformational analysis.

Buckingham, Marzilli, and Sargeson[26] separated the conformational isomers of complexes of the type *trans-*$[Co(trien)X_2]^{n+}$ and reported evidence for the existence of conformational isomers of β-$[Co(trien)(H_2O_2)]^{3+}$ in solution, but the isomers were not isolated. One of the latter isomers undergoes mutarotation to give the more stable conformation of trien.

Lin[23,27] isolated complexes identical to β_1- and β_2-$[Co(trien)gly]^{2+}$ and also corresponding isomers of a series of optically active amino acids. All attempts to prepare the corresponding α-isomers were unsuccessful. Two optical isomers (Δ and Λ) were isolated for the β_1-isomer of each of the complexes formed with glycine, alanine, methonine, and proline. In each

*SS (RR, RS, or SR) refers to the absolute configurations of the nitrogens of the secondary amines— these are fixed by the chelate ring conformations. The absolute configurations of amino acids are designated S (for L) or R (for D).

case the β_1-isomer was identified from the visible, infrared, circular dichroism (CD), and pmr spectra. No evidence was found for more than one conformational isomer of trien for this case. Following the discussion of Marzilli and Buckingham,[24] the β_1 complexes isolated were assumed to have the RR or SS conformation of the trien chelate rings. Scale models reveal that the Λ-β_1-SS- and Δ-β_1-RR-isomers should be much more favorable than the other β_1-isomers. CD curves are shown in Fig. 4 for β_1-[Co(trien)-gly]I_2 and β_1-[Co(trien)S-ala]$I_2 \cdot H_2O$. The spectra for the β_1 complex with S-methionine are very similar to those for the S-alanine complex.

Two series of β_2 isomers were isolated with glycine, S-alanine, and S-methionine. The complexes all had visible and infrared spectra expected of β_2 isomers,[24] but the two series differed in CD spectra. They are interpreted as conformational isomers, as represented in Fig. 3. Thus far, pmr spectra have been inconclusive in distinguishing the conformational isomers. The CD and absorption spectra of the six β-isomers of [Co(trien)(S-ala)]$I_2 \cdot H_2O$ are compared in Fig. 5. The absorption curves are the same for both β_2-isomers.

The SS or RR conformation assigned to the complexes reported by Marzilli and Buckingham[24] is supported by the crystal structure analysis of the β_2-[Co(trien)(gly-gly-O-Et)]$^{3+}$ (gly-gly-O-Et = glycylglycine-O-ethyl ester) and β_2-[Co(trien)Cl(H$_2$O)]$^{2+}$ ions.[28] The corresponding isomers, designated Λ-SS and Δ-RR, (see footnotes on p. 29 and 32) were prepared by the reaction of cis-α-[Co(trien)Cl$_2$]Cl with the sodium salt of the amino acid. The complexes were resolved using silver antimonyl-(+)-tartrate as resolving agent. Only one isomer was obtained by this procedure for S-proline.

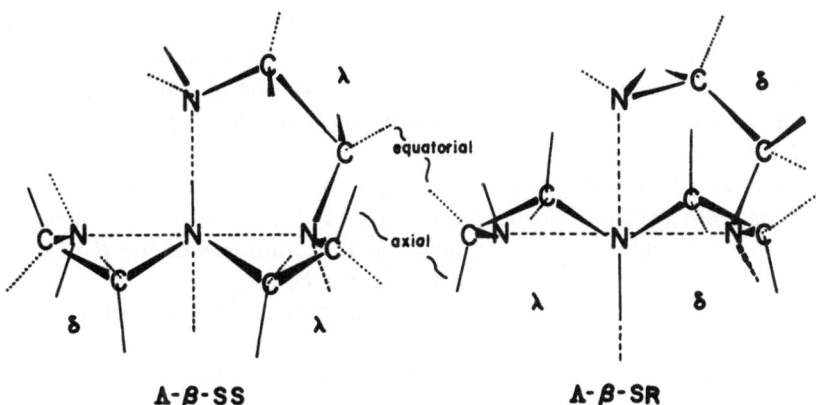

Λ-β-SS $\qquad\qquad$ Λ-β-SR

Fig. 3. Conformational isomers of β-triethylenetetraamine complexes.

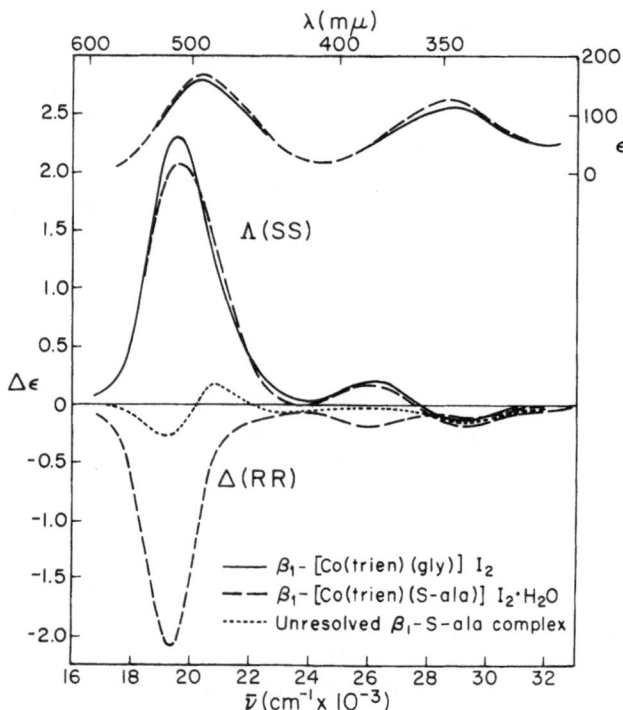

Fig. 4. Circular dichroism and electronic absorption spectra for
β_1-(RR and SS)-[Co(trien)S-ala]I$_2$·H$_2$O.

The other series of β_2-isomers, designated Λ-SR or Δ-RS, were prepared
by converting α-[Co(trien)Cl$_2$]Cl to β-[Co(trien)(OH)$_2$]$^+$ in solution, followed
by the addition of the amino acid, and were resolved as above. Two isomers
were isolated for proline by the second method, one of which was identical
to that from the first method, and both of which were interpreted as belonging
to the RR-SS series.

The complexes assigned to the Λ-(+)-β_1-SS- or Δ-(−)-β_1-RR-confor-
mations were prepared from [Co(trien)(OH)$_2$]$^+$, as above, but using two
molar equivalents of the amino acid. Smaller amounts of the β_2-RR- and
β_2-SS-isomers were also formed by this procedure.

The isolation of only a β_2-complex (no β_1) of sarcosine and the known
stereospecificity of sarcosine[25] were used in the characterization of the β_1
and β_2 complexes by Marzilli and Buckingham.[24] The stereoselectivity of
the cyclic amino acid proline might be expected to be greater than that of
sarcosine. A β_1-isomer was isolated for proline, but only using one of several

preparative procedures tried. Models reveal that the repulsive interactions are at a minimum in the complex ion Δ-β_2-RR-[Co(trien)(S-proline]$^{2+}$ as compared to the other β_2-isomers. Only one isomer was obtained in the preparation by the first method. The assignment of this as the most favorable isomer (Δ-β_2-RR) is consistent with the interpretation of CD curves and the earlier work.[9,24] The less favorable Λ-β_2-SS-isomer was obtained in smaller yields by the second preparative method, but no evidence was found for the formation of the still less favorable Δ-β_2-SR- and Δ-β_2-RS-isomers of the proline complex ion.

RING CONFORMATIONAL EFFECTS IN TRIETHYLENETETRAAMINE COMPLEXES

Although there are significant differences among the CD curves for the isomers shown in Fig. 5, they are of the same form as those of the [Co(en)$_2$aa]$^{2+}$

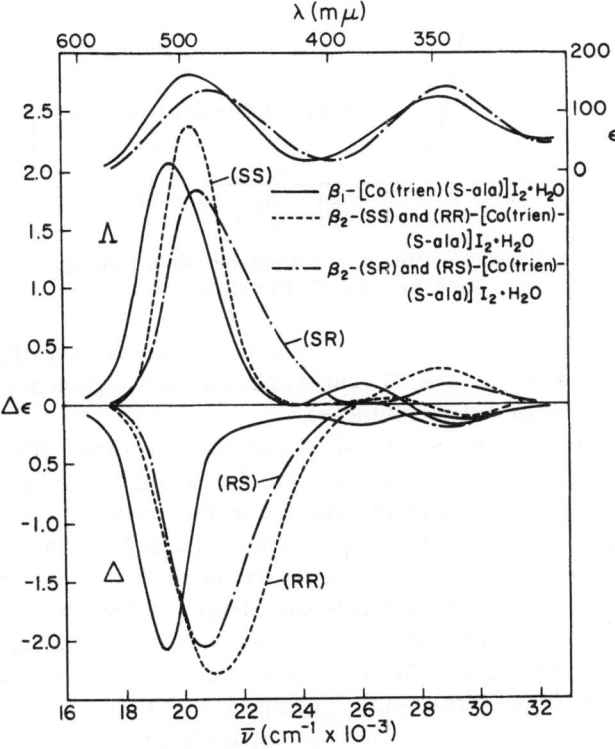

Fig. 5. Circular dichroism and electronic absorption spectra for β_1- and β_2-[Co(trien)S-ala]I$_2$·H$_2$O.

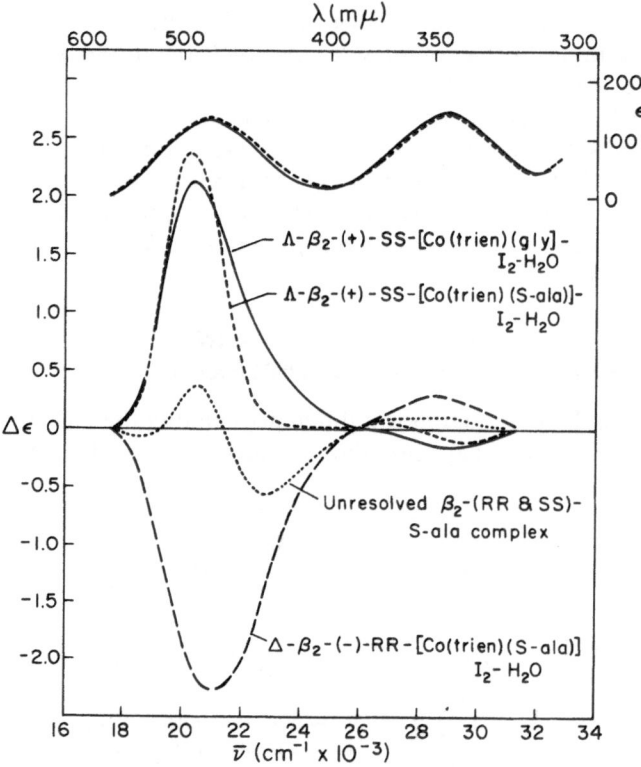

Fig. 6. Circular dichroism and electronic absorption spectra for
β_2-(RR and SS)-[Co(trien)S-ala]$I_2 \cdot H_2O$.

complex ions (Fig. 1). The absolute configurations are known for the Δ-$(-)_D$-
[Co(en)$_2$aa]$^{2+}$ ions[29] so the CD curves permit the assignments of absolute
configurations to the trien complexes.

The CD curves for Δ-$(-)$-β_2-RR- and Λ-$(+)$-β_2-SS-[Co(trien)S-ala]$^{2+}$
are shown in Fig. 6 along with the curve for the unresolved complex. The
latter curve agrees well with the average of the curves for the Δ- and Λ-
isomers, verifying the complete separation of the isomers and the presence
of equimolar proportions of the two isomers in the unresolved product.
Adding the two CD curves results in cancellation of the contributions from
the opposite chelate ring spirals. It leaves conformational contributions
that do not cancel and any "true" vicinal effect. The resultant curve is very
similar to that for [Co(en)$_2$S-aa]$^{2+}$ complex ions (Fig. 1), except that the
peaks are somewhat more intense for the trien case. If one subtracts the
curve for the unresolved complex from that for either isomer, one obtains a
curve in fairly good agreement with that for the corresponding glycine

complex. These results suggest that the contributions from trien largely cancel when the curves for the two isomers are added so that the curve for the unresolved complex is primarily the contribution of S-alanine. If the S-alaninato chelate ring is more puckered in the trien complex as compared to $[Co(en)_2S\text{-}ala]^{2+}$, one would expect more intense CD peaks from the conformational contribution.

Although the CD curves are significantly different for the β_1(RR and SS)- and β_2(RR and SS)-isomers (Fig. 4 and 5), the curves for the unresolved complexes are similar. They differ in the relative intensities of the three peaks in the first band region, and there is generally lower intensity for the β_1 complex. Nevertheless, the curve for the unresolved complex is related to those for the Δ- and Λ-isomers as for the β_2 case. The curve for either isomer minus that for the unresolved complex gives a curve which agrees well with that for the corresponding glycine complex (Fig. 4).

The CD curves for the isomers Δ-(+)-β_2-SR- and Λ-(−)-β_2-RS-$[Co(trien)(S\text{-}ala)]^{2+}$ are almost mirror images (Fig. 7). The CD curve for the unresolved complex has only one broad, very weak band. This curve can account for the slight differences between the CD curves for the Δ- and Λ- isomers. It also accounts for the differences between the curves for the

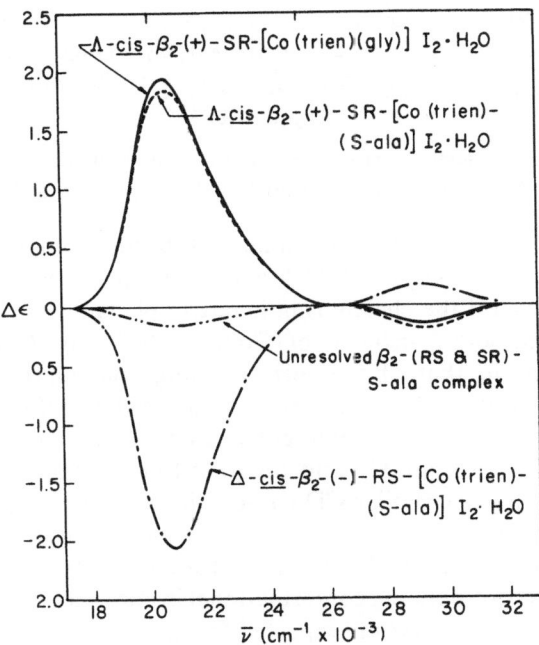

Fig. 7. Circular dichroism and electronic absorption spectra for β_2-(RS and SR)-[Co(trien)S-ala]$I_2 \cdot H_2O$.

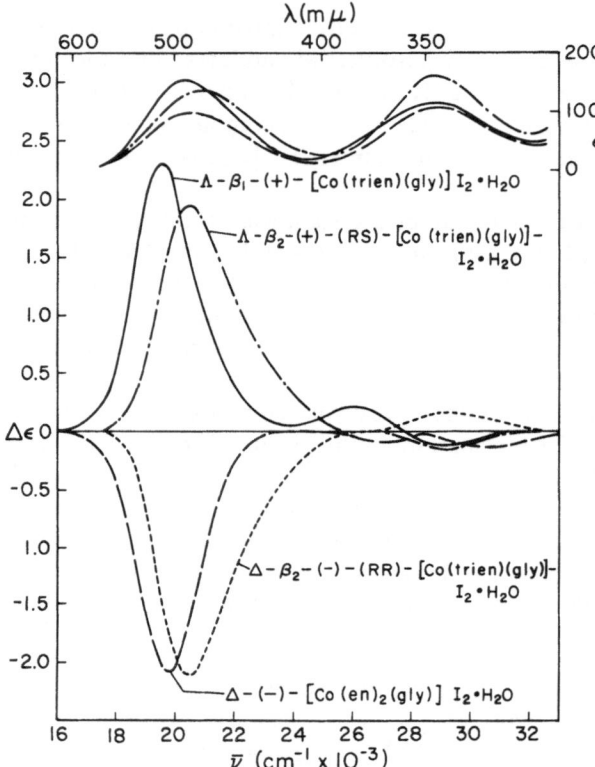

Fig. 8. Circular dichroism and electronic absorption spectra for
the isomers of β-[Co(trien)gly]I$_2$ and of $(-)$-[Co(en)$_2$gly]I$_2$.

corresponding isomers of the S-alanine and glycine complexes. The small
effect observed here suggests that S-alanine makes little contribution, as
might be expected if the chelate ring is nearly planar.

The β_1- and both β_2-isomers of [Co(trien)gly]$^{2+}$ give CD curves which
are very similar to that of the corresponding configurational isomer of
[Co(en)$_2$gly]$^{2+}$ (Fig. 8). Thus, the trien in any of the conformations makes
no contribution significantly different from that of two ethylenediamine
chelate rings. The RR (or SS) and SR (or RS) conformational isomers of
β_2-[Co(trien)gly]I$_2$ give similar CD curves which differ only slightly in peak
intensities. Great care was taken to assure complete resolution in each case.
These isomers were identified by the preparative method used, i.e., the method
which gave the corresponding conformational isomer with other amino acids.
This similarity in CD curves is expected for the conformational isomers in
the case of the glycine complexes, since the general results indicate that the
differences observed for the isomers of optically active amino acids are

caused by contributions of the amino acids. They are not caused by a direct contribution from the asymmetric nitrogen atoms or from the conformations of the trien chelate rings.

The nonbonding interactions between trien and proline presumably prevent the formation of one pair of isomers (β_2-RS and β_2-SR). The great differences among the CD curves for the other isomers (Fig. 9) indicate more interaction between proline and trien than for the noncyclic amino acids. It is of interest that the form of the average CD curve for the β_2-isomers is the same as that for the other optically active amino acids.

The small contributions of the fixed conformations of trien are inconsistent with the explanations of the appearance of one or two CD peaks in the T_{1g} band region for tris(diamine) complexes given by Woldbye,[16] Dingle and Ballhausen,[15] and Mason,[18] *vide ante.* These small effects do not lead one to expect that the two peaks observed for $[Co(en)_3]^{3+}$ could arise from the presence of two conformational isomers.[15,16] Nor should one expect that there would be a significant change in frequency interval

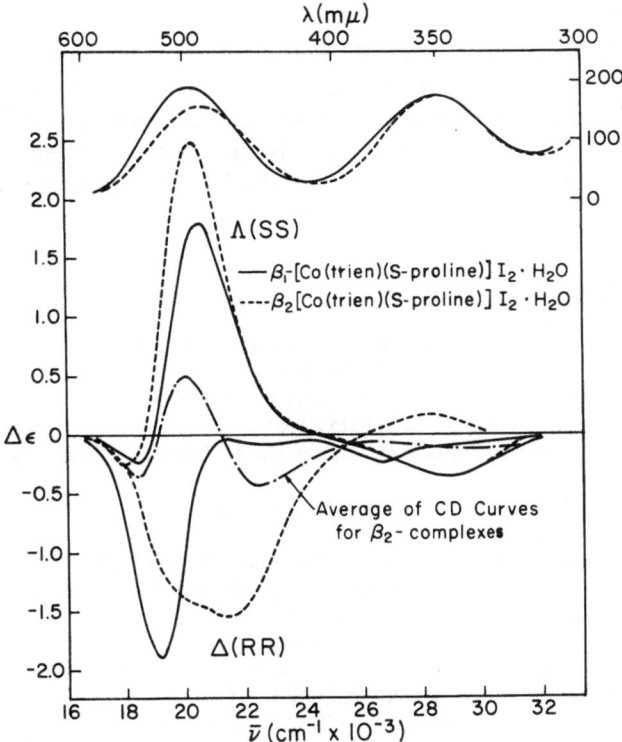

Fig. 9. Circular dichroism and electronic absorption spectra for the isomers of β-(RR and SS)-[Co(trien)S-prol]$I_2 \cdot H_2O$.

or splitting parameter through the series of $[Co(pn)_3]^{3+}$ conformational isomers to account for the disappearance of the second peak in some cases.[18]

It seems reasonable to conclude that changes in chelate ring conformations do not cause significant changes either in the energies of electronic transitions or in the splitting parameters. One would expect some contribution to the intensity of each CD peak from the configurational and conformational effects. The contribution to each peak might differ in magnitude or sign for these effects, altering the shape of the CD curve. Thus, we might expect to obtain detailed information concerning electronic energy levels from the "vicinal" effect where all of the contributions are weak and peaks are less likely to be masked by stronger neighboring peaks.

The origin of optical activity in metal chelates is not well understood. Circular dichroism spectra, together with other physical methods, should give detailed stereochemical and conformational information which might provide a basis for theoretical advancements.

ACKNOWLEDGMENT

The author wishes to thank Prof. Bailar for his inspirational example and his associates whose contributions, in the form of ground-work and stimulation, have not been cited. Dr. C.-Y. Lin has been responsible for all the careful work on conformational isomers of triethylenetetraamine complexes in our laboratory. Support of the National Institutes of Health (GM-10829) is gratefully acknowledged.

REFERENCES

1. E. J. Corey and J. C. Bailar, Jr., *J. Am. Chem. Soc.* **81**: 2620 (1959).
2. J. I. Legg and B. E. Douglas, *ibid.* **88**: 2697 (1966).
3. C. J. Hawkins and E. Larsen, *Acta Chem. Scand.* **19**: 185 (1965).
4. Y. Shimura, *Bull. Chem. Soc. Japan* **31**: 315 (1958).
5. P. Pfeiffer, W. Christeleit, T. Hesse, H. Pfitzner, and H. Thielert, *J. Prakt. Chem.* **150**: 261 (1938).
6. C. T. Liu and B. E. Douglas, *Inorg. Chem.* **3**: 1356 (1964).
7. B. E. Douglas, *ibid.* **4**: 1813 (1965).
8. K. Ogino, K. Murano and J. Fujita, *Inorg. Nucl. Chem. Letters* **4**: 351 (1968).
9. S. Larsen, K. J. Watson, A. M. Sargeson, and K. R. Turnbull, *Chem. Commun.* 847 (1968).
10. S. K. Hall and B. E. Douglas, *Inorg. Chem.* **8**: 372 (1969).
11. B. E. Douglas and S. Yamada, *ibid.* **4**: 1561 (1965).
12. J. Fujita, T. Yasui, and Y. Shimura, *Bull. Chem. Soc. Japan.* **38**: 654 (1965).
13. T. Yasui, J. Hidaka, and Y. Shimura, *ibid.* **39**: 2417 (1966).
14. D. C. Bhatnagar and S. Kirschner, *Inorg. Chem.* **3**: 1256 (1964).
15. R. Dingle and C. J. Ballhausen, *Mat. Fys. Medd. Dan. Vid. Selsk.* **35**: 1 (1967).
16. F. Woldbye, *Rec. Chem. Progr.* **24**: 197 (1963).
17. A. J. McCaffery and S. F. Mason, *Mol. Phys.* **6**: 359 (1963).

18. A. J. McCaffery, S. F. Mason, B. J. Norman, and A. M. Sargeson, *J. Chem. Soc.* A, 1304 (1968).
19. C. J. Hawkins, E. Larsen, and I. Olsen, *Acta Chem. Scand.* **19**: 1915 (1965).
20. H. L. Smith and B. E. Douglas, *J. Am. Chem. Soc.* **86**: 3885 (1964); *Inorg. Chem.* **5**: 784 (1966).
21. S. F. Mason and B. J. Norman, *Proc. Chem. Soc.* 339 (1964); *J. Chem. Soc.* A, 307 (1966); R. Larsson, S. F. Mason, and B. J. Norman, *ibid.* 301 (1966).
22. A. M. Sargeson and G. H. Searle, *Inorg. Chem.* **4**: 45 (1965); **6**: 787 (1967).
23. C.-Y. Lin, Ph.D. Thesis, University of Pittsburgh, 1968.
24. L. G. Marzilli and D. A. Buckingham, *Inorg. Chem.* **6**: 1042 (1967).
25. D. A. Buckingham, S. F. Mason, A. M. Sargeson, and K. R. Turnbull, *ibid.* **5**: 1649 (1966).
26. D. A. Buckingham, P. A. Marzilli and A. M. Sargeson, *Inorg. Chem.* **6**: 1032 (1967).
27. C.-Y. Lin and B. E. Douglas, *Inorg. Nucl. Chem. Letts.* **4**: 15 (1968).
28. D. A. Buckingham, P. A. Marzilli, I. E. Maxwell, A. M. Sargeson, M. Fehlman, and H. C. Freeman, *Chem. Commun.* 488 (1968).
29. J. F. Blount, H. C. Freeman, A. M. Sargeson, and K. R. Turnbull, *Chem. Commun.* 324 (1967).

OPTICAL ROTATORY DISPERSION, CIRCULAR DICHROISM, AND THE PFEIFFER EFFECT IN COORDINATION COMPOUNDS

S. Kirschner and N. Ahmad

Wayne State University
Detroit, Michigan

INTRODUCTION

During the past decade there has been a considerable resurgence of interest in the optical rotatory dispersion and circular dichroism of co-ordination compounds.[1] In addition, there also has been a considerable renewal of interest in the Pfeiffer effect during this same period of time.[2-8] This coincidental renewal of interest in these fields is most fortunate, because it is possible to apply optical rotatory dispersion and circular dichroism techniques to the study of the Pfeiffer effect with fruitful results (*vide infra*).

OPTICAL ROTATORY DISPERSION (ORD)

This is the wavelength dependence of optical rotation. Figure 1 shows the optical rotatory dispersion of sucrose in the visible region. This type of curve is known as a plain or normal, optical rotatory dispersion curve, and is typical of ORD curves of optically active substances in wavelength regions outside their absorption bands. Several excellent reviews of the application of optical rotatory dispersion and circular dichroism techniques to coordination compounds have been published in recent years.[9-13,1] Drude[14,15] developed an equation to describe the optical rotatory dispersion of a substance in the wavelength region outside the optically active absorption band, and it is usually given as

$$[\alpha] = [(c_m)/(\lambda^2 - \lambda_m^2)] \tag{1}$$

where c_m is a constant which depends upon the number of absorption peaks and other factors present in a given medium, λ refers to the wavelength of the optical rotation under study, and λ_m refers to the wavelengths

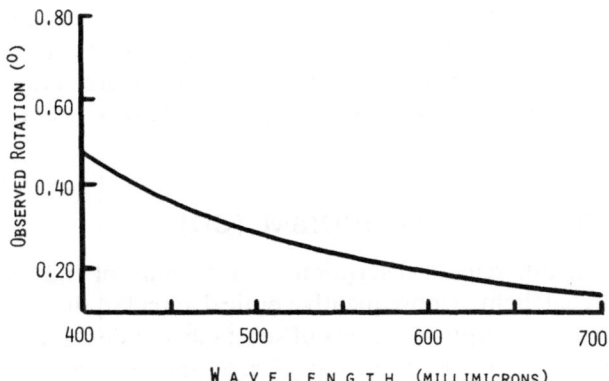

Fig. 1. The optical rotatory dispersion of d-sucrose at 20°C
(0.301 g of d-sucrose in 100 ml of water).

of the peaks of the absorption bands nearest λ. The equation is usually
expressed with two terms as

$$[\alpha] = \frac{C_1}{\lambda^2 - \lambda_0^2} + \frac{C_2}{\lambda^2 - \lambda_1^2} \tag{2}$$

where λ_0 and λ_1 are the wavelengths of the absorption peaks nearest
λ ($\lambda_0 > \lambda_1$). Heller[16,17] has developed a technique for determining whether
a one-term Drude equation adequately describes the optical rotatory dis-
persion of a system outside its absorption band. In organic chemistry it is
usual in the case of ORD curves to plot wavelength against *specific rotation*
$[\alpha]_\lambda^t$:

$$[\alpha]_\lambda^t = \alpha_{\lambda\,\text{obs.}}^t / c l_d \tag{3}$$

or against *molar rotation*, $[M]_\lambda^t$:

$$[M]_\lambda^t = [\alpha]_\lambda^t \left(\frac{M}{100} \right) = \frac{\alpha_{\lambda\,\text{obs.}}^t}{[C] l_m} \tag{4}$$

where c is the concentration of the solute in grams per millileter of solvent
(or density, for a pure substance), l_d is the path length in decimeters, M is
the molecular weight of the solute, $[C]$ is the molar concentration of the solute,
and l_m is the path length in meters. However, for coordination compounds,
it is preferable to plot wavelength against *equivalent rotation*, $[E]_\lambda^t$, where

$$[E]_\lambda^t = \frac{[M]_\lambda^t}{\bar{n}} \tag{5}$$

and where $\bar{n}$ is the number of optically active units per molecule or formula unit. This eliminates non-real differences in the calculation of optical rotation of coordination compounds containing different numbers of the same optically active complex per molecular unit (e.g., $[Co(en)_3]Cl_3$ and $[Co(en)_3]_2$-$(SO_4)_3$).

OPTICAL CIRCULAR DICHROISM (CD)

This is the differential absorption by a medium of right and left circularly polarized light. Consequently, optical circular dichroism spectra occur only inside absorption regions of substances under study. Substances exhibiting this differential absorption of right and left circularly polarized light are said to be *circularly dichroic*. Further, it should be mentioned that incident *linear* (plane) polarized light (which may be considered to be the resultant of identical beams of right and left circularly polarized light), having a wavelength inside the absorption band of a circularly dichroic medium, leaves that medium both with its line (plane) of polarization having been rotated and with itself having been converted into *elliptically* polarized light. This type of dichroism results from the different electronic transition probabilities, which arise when optically active molecules are excited by right and left circularly polarized light inside an optically active absorption band.[18,19]

Figure 2 describes how linear polarized light is rotated and converted into elliptically polarized light as it passes through a medium which exhibits both differential absorption (circular dichroism) and differential index of refraction (optical rotation). In this figure the greater absorption of the left component of circularly polarized light is indicated by the smaller length of $O - E_L$, and its smaller velocity is indicated by an angle IOE_L which is smaller than IOE_R. The angle of ellipticity (ψ, the degree of "flatness" of the ellipse) is defined as the angle formed by the major axis (OB) and the line (OD) drawn from the origin through the end of the minor axis (which has been displaced to the tip of the ellipse, BD).

Just as it is possible to define specific rotation and molar rotation (equations 3 and 4), it is also possible to define specific ellipticity, $[\psi]_\lambda^t$ and molar ellipticity $[\theta]_\lambda^t$, as:

$$[\psi]_\lambda^t = \frac{\psi_\lambda^t}{cl_d} \tag{6}$$

$$[\theta]_\lambda^t = [\psi]_\lambda^t (M/100) = \frac{\psi_\lambda^t}{Cl_m} \tag{7}$$

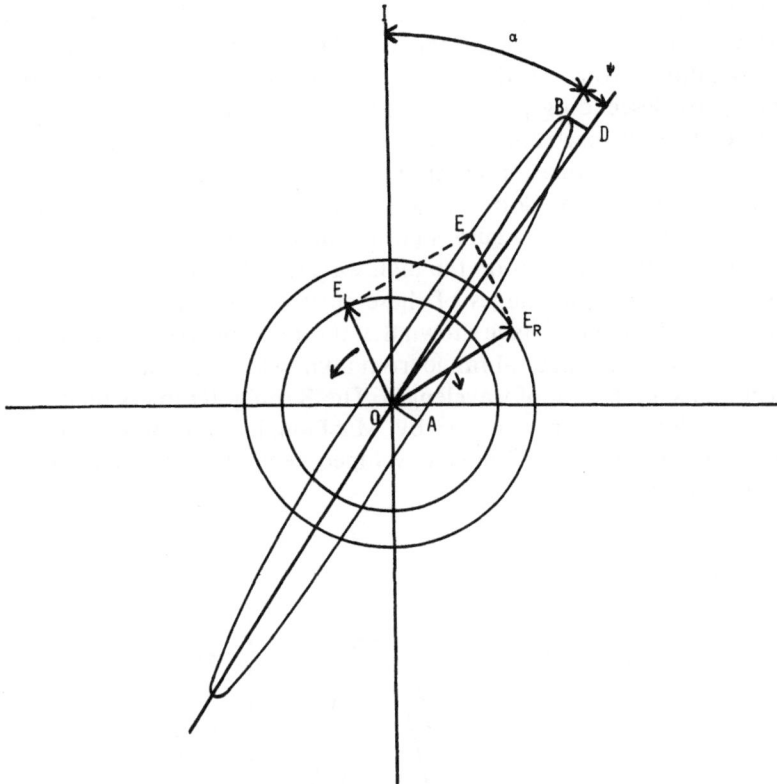

Fig. 2. Incident linear polarized light (I) rotated by an angle α and converted into elliptically polarized light having an angle of ellipticity ψ.

Optical circular dichroism spectra are usually depicted as plots of molar ellipticity or circular dichroic absorption ($\varepsilon_L - \varepsilon_R$ or $\Delta\varepsilon$) *vs.* wavelength. The reason for the use of the latter term is that, in practice, a series of equations have been described[1] which lead to equation 8:

$$[\theta]_\lambda^t = 3300(\varepsilon_L - \varepsilon_R) \qquad \text{(in deg-cm}^{-1}\text{ mole}^{-1}\text{ ml} \times 10^1) \qquad (8)$$

and, since circular dichroic absorption ($\Delta\varepsilon$) is related to circular dichroic optical density ($D_L - D_R$, or circular dichroic absorbance, $A_M - A_R$), it is relatively easy to determine circular dichroic absorption experimentally and to plot it directly.

The Cotton Effect

Inside an absorption region, the appearance of a circular dichroism spectrum, coupled with the conversion of linear polarized light into elliptically polarized light and the appearance of an "S"-shaped ORD curve is known as the Cotton effect.[20] The S-shaped ORD curves are often referred to as "anomalous" rotatory dispersion curves (compared to the smoothly ascending or descending plain or normal curves). Figure 3 shows both a circular dichroism spectrum and an optical rotatory dispersion curve for an optically active coordination compound inside an optically active absorption band. The sign of the Cotton effect in Fig. 3 is positive. It is common practice to refer to positive CD curves (and S-shaped ORD curves which have their most positive rotation on the high wavelength side) as *positive* Cotton effects, and the mirror images of these curves as *negative* Cotton effects. Kronig[21] and Kramers[22] have shown that it is possible to derive a dispersion curve over the entire spectral range from a knowledge of the corresponding absorption as a function of wavelength. The Kronig–Kramers relationships have been described in a paper by MacDonald and Brachman,[23] and Moffit and Moscowitz[24–28] have developed general relationships between CD and ORD curves.

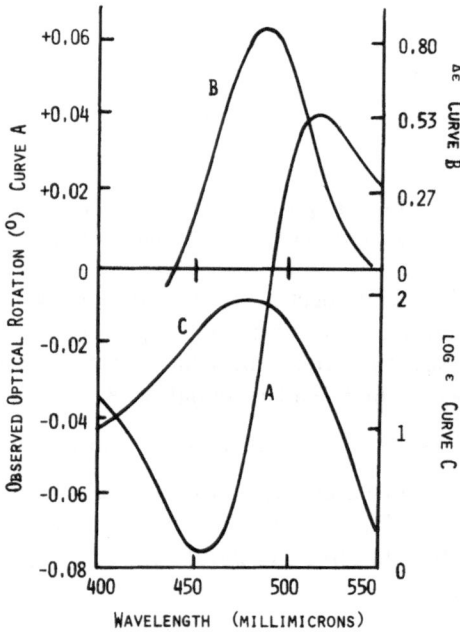

Fig. 3. The optical rotatory dispersion (A), circular dichroism (B), and absorption spectrum (C) of $(+)_D[Co(en)_3]^{+3}$ in water at 20°C.

THE APPLICATION OF ORD AND CD TECHNIQUES TO COORDINATION COMPOUNDS—ENVIRONMENTAL EFFECTS

Optically Inactive Ion Association

Kirschner and his co-workers[29-32] and McReynolds and Witmeyer[33] have studied the effects of optically inactive ions on optically active dissymmetric complex ions. The former authors have indicated that the major effect is an alteration of the rotatory strength of the optically active complex at the peak and trough wavelengths of the ORD curve inside the absorption region, and that the magnitude of this effect depends upon the extent of association of the oppositely charged ions. Smith and Douglas[34] have described the effects of optically inactive ions on the CD of complex ions, and they also have observed that the magnitude of the effect is quite sensitive to changes in the symmetry of the complex ion, as a result of interactions between optically active and inactive ions of opposite charge. Mason and Norman[35] have determined that "outer-sphere" complex formation between an optically active, colorless, organic compound and an optically inactive, colored complex cation is strong enough to induce the appearance of a CD peak in the absorption band of the optically inactive complex ion. It is proposed that hydrogen bonds form between the ligands on the complex ion and the optically active organic compound.

Solvent Effects

In many cases the nature of the solvent in which an ORD or CD spectrum is taken has been found to exert a considerable effect on the curves obtained.[18,35] Theoretical approaches to this problem have been undertaken by Kauzmann, Walter, and Eyring,[26] Weigang,[36] and Condon.[26] Kirschner and his co-workers[30] have studied the ORD in the visible region of $[Ru(o\text{-phenanthroline})_3](ClO_4)_2$ in water, methanol, ethanol, acetone, and chloroform. They found that although the wavelengths of the peaks and troughs of the ORD curves do not change, the rotatory strengths of these peaks and troughs depend markedly on the solvent. In particular, it was noted that as the dielectric constant of the solvent decreased from ethanol to chloroform, the increasing extent of ion-association resulted in more positive rotations at the ORD trough. However, it was also noted that in the high dielectric constant solvents (water and methanol), there was an even greater shift to more positive rotations at this same ORD trough. This is postulated as being due to a strong distortion of the dissymmetric electron cloud by the highly polar solvent, and to relatively little association between the perchlorate anion and the complex cation in these polar solvents.

Similar solvent effects have been found in organic systems.[37] Bosnich[38] observed an interaction between an optically active, colorless, solvent,

$d(-)$-2,3-butanediol and the $[PtCl_4]^{--}$ anion, as evidenced by the appearance of a negative CD peak in the visible region for this system. He and Watts[39] have also studied the energetics of the dissymmetric interactions between this same solvent and various enantiomers of cis-$[Co(en)_2Cl_2]ClO_4$ through the differential solubilities of these enantiomers in this solvent. Similar studies have been carried out by Premuzic and Scott[40] and Haines and Smith[41] with different systems.

"Loading Effect"

Kirschner and his co-workers[30] have investigated the ion-association of optically active complex ions with optically inactive anions. They have first examined the optical rotatory dispersion of a complex such as $(+)_D$-[Co-$(en)_3]Cl_3$, which contains three chloride ions for each complex ion. They then proceeded to add additional chloride (as NaCl) and studied the optical rotatory dispersion at Cl:complex ion ratios of 3:1, 30:1, 300:1, and 3000:1.[32] They noticed an enhancement in the rotatory strength at the peak and troughs of the optical rotatory dispersion curves of these complexes with increasing concentration of the optically inactive anion. Further, using phosphate anion instead of chloride,[30] the authors noted that this effect is very much enhanced, which is to be expected on the basis of the increased ion-pairing ability of phosphate over chloride.

The Determination of Coordination of Optically Active Ligands

In 1964 Bhatnagar and Kirschner studied the optical rotatory dispersion and absorption of the following compounds:

(I) $[Co(NH_3)_4(d\text{-tartrate})]ClO_4$	(IV) d-H_2tartrate	
(II) $[Co(NH_3)_5(d\text{-tartrate})]ClO_4$	(V) d-NaH tartrate	
(III) $[Co(NH_3)_6]_2(d\text{-tartrate})_3$	(VI) d-Na_2tartrate	

These investigators[42] noted the appearance of a Cotton effect in the visible region of the optical rotatory dispersion curve of compounds (I) and (II) as shown in Fig. 4. However, the ORD of compound (III) is essentially identical to that of compound (VI) in the visible region. These investigators have postulated that ORD (and CD) techniques are suitable for determining whether an optically active ligand is actually coordinated. In addition, they have postulated that it may be possible to determine whether an optically active compound is acting as a bidentate or monodentate ligand, assuming it has the ability to act in both ways. It is possible to do this because of the significantly stronger rotatory strength (in the Cotton-effect region) of the complex when the ligand is behaving as a bidentate group instead of a monodentate one.

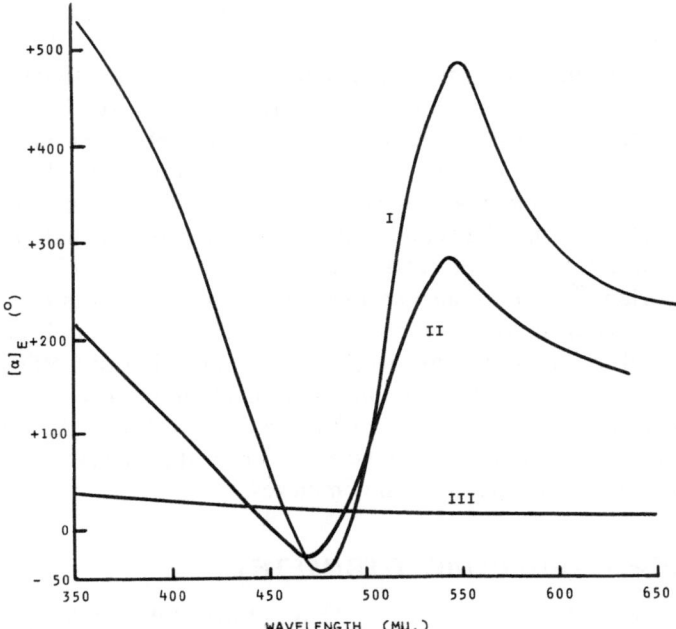

Fig. 4. The optical rotatory dispersion of $[Co(NH_3)_4(d\text{-tart})]^+$ (I), $[Co(NH_3)_5(d\text{-tart})]^+$ (II), and $[Co(NH_3)_6]_2(d\text{-tart})_3$ (III) in water at 25°C.

Kirschner and Pearson[43] utilized both ORD and CD techniques to demonstrate that it is possible for an optically active monodentate ligand to induce a Cotton effect in the absorption region of a complex, even if the dissymmetric center of the ligand is as many as five atoms removed from the metal ion.

ABSOLUTE CONFIGURATION AND CONFORMATION

A considerable amount of work has been put forth by many research investigators in an effort to arrive at the absolute configuration and conformation of optically active complexes on the basis of ORD and CD data. The basic technique involved is to compare the ORD and CD spectra of a complex whose absolute configuration is not known with those of a very similar complex of known absolute configuration. The absolute configurations (and/or conformations) of more than 15 compounds have been determined by x-ray studies.[44–57] Several investigators including Mason and co-workers,[47,58–62] Mathieu,[63] Gillard and co-workers,[64–66] Michelsen,[67] Bosnich,[68,69] K. Saito and co-workers,[70,71] Jaeger and Bijkerk,[72] and

Douglas and co-workers[73–76] have done absolute configuration studies of this type.

In addition to being used for absolute configuration studies, optical rotatory dispersion and circular dichroism techniques have also been used for the determination of absolute conformation. In 1959, Corey and Bailar[77] in a very significant paper, predicted the conformations of ethylenediamine when it behaves as a bidentate chelating agent in coordination compounds. Gollogly and Hawkins[78] have undertaken a conformational analysis of such coordination compounds of cobalt(III), and Sargeson, Buckingham, and their co-workers[79–85] have also carried out important studies on the conformations of coordination compounds. Several strong research groups in Japan have also done significant work in the application of ORD and CD to coordination compounds including those of Yasui and his co-workers,[86] Shibata and his co-workers,[87] Misumi and his co-workers,[88] Shimura and his co-workers.[89] Bailar and his co-workers[90] have applied ORD techniques to the study of optical inversions in complexes.

SPECTROPOLARIMETRIC TITRIMETRY

Kirschner and his co-workers[91–93] have utilized ORD techniques in the titrimetric determination of optically active and inactive strong and weak acids and bases, optically active and inactive ligands, metal ions, and n—the number of ligands bound to a metal ion in a complex. The general technique for determining a metal ion or an optically inactive ligand is to titrate one with the other using an added optically active indicator which has a stability constant that is at least 100 times smaller than that of the metal-ligand pair being used in the titration. An example is given in Fig. 5, which shows a spectropolarimetric titration of disodium dihydrogen ethylenediamine-tetraacetate (Na_2H_2edta) and zinc nitrate using *levo*-histidine as the optically active indicator. It can be seen that during the course of titration there is very little change in the optical rotation of the solution as the zinc-edta complex is being formed. After the 1:1 complex has formed essentially completely, the *levo*-histidine begins to complex the additional zinc ion added to the solution, which results in a marked change in the optical rotation of the system. The determination of the optical rotation at three or four points both before and after the end-point allows the end-point to be determined by drawing two straight lines and locating their intersection. Pecsok and Juvet[94] have studied the change of optical rotation of certain lead complexes containing an optically active ligand, as a function of a change in pH. They refer to these studies as "rotometric titrations."

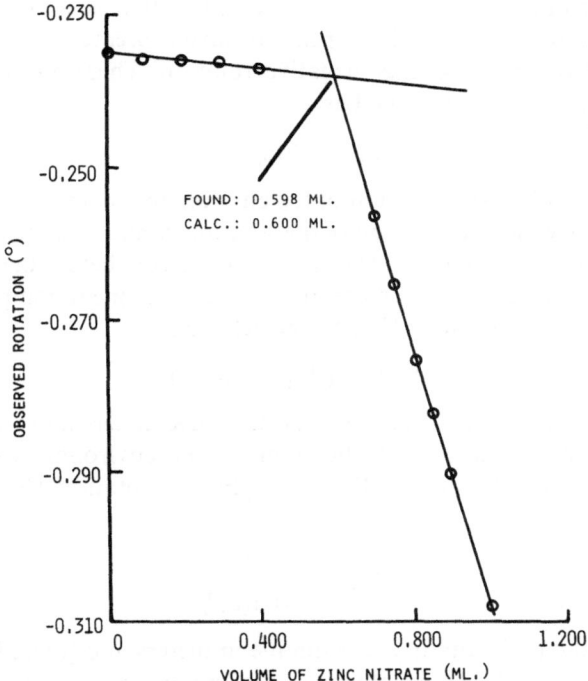

Fig. 5. Spectropolarimetric titration of aqueous zinc(II) nitrate and disodium dihydrogen ethylenediaminetetraacetate at 350 mμ using *levo*-histidine as an optically active indicator.

ORD, CD, AND THE PFEIFFER EFFECT

The Pfeiffer Effect

In 1931, Pfeiffer and Quehl[95] observed that the optical rotation of a solution of an optically active compound (e.g., ammonium d-α-bromo-camphor-π-sulfonate, hereinafter called an "environment" compound) changes significantly upon the addition of solutions of racemic mixtures of certain coordination compounds (e.g., D,L-[Ni(o-phen)$_3$]Cl$_2$). Brasted and his students,[2-5,96] and Dwyer and his co-workers[97] have studied this effect in some detail with particular reference to the source and nature of the effect. Kirschner and his co-workers[6-8] have also studied the effect and have found considerable evidence in support of the mechanism for this effect described by Dwyer and his co-workers.[97] Kirschner and Magnell[4] have developed quantitative expressions for the rotation change due to the Pfeiffer effect, and have defined a *positive Pfeiffer rotation* as an enhancement of optical rotation upon addition of the racemic complex; that is, a positive rotating environment substance would appear to have a greater positive rotation, and a negative one, a greater negative rotation. Therefore, the observed Pfeiffer rotation $(P)_{obs.}$ is defined as:

$$(P)_{obs.} = \pm(\alpha_{e+c} - \alpha_e) \qquad (9)$$

where α_e is the observed rotation of the environment compound in solution and α_{e+c} is the observed rotation of the solution containing the environment and the complex compounds. The sign to be inserted before the parentheses is the same as the sign of the solution of the environment compound alone. The specific Pfeiffer rotation $[P]_\lambda^t$ can be defined as:

$$[P]_\lambda^t = (P)_{obs.}/(c)(e)(l_d) \qquad (10)$$

where l_d is the path length of the solution in decimeters and where c and e are concentrations (in g/ml) of the complex and environment compounds, respectively [see equation (3)]. The Molar Pfeiffer rotation, $[P_M]_\lambda^t$, therefore is defined as:

$$[P_M]_\lambda^t = \frac{(P)_{obs.}}{[C][e]d_m} \qquad (11)$$

where d_m is the path length of the sample in meters and $[e]$ and $[C]$ are the concentrations of the environment and complex compounds, respectively [see equation (4)].

Pfeiffer-Active Systems

Several new systems exhibiting the Pfeiffer effect have been described by Kirschner and his co-workers.[6-8] Additionally, Table 1 gives some new Pfeiffer-active systems which are reported here for the first time.

Solvents and Environmental Compounds

The authors have observed the Pfeiffer effect in water, glacial acetic acid, and N,N-dimethylformamide. With regard to environment compounds, several have been reported,[2-8] and there seems to be a wide variety of compounds which can act in this capacity.

Nature of the Pfeiffer Effect

Gyarfas and Dwyer[97] postulate that those complexes which will exhibit the Pfeiffer effect are relatively optically labile, and, therefore, an equilibrium of the following type exists in such systems:

$$D\text{-complex} \overset{e^*}{\rightleftharpoons} L\text{-complex} \qquad (12)$$

$*e$ = the optically active environment solution

Table 1

Molar Pfeiffer Rotation, Absorption, and Circular Dichroism of Metal Oxalate Complexes

Complex[a]	Environment	Molar Pfeiffer rotation	Absorption	Circular dichroism	
		$[P_M]_{589}^{23°}$	λ_{max} (mμ)	λ_{max} (mμ)	$\Delta\varepsilon \times 10^4$
$K_3[Co(ox)_3]\cdot3H_2O$	d-Cinchonine Hydrochloride in Water	$-6500°$	605 425	618	$+768$
$K_3[Cr(ox)_3]\cdot3H_2O$	d-Cinchonine Hydrochloride in Water	$+4833°$	570 420	640 556 428	-92 $+820$ -152
$K_3[V(ox)_3]\cdot3H_2O$	d-Cinchonine Hydrochloride in Water	$-1533°$	605	600 440	$+182$ -786.7
$K_3[Fe(ox)_3]\cdot3H_2O$	d-Cinchonine Hydrochloride in Water	$-640°$	670	643 620	-1.2 -1.6
$K_3[Al(ox)_3]\cdot3H_2O$	d-Cinchonine Hydrochloride in Water	$-534°$	340	—	—

[a] ox = oxalate anion.

As yet, it is not possible to predict with absolute certainty which complexes will exhibit the Pfeiffer effect and which will not. One characteristic common to almost all of the complexes which exhibit this effect is the presence of unsaturation (such as benzene rings, double bonds, or partial double bonds) in the ligand or the ligand-metal system. Although it might be expected that the Pfeiffer effect would occur with a complex cation when the environment compound is a negative ion or a neutral molecule (and vice versa for negative ion complexes), it has been stated[98] that the Pfeiffer effect can occur for a racemic complex cation in a system in which the optically active environment substance is also a positive ion.

There is now considerable evidence to support the idea that an equilibrium between the two enantiomers of the optically labile complex does indeed exist, and that the equilibrium constant in an optically inactive environment is 1, but that in an optically active environment it is something other than 1. This comes about as the result of the conversion of one enantiomer into another with a resultant change in the optical rotation (and in the circular dichroism inside an optically active absorption band) of the system. Some evidence in support of this view follows.

Stability Constant and Pfeiffer Effect Changes

Kirschner and Magnell[6] have reported that the magnitude of the Pfeiffer effect increases for the series of tris(*ortho*-phenanthroline) complexes of zinc(II), cadmium(II), and mercury(II). The stability constants of these complexes, however, have been shown to be in the reverse order, indicating that the mercury complex is the most labile of the three, and, therefore, would be expected to be able to undergo the Pfeiffer effect with the greatest facility and to the greatest extent.

The Racemization of Pfeiffer-Active Complexes

During recent work on the Pfeiffer effect in aqueous solution, the authors have found a method of studying the racemization of Pfeiffer-active complexes without going through a resolution procedure involving diastereoisomer formation and fractional crystallizations. Figure 6 shows that malic acid does not change its optical rotation in solution over several days. This figure also shows the appearance of the Pfeiffer effect for the racemic nickel(II) tris(*ortho*-phenanthroline) complex over a period of 120 hr in the presence of *levo*-malic acid. At that point in time, addition of the same quantity of *dextro*-malic acid results in an immediate change in rotation (positive direction) equal in magnitude to the negative rotation of the original *levo*-malic acid. It also results in the displaced equilibrium described in equation (12) now finding itself in an optically inactive, racemic environment. Consequently, if the *levo* enantiomer of the nickel complex is actually in excess, it should begin to undergo racemization. This racemization is observed as shown by the curve (D) in Fig. 6, and it is identical to the racemization curve for the *levo* enantiometer of the nickel(II) tris(*ortho*-phenanthroline) complex cation, which is resolvable by conventional techniques.[99,100] This supports the postulate of Dwyer that a displacement of an equilibrium such as described in equation (12) actually occurs.

The Resolution of Pfeiffer-Active Complexes

Kirschner and Ahmad[101] have applied the Pfeiffer effect to the resolution of complex inorganic compounds. These authors allowed a Pfeiffer effect to develop in the initially *racemic* nickel(II) tris(2,2'-dipyridyl) complex in *levo*-malic acid, and then succeeded in freezing out the non-equimolar mixture of enantiomers from the system by means of precipitation of the enantiomers as the relatively insoluble perchlorates. These insoluble perchlorates were then redissolved by allowing an aqueous suspension of them to come in contact with Amberlite IRA-400 anion exchange resin in the chloride form, and the optical rotation of a solution of the soluble complexes (as chlorides) was then read on a photoelectric polarimeter. Table 2 shows the optical rotations of this complex, and Fig. 7 shows a comparison of the

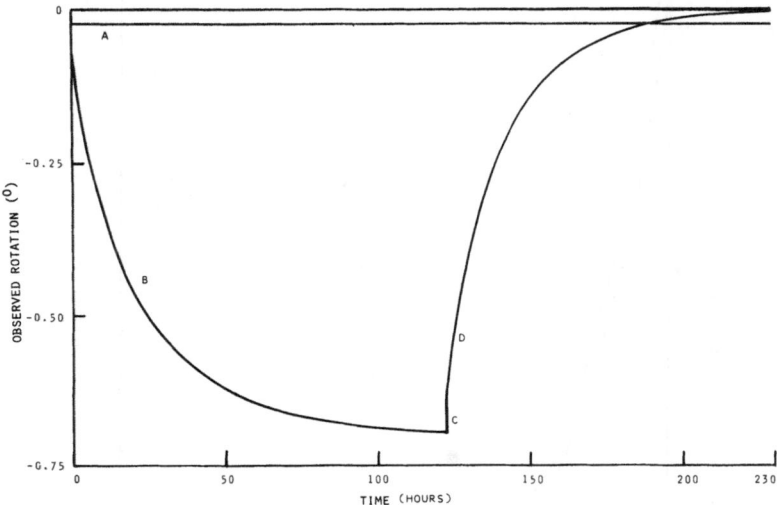

Fig. 6. The Pfeiffer effect and racemization in water of $[Ni(o\text{-}phen)_3]Cl_2$ with *levo*-malic acid as the environment compound. Optical rotation as a function of time for A: *levo*-malic acid; B: *levo*-malic acid + $D,L\text{-}[Ni(o\text{-}phen)_3]Cl_2$; C: system B + *dextro*-malic acid added after 123 hr; D: racemization of the $L\text{-}[Ni(o\text{-}phen)_3]Cl_2$ produced by the Pfeiffer effect (system C allowed to decay).

racemization curve of the *levo* complex obtained by the Pfeiffer effect resolution procedure, compared with the racemization curve for the same complex obtained by conventional resolution techniques.[102] The authors of this paper also have succeeded in achieving a partial resolution of the racemic tris(oxalto) complexes of cobalt(III) and chromium(III) by allowing the Pfeiffer effect to develop in each of these complexes in the presence of

Table 2
Resolution of $D, L\text{-}[Ni(dipy)_3]^{++}$ by the Pfeiffer Effect

Initial racemic complex	Environment compound	Pfeiffer rotation $(P_{obs.})^a$ (deg.)	Conc. of complex, M/liter	Observed rotation,[b] $\alpha_{obs.}$	$[M]^a$	% resolution
$[Ni(dipy)_3]^{++}$ (0.05 M)[c]	l-Malic Acid (0.2 M)	0.396°	0.048	−0.151°	−31.46°	1.5%

[a]25°C; 589 mμ.
[b]After resolution.
[c]dipy = 2,2'-dipyridyl.

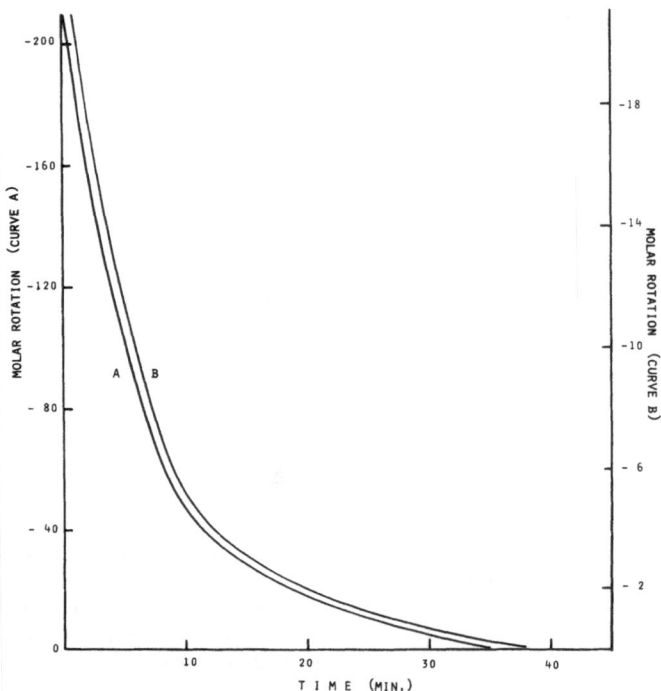

Fig. 7. The racemizations (at the sodium D line) in aqueous solution at
25°C. of (a) L-[Ni(dipy)$_3$]$^{++}$ (0.02 M) and (B) L-[Ni(dipy)$_3$]$^{++}$ partially
resolved by the Pfeiffer effect technique (0.048 M).

dextro-cinchonine hydrochloride. Then the optically active environment
(*d*-cinchonine hydrochloride) was completely removed from solution by
precipitation with the stoichiometric quantity of base, leaving the displaced
equilibrium mixture in the presence of an optically inactive environment
(water plus NaCl). The optical rotations of these partially resolved optically
active systems were then determined with a photoelectric polarimeter, and
were compared with the known rotations of the pure enantiomers, which
are capable of being resolved by conventional means.[103,104] Table 3 shows
the rotations of these partially resolved complexes.

Racemization Rate Studies of Pfeiffer-Active Systems

Kirschner and Ahmad[105] studied the rate of racemization of the nickel-
(II) tris(*ortho*-phenanthroline) complex in: (a) water; (b) water containing
levo-malic acid; and (c) water containing *dextro*-malic acid. If Dwyer and
Gyarfas' postulate concerning the existence of an equilibrium which is
shifted (in the presence of *levo*-malic acid, for example) in the direction

Table 3
Partial Resolution of Racemic Complexes by the Pfeiffer Effect

Initial racemic compound	Environment compound	Observed Pfeiffer rotation[a]	Conc. of complex,[b] M/liter	Observed rotation,[b] $\alpha_{obs.}$	$[M]$[a,b]	Percentage resolution
$K_3[Cr(ox)_3]\cdot3H_2O$ (0.05 M)[c]	d-Cinchonine Hydrochloride	$+0.287°$	0.012	$+0.023°$	$+190°$	2.6
$K_3[Co(ox)_3]\cdot3H_2O$ (0.05 M)	d-Cinchonine Hydrochloride	$-0.388°$	0.012	$-0.016°$	$-137°$	1.4

[a] 25°C.; 589 mμ.
[b] After resolution.
[c] ox = oxalate anion.

indicated by equation (12), then the racemization rate of the *levo* complex in the presence of this *levo* acid should be greater than that in pure water. Additionally, the racemization rate of the *dextro* complex in the presence of the levo form of the acid should be smaller than that in pure water. Figure 8

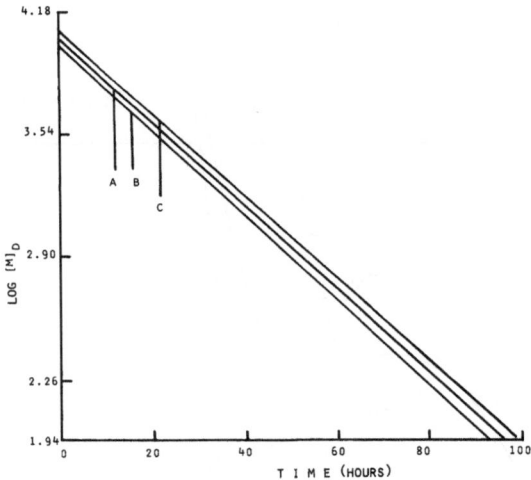

Fig. 8. The racemization of the enantiomers of [Ni(o-phen)$_3$]Cl$_2$ in aqueous *levo*-malic acid (conc. of Complex: 10^{-3}M; Conc. of *levo*-malic acid: 3×10^{-3}M) (A) L-[Ni(o-phen)$_3$]Cl$_2$ in water (no malic acid) ($t_{1/2}$: 14.25 hr; k: 13.51×10^{-6} sec.$^{-1}$); (B) D-[Ni(o-phen)$_3$]Cl$_2$ + Aqueous *levo*-malic acid ($t_{1/2}$: 13.75 hr; k: 14.00×10^{-6} sec.$^{-1}$); (C) L-[Ni(o-phen)$_3$]Cl$_2$ + aqueous *levo*-malic acid ($t_{1/2}$: 14.75 hr; k: 13.05×10^{-6} sec.$^{-1}$).

shows that these are the results obtained experimentally for this system, which offers additional support for the proposal of Dwyer and Gyarfas.[97]

ORD and CD Studies of Pfeiffer-Active Systems

If the Dwyer and Gyarfas proposal for the existence of an equilibrium between optically labile enantiomers exists, and if this equilibrium is displaced in the presence of an optically active environment, then a study of the optical rotatory dispersion and circular dichroism in the visible region of a Pfeiffer-active colored complex (in the presence of a colorless, optically active environment) should yield ORD and CD curves which are similar to those obtained for the pure, resolved enantiomers of these complexes. That this is actually the case is shown in Figs. 9 and 10, which give the

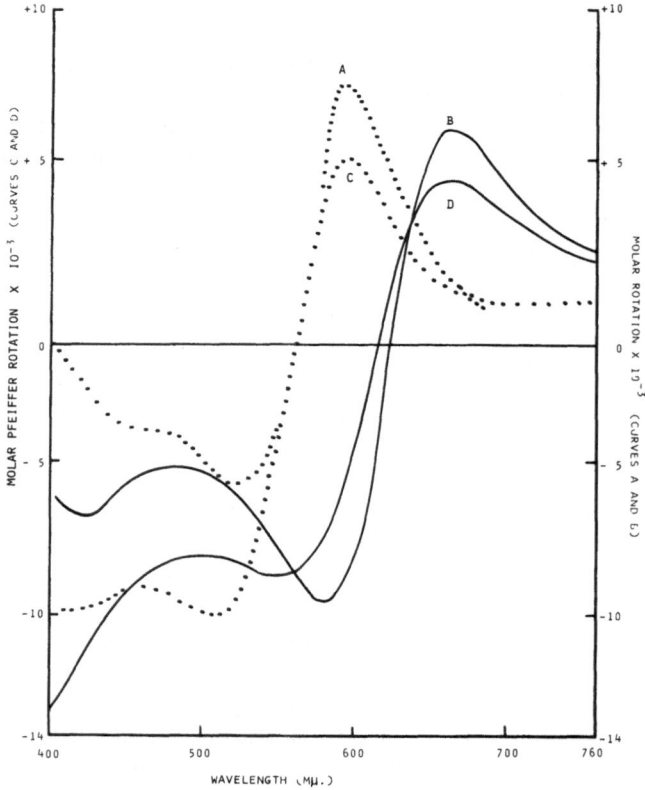

Fig. 9. Optical rotatory dispersions in water at 23°C. of (A) $(+)_D$[Cr-$(C_2O_4)_3]^{-3}$, (B) $(-)_D$[Co($C_2O_4)_3]^{-3}$ and Pfeiffer rotatory dispersions in water containing d-cinchonine hydrochloride at 23°C. of (C) D,L-[Cr($C_2O_4)_3]^{-3}$, (D) D,L-[Co($C_2O_4)_3]^{-3}$.

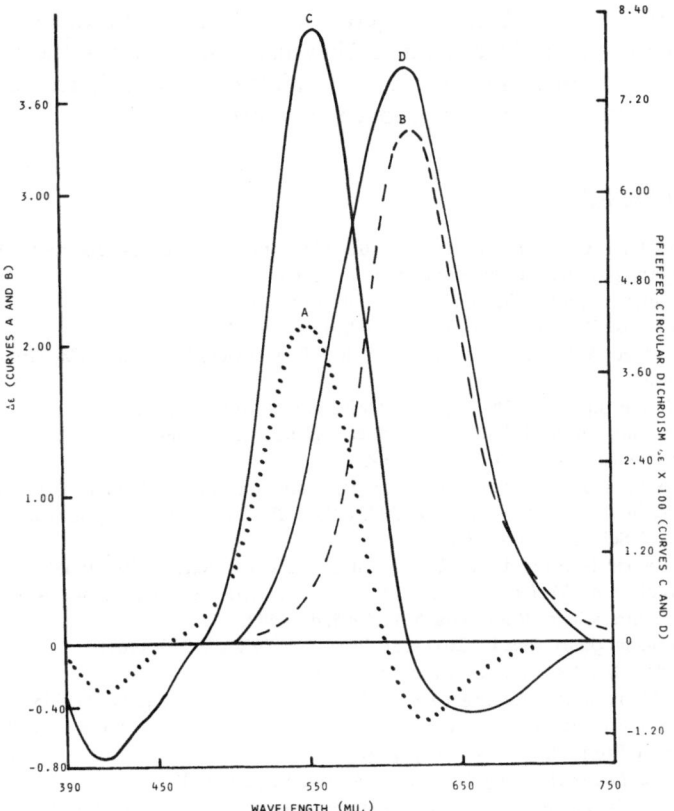

Fig. 10. Circular dichroism spectra in water at 23°C of (A) $(+)_D[Cr(C_2O_4)_3]^{-3}$, (B) $(-)_D[Co(C_2O_4)_3]^{-3}$ and Pfeiffer circular dichroism spectra in water containing d-Cinchonine hydrochloride at 23°C of (C) D,L-$[Cr(C_2O_4)_3]^{-3}$, (D) D,L-$[Co(C_2O_4)_3]^{-3}$.

ORD and CD curves, respectively, of the oxalate complexes described above, both as pure enantiomers resolved by conventional means, and as partially resolved enantiomers existing in Pfeiffer-active solutions. These results confirm the presence of an excess of one enantiomer of the metal complex in the Pfeiffer-active systems. The ORD and CD curves were taken with a Cary-60 Spectropolarimeter with a CD attachment.

ACKNOWLEDGMENTS

The authors wish to express their appreciation to the National Science Foundation for a research grant (GP-8450) and for a research equipment grant to Wayne State University (GP-6936) which contributed significantly to the progress of this investigation. They also wish to express their appreciation to Prof. John C. Bailar, Jr., for inspiration and guidance many years ago, which ultimately helped produce this work.

REFERENCES

1. S. Kirschner, *Coord. Chem. Revs.* **2**: 461 (1967); many references relating to these topics during the past decade are included in this review.
2. E. J. Kuhajek, Ph.D. Thesis, University of Minnesota, 1962.
3. V. J. Landis, Ph.D. Thesis, University of Minnesota, 1957.
4. F. R. Longo, A. Ventresca, Jr., J. E. Drach, J. E. McBride, R. F. Suers, *Chemist-Analyst* **54**: 101 (1965).
5. P. E. R. Nordquist, Jr., Ph.D. Thesis, University of Minnesota, 1964.
6. S. Kirschner and K. R. Magnell, Advances in Chemistry Series No. 62, American Chemical Society, Washington, D.C., 1966, p. 366.
7. S. Kirschner, K. Magnell, and K. H. Pearson, *Rev. Chim.* (Bucharest) **17**: 588 (1966); S. Kirschner, *J. Am. Chem. Soc.* **78**: 2372 (1956); *Proc. IIIrd Nat. Conf. for Chem.*, Roumanian Acad. Sci. Timisoara 1966, p. 35.
8. S. Kirschner, N. Ahmad, and K. Magnell, *Coord. Chem. Revs.* **3**: 201 (1968).
9. F. Woldbye, in: *Technique of Inorganic Chemistry*, (eds. H. Jonassen and A. Weissberger), Vol. IV, Interscience Publishers, N.Y., 1965, pp. 249 ff.
10. S. F. Mason, *Quart. Rev.* **17**: 20 (1963); *Proc. Roy. Soc.* A **297**: 3 (1967).
11. W. Kuhn, *Ann. Revs. Phys. Chem.* **9**: 417 (1958).
12. R. D. Gillard, in *Progr. Inorg. Chem.* **7**: 215 (1966); *Chem. Revs.* **65**: 603 (1965).
13. A. M. Sargeson, in *Chelating Agents and Metal Chelates*, (eds. F. P. Dwyer and D. P. Mellor) Academic Press, New York, 1964, pp. 184 ff.
14. P. Drude, *Lehrbuch der Optik*, S. Hirzel Verlag, Leipzig, 1906.
15. T. Lowry, *Optical Rotatory Power*, Longmans, Green and Co., London, 1935.
16. W. Heller, *J. Phys. Chem.* **62**: 1569 (1958).
17. W. Heller, in: *Physical Methods of Organic Chemistry* (ed. A. Weissberger) Vol. I, Part III, Interscience Publishers, New York, 1960, Chapter XXXIII, p. 2152.
18. A. Moscowitz in: *Advances in Chemical Physics* (ed. I. Prigogine) Vol. IV, Interscience Publishers, New York, 1962, pp. 67 ff.
19. E. U. Condon, W. Altar, and H. Eyring, *J. Chem. Phys.* **5**: 753 (1937).
20. A. Cotton, *Compt. Rend.* 120, 989, 1044 (1895); *Ann. Chim. Phys.* **18**: 347 (1896).
21. R. De L. Kronig, *J. Opt. Soc. Am.* **12**: 547 (1926).
22. H. A. Kramers, *Atti Congr. Intern. Fisici, Como* **2**: 545 (1927).
23. J. R. Macdonald and M. K. Brachman, *Rev. Mod. Phys.* **28**: 393 (1956).
24. A. Moscowitz, *Tetrahedron* **13**: 48 (1961).
25. A. Moscowitz, in: *Optical Rotatory Dispersion*, (ed. C. Djerassi) McGraw–Hill Book Co., Inc. New York, 1960, Chapter 12.
26. W. Kauzmann, J. Walter, and H. Eyring, *Chem. Revs.* **26**: 339 (1940); E. Condon, *Rev. Mod. Phys.* **9**: 432 (1937); W. Kuhn, *Ann. Rev. Phys. Chem.* **9**: 417 (1958); W. Moffit, *J. Chem. Phys.* **25**: 1189 (1956); H. M. Powell, *Endeavor* **15**: 20 (1956); M. Vol'kenshtein, *Zh. Exsperim. i Teor. Fiz. USSR.* **20**: 342 (1950); R. Servant, *J. Phys. Radium* [8] **3**: 90 (1942); J. Kirkwood, *J. Chem. Phys.* **5**: 479 (1937); M. Born, *Proc. Roy. Soc.* (*London*) A **150**: 84 (1935); A **153**: 339 (1936); M. Betti, *Trans. Faraday Soc.* **26**: 337 (1930); C. Djerassi, *Optical Rotatory Dispersion*, McGraw–Hill Book Co., Inc., New York, 1960.

27. W. Moffitt and A. Moscowitz, *J. Chem. Phys.* **30**: 648 (1959); D. F. Detar, *Biophys. J.* **6**: 505 (1966).
28. A. Moscowitz, Ph.D. Dissertation, Harvard University, 1957.
29. M. J. Albinak, D. C. Bhatnagar, S. Kirschner, and A. J. Sonnessa, in *Advances in: The Chemistry of the Coordination Compounds* (ed. S. Kirschner) The Macmillan Co., New York, 1961, pp. 154 ff.
30. S. Kirschner, D. C. Bhatnagat, M. J. Albinak, and A. J. Sonnessa, in: *Theory and Structure of Complex Compounds* (ed. B. J. Trzbiatowska) Pergamon Press, Oxford, 1964, pp. 63 ff.
31. M. J. Albinak, D. C. Bhatnagar, S. Kirschner, and A. J. Sonnessa, *Can. J. Chem.* **39**: 2360 (1961).
32. S. Kirschner and D. C. Bhatnagar, in: *Proceedings of the Symposium on Coordination Chemistry, Tihany, Hungary, 1964*, Akademiai Kiato, Publishing House of the Hungarian Academy of Sciences, Budapest, 1965, pp. 23 ff.
33. J. P. McReynolds and J. R. Witmeyer, *J. Am. Chem. Soc.* **62**: 3148 (1940).
34. H. L. Smith and B. E. Douglas, *J. Am. Chem. Soc.* **86**: 3885 (1964); *Inorg. Chem.* **5**: 784 (1966).
35. S. F. Mason and B. J. Norman, *Proc. Chem. Soc. (London)* **339**: (1964); *Chem. Comm.* 73 (1965); W. Kuhn, *Z. Physik. Chem. (Leipzig)* **B 30**: 356 (1935).
36. O. Weigang, *J. Chem. Phys.* **41**: 1435 (1964).
37. S. Yamada, K. Ishikawa, and K. Achiwa, *Chem. Pharm. Bull. Japan* **13** (7): 892 (1965).
38. B. Bosnich, *J. Am. Chem. Soc.* **88**: 2606 (1966).
39. B. Bosnich and D. W. Watts, *J. Am. Chem. Soc.* **90**: 6228 (1968).
40. E. Premuzic and A. I. Scott, *Chem. Comm.* 1078 (1967).
41. R. A. Haines and A. A. Smith, *Can. J. Chem.* **46**: 1444 (1968).
42. D. C. Bhatnagar and S. Kirschner, *Inorg. Chem.* **3**: 1256 (1964).
43. S. Kirschner and K. H. Pearson, *Inorg. Chem.* **5**: 1614 (1966).
44. Y. Saito, K. Makatsu, M. Shiro, and H. Kuroya, *Acta Cryst.* **7**: 636 (1954); **8**: 729 (1955); K. Nakatsu, M. Shiro, Y. Saito and H. Kuroya, *Bull. Chem. Soc. Japan* **30**: 158, 795 (1957); K. Nakatsu, *ibid.* **35**: 832 (1962). Also see J. M. Bijvoet, *Endeavour*, **14**: 71 (1955).
45. R. Pepinsky, *Rec. Chem. Progr.* **17**: 145 (1956).
46. Y. Saito, H. Iwasaki, and H. Ota, *Bull. Chem. Soc. Japan* **36**: 1543 (1963); K. Nakatsu, *ibid.* **35**: 832 (1962).
47. A. J. McCaffery, S. F. Mason, and B. J. Norman, *Chem. Comm.* 49 (1965).
48. Y. Saito, *Proc. 10th Internat. Conf. on Coord. Chem., Tokyo and Nikko, Japan*, Butterworths, London, 1967.
49. O. L. Carter, A. T. McPhail, and G. A. Sim., *J. Chem. Soc.* 365 (1967).
50. H. Scouloudi and C. H. Carlistle, *Acta Cryst.*, **6**: 651 (1953); *Nature* **166**: 357 (1950); B. W. Brown and E. C. Lingafelter, *Acta Cryst.* **17**: 254 (1964).
51. A. Nakahara, Y. Saito, and H. Kuroya, *Bull. Chem. Soc. Japan* **25**: 331 (1952).
52. S. Ooi, Y. Komiyama, Y. Saito, and H. Kuroya, *ibid.* **32**: 263 (1959).
53. Y. Saito and H. Iwasaki, in: *Advances in the Chemistry of the Coordination Compounds* (ed. S. Kirschner,) The Macmillan Co., New York, 1961; *Bull. Chem. Soc. Japan* **35**: 1131 (1962).
54. H. Reihlen, E. Weinbrenner, and G. v. Hessling, *Ann.* **494**: 143 (1932).
55. J. N. Van Niekerk and F. R. L. Scheoning, *Acta Cryst.* **5**: 196, 475, 499 (1952).
56. H. Kuroya, K. Matsumoto, and Y. Kushi, *Proc. 10th Internat. Conf. on Coord. Chem., Tokyo and Nikko, Japan* (ed. K. Yamasaki) Chem. Soc. Japan, 90 (1967).
57. Y. Saito, *Nippon Kagaku Zasshi* **88**: 367 (1967).
58. S. F. Mason, in: *Optical Rotatory Dispersion and Circular Dichroism in Organic Chemistry* (ed. G. Snatzke) Heyden and Son. Ltd., London, 1967, pp. 116 ff.

59. R. E. Ballard, A. J. McCaffery, and S. F. Mason, *Proc. Chem. Soc.* 331 (1962); A. J. McCaffery and S. F. Mason, *Mol. Phys.* **6**: 359 (1963); *Proc. Chem Soc.* 388 (1962).
60. S. F. Mason and B. J. Norman, *Inorg. Nucl. Chem. Letters* **3**: 285 (1967).
61. S. F. Mason, *Proc. Chem. Soc.* **137** (1962); A. J. McCaffery and S. F. Mason, *Trans. Far. Soc.* 481, 1 (1963); *Mol. Phys.* **6**: 359 (1963).
62. D. H. Templeton, A. Zalkin, and T. Ueki, *Acta Cryst.* **21** (1966) A 154 (supplement).
63. J. P. Mathieu, *Ann. Phys.* **19**: 335 (1944); *Bull. Soc. Chim. France*, (5) 3, 476 (1936); (5) 3, 476 (1926); *J. Chim. Phys.* **33**: 78 (1936).
64. J. H. Dunlop and R. D. Gillard, *J. Inorg. Nucl. Chem.* **27**: 361 (1965); M. G. B. Drew, J. H. Dunlop, R. D. Gillard, and D. Rogers, *Chem. Comm.*, 42 (1966); J. H. Dunlop, R. D. Gillard and G. Wilkinson, *J. Chem. Soc.* 3160 (1964); K. Garbett and R. D. Gillard, *J. Chem. Soc.* 6084 (1965); *Coord. Chem. Rev.* **1**: 179 (1966); B. J. Norman, *Inorg. Chim. Acta* **1**: 177 (1967).
65. K. Garbett and R. D. Gillard, *Chem. Comm.* 76 (1965).
66. J. H. Dunlop, R. D. Gillard, and N. C. Payne, *J. Chem. Soc.* 1469 (1967); *Progress in Inorganic Chemistry* (ed. F. A. Cotton) **7**: 215 (1966); *Physical Methods in Advanced Inorganic Chemistry* (eds. H. Hill and P. Day) Interscience Publishers, New York, 1968, pp. 167 ff.
67. K. Michelsen, *Acta. Chem. Cand.* **19**: 1175 (1965).
68. B. Bosnich, *Inorg. Chem.* **7**: 178 (1968).
69. B. Bosnich, *J. Am. Chem. Soc.* **90**: 627 (1968).
70. H. Ito, J. Fugita, and K. Saito, *Bull. Chem. Soc. Japan* **40**: 2584 (1967).
71. J. Fugita, K. Ohashi, and K. Saito, *ibid.* **40**: 2986 (1967).
72. F. M. Jaeger and L. Bijkerk, *Proc. Acad. Sci. Amsterdam* **40**: 246, 254 (1937).
73. J. I. Legg and B. E. Douglas, *J. Am. Chem. Soc.* **88**: 2697 (1966).
74. C. T. Liu and B. E. Douglas, *Inorg. Chem.* **3**: 1356 (1964).
75. H. L. Smith and B. E. Douglas, *J. Am. Chem. Soc.* **86**: 3885 (1964), *Inorg. Chem.* **5**: 784 (1966).
76. C. Y. Lin and B. E. Douglas, *Inorg. Nucl. Chem. Letters* **4**: 15 (1968).
77. E. J. Corey and J. C. Bailar, Jr., *J. Am. Chem. Soc.* **81**: 2620 (1959).
78. J. R. Gollogly and C. J. Hawkins, *Aust. J. Chem.* **20**: 2395 (1967).
79. A. M. Sargeson and G. A. Searle, *Inorg. Chem.* **4**: 45 (1965); **6**: 787 (1967).
80. L. G. Marzilli and D. A. Buckingham, *Inorg. Chem.* **6**: 1042 (1967).
81. D. A. Buckingham, S. Mason, A. M. Sargeson, and K. R. Turnbull, Inorg. Chem. **5**: 1649 (1966).
82. D. A. Buckingham, P. A. Marzilli, and A. M. Sargeson, *ibid.* **6**: 1032 (1967).
83. D. A. Buckingham, P. A. Marzilli, I. E. Maxwell, A. M. Sargeson, M. Fehlman, and H. C. Freeman, *Chem. Comm.* 488 (1968).
84. J. F. Blount, H. C. Freeman, A. M. Sargeson, and K. R. Turnbull, *Chem. Commun.* 324 (1967).
85. A. M. Sargeson, in: *Transition Metal Chemistry* (ed. R. Carlin) Marcel Dekker, Inc., 1966, Vol. 3, pp. 303 ff.; in: *Chelating Agents and Metal Chelates* (eds. F. P. Dwyer and D. P. Mellor) Academic Press, New York (1964).
86. T. Yasui, J. Hidaka and Y. Shimura, *J. Am. Chem. Soc.* **87**: 2762 (1965); *Bull. Chem. Soc. Japan* **39**: 2417 (1966); **38**: 654 (1965); T. Yasui, *ibid.* **38**: 1746 (1965).
87. M. Shibata, H. Nishikawa, and K. Hosaka, *ibid.* **40**: 236 (1967); M. Shibata, H. Nishikawa, and Y. Nishida, *ibid.* **39**: 2310 (1966).
88. S. Kida, T. Isobe, and S. Misumi, *ibid.* **39**: 2786 (1966).
89. K. Ohkawa, J. Fujita, and Y. Shimura, *ibid.* **38**: 67, 654 (1965).
90. J. C. Bailar, Jr., *Rev. Pure Appl. Chem.* **16**: 91 (1966); J. C. Bailar, Jr. and R. W. Auten, *J. Am. Chem. Soc.* **56**: 774 (1934); E. Kyuno and J. C. Bailar, Jr., *ibid.* **88**: 5447, 1120 (1966); E. Kyuno, L. J. Boucher, and J. C. Bailar, Jr., *ibid.* **87**: 4458 (1965); L. J. Boucher, E. Kyuno, and J. C. Bailar, Jr., *ibid.*, **86**: 3656 (1964).

91. S. Kirschner and D. C. Bhatnagar, *Anal. Chem.* **35**: 1069 (1963).
92. S. Kirschner and K. Pearson (in press).
93. K. H. Pearson, Ph.D. Dissertation, Wayne State University, 1966.
94. R. L. Pecsok and R. S. Juvet, *J. Am. Chem. Soc.* **78**: 3967 (1956); R. S. Juvet, *ibid.* **81**: 1796 (1959).
95. P. Pfeiffer and K. Quehl, *Ber.* **64**: 2667 (1931); **65**: 560 (1932), P. Pfeiffer and Y. Nakasuka, *ibid.* **66**: 410 (1933).
96. R. Brasted, Ph.D. Thesis, University of Illinois, 1942.
97. E. C. Gyarfas and F. P. Dwyer, *Rev. Pure Appl. Chem.* **4**(1): 73 (1954); *J. Am. Chem. Soc.* **73**: 2332 (1951); *J. Chem. Soc.* 953 (1950); *Nature* **167**: 1036 (1951); *ibid.* **168**: 29 (1951); with G. Barnes and J. Backhous, *ibid.* **89**: 151 (1956).
98. E. C. Gyarfas, private communication.
99. T. R. Harkins, Jr., J. L. Walter, O. F. Harris, and H. Frieser, *J. Am. Chem. Soc.* **78**: 260 (1956).
100. F. P. Dwyer and E. C. Gyarfas, *J. Proc. Roy. Soc. N. S. Wales* **83**: 232 (1949).
101. S. Kirschner and N. Ahmad, *J. Am. Chem. Soc.* **90**; 1910 (1968).
102. G. Morgan and F. Burstall, *J. Chem. Soc.* 2213 (1931).
103. G. B. Kauffman and L. T. Takahashi, *J. Chem. Educ.* **39**: 481 (1962).
104. J. W. Vaughen and V. E. Magnuson, *Inorganic Syntheses*, 1969 (in press).
105. S. Kirschner and N. Ahmad, in: *Progress in Coordination Chemistry* (ed. M. Cais) Elsevier Publishing Co., New York, 1968, pp. 582 ff.

ENVIRONMENTAL ASPECTS OF THE PFEIFFER EFFECT

R. C. Brasted, V. J. Landis,* E. J. Kuhajek,†
P. E. R. Nordquist,‡ and L. Mayer

University of Minnesota
Minneapolis, Minnesota

INTRODUCTION

Pfeiffer and co-workers[1-3] in an attempt to prepare and resolve complexes of the zinc ion using bidentate ligands encountered an optical anomaly which since has been named the Pfeiffer effect.§ Pfeiffer and Quehl determined the rotation of a 0.04 M solution of zinc camphor sulfonate, $Zn(CS)_2 \cdot 6H_2O$, to be 0.92°. On addition of a three mole quantity of phenanthroline (phen) the rotation diminished to 0.08°. In a similar experiment using the brom-camphor sulfonate (BCS^-) salt of zinc (in place of camphor sulfonate), there was noted an increase of the rotation from 4.55° to 8.44° at Na-D line. It has long been recognized that these two resolving agents (as NH_4CS and NH_4BCS) do give opposite enantiomers when used in resolutions of complexes. It was also reported by the same authors that a solution of zinc sulfate, NH_4BCS and phen in the ratio $1:2:3$ gave the same optical rotation as a solution containing the same concentrations of the complex represented as $[Zn(phen)_3](BCS)_2$. The present authors have investigated this claim further and the experiments will be discussed later.

When phen is replaced by the ligand 2,2′-dipyridyl (dipy) an anologous effect is noted, but diminished in rotational intensity by a factor of about one tenth. From all of the solutions, regardless of the ligand, it was possible to regain the stable, optically active species (e.g., CS^- or BCS^-) without change in the original rotation. The complex itself is also precipitatable as the bromide or iodide without optical activity. It is evident that no resolution of the complex has been achieved.

*Present address: Dept. of Chemistry, San Diego State University, San Diego, Calif.

†Present address: Morton Chemical Co., Chicago, Ill.

‡Present address: Monsanto Chemical Co., St. Louis, Mo.

§The term "Pfeiffer effect" was first applied to these systems during discussions with the late Prof. F. P. Dwyer and one of us (RCB) at the time of a symposium on coordination chemistry at the Indiana University in 1954.

Even in the earliest work of Pfeiffer[1-3] it was noted that the observed rotation was a function of the wavelength of the plane polarized light. For solutions 0.08 M in NH_4BCS and 0.04 M in Zn^{+2}, the rotation varied from 3.69° at 656.3 mμ to 6.05° at 546 mμ. When the solutions were 0.12 M (a three to one ratio) in phen, the α obs. values were 6.65° at 656.3 mμ and 10.62 at 546 mμ. Properly treated data using the Drude equations have been found useful by the present authors in investigating the source or sources of the Pfeiffer rotation.

The changes in rotation have been noted with resolved species that are positively charged, negatively charged, and neutral. A number of alkaloids have been used. Cinchonine hydrochloride, $C_{19}H_{22}N_2O \cdot HCl \cdot 2H_2O$, as a 0.04 M solution in $ZnSO_4$ of the same concentration exhibits a rotation of 5.29°. When a three-fold molar quantity of phen is added the rotation decreases to $-1.89°$ at "zero time" with a drop to $-2.46°$ at equilibrium. The alkaloid is isolatable from the solution unchanged, as is also the complex (*vide supra*). Strychnine acidsulfate, $C_{21}H_{22}N_2O_2 \cdot H_2SO_4 \cdot H_2O$, gives a positively charged species in solution showing the effect. The fact that both negatively (e.g., CS^- and BCS^-) and positively charged active species are operative in causing the Pfeiffer rotation suggests a more comprehensive study on such environmental factors as pH and ionic strength. Nicotine is still another of the alkaloids with efficacy, yet being neutral in charge. Inconsistent time data are found in the literature using alkaloids (*vide infra*).[4] Another inconsistency in the use of alkaloids is found when a resolved species cinchonine monochloromethylate is used. This substance is Pfeiffer inactive according to Pfeiffer and Quehl. They assumed from the experiments that a "free amine nitrogen" must be present in the alkaloid structure if the anomaly is to be observed. They stated that the alkaloid could not have entered into the metal ion coordination sphere of the metal complex, since the complex is recoverable without the alkaloid.[2] That this deduction is unsound is obvious because in any system the lower energy species is more likely to be formed and is isolatable. It is not expected that an alkaloid would compete with a strong field ligand such as phen.

Pfeiffer and Nakatsuka[3] extended time dependence studies from the zinc system to nickel. The rotation of the system $[Ni(phen)_3]SO_4$ (0.04 M) and NH_4BCS (0.08 M) was followed over a two-week period. The zero time rotation of 4.96° changed slowly to 6.35° after 13 days. The rotation accountable to the NH_4BCS alone was 3.78°. It is evident that the initial interaction gave rise to a substantial fraction of the total change. The rotation after 3 hr was only 0.09° greater than that at zero time. Further mention is made of this time function.

The Pfeiffer rotation as a function of the concentration of the ligand (e.g., phenanthroline and dipyridyl) was originally observed by Pfeiffer and

Quehl.[2] The exhaltation of the rotation was attributed to the formation of small quantities of the tris complex although only a 1:1 ratio of metal ion to the ligand existed in solution. The present investigators, disagreeing with this deduction, have studied the Zn^{+2}:phen system in detail (*vide infra*) with several active species.

Throughout their studies Pfeiffer and his co-workers have suggested only one reason for the anomalous rotation arising from a resolved species in the presence of a resolvable metal complex. This change in rotation was attributed to a shift in an equilibrium rather than the formation of any new asymmetric center.

$$d\text{-(complex)} \underset{\text{species}}{\overset{\text{resolved}}{\rightleftharpoons}} l\text{-(complex)}$$

According to this explanation, a resolving agent such as d-camphor sulfonate would shift this equilibrium to the right, whereas, d-bromcamphor sulfonate would shift the equilibrium in the opposite direction. By a not clearly defined mechanism one or the other of the antipodes would increase in thermodynamic activity with a resulting change in optical activity.

Turner and Harris[5] have discussed the more general problem of rotational changes when a species, stable with respect to racemization, is added to a solution containing an optically labile species (not necessarily a metal complex). When new rotations evolve, the authors describe the changes as occurring through asymmetric transformations. Using an equilibrium equation, in which the resolved stable species (base) is B and the racemic species (acid) is A, the fraction distribution (x) could be represented as

$$d, l - (A) + dB \rightleftharpoons d(A)d(B) + l(A)d(B)$$

$$(1 - x) \qquad (x)$$

Whenever x is greater than or less than 0.5, a rotation will be observed. Although the Pfeiffer and the Harris explanations appear to be identical, Harris contends, in opposition to Pfeiffer but in agreement with the present authors, that an A–B association of some kind is essential to the development of the Pfeiffer-type activity. The term "first-order asymmetric transformation" is used for a system in which no optically active phase other than that originally introduced can be isolated. If such a new phase is isolatable, the transition is called "second order." Turner and Harris have used the former term (first order) to describe the observations of Pfeiffer, Dwyer, and others.[6]

A case in point which does not involve a metal complex (excluding for purposes of discussion the proton as a pseudo metal ion) is the Kuhn effect. The late R. Kuhn and Albrecht[7] showed that the quinine salt of 4,4'-dinitrodiphenic acid is dextro rotatory in chloroform. The quinine salts of two structurally similar acids, phthalic and *m*-benzoic exhibit *levo* rotation. Other studies show[8-17] similar rotations when the cinchona alkaloids are used with various other diphenic acids. It has been suggested by Kuhn that a stereospecificity exists with these alkaloids in causing rotation of one phenyl group about the biphenyl bond leading to an unsymmetrical "complex." Since no active phase (other than the original alkaloid) is isolable the system may be classified as a first order asymmetric transformation.

Biswas[17,18] and others have reported the effect of *d*-tartaric acid on molybdates in solution. Although no mechanism is proven and no active complex is isolated, it is assumed that in solution the molybdate and *d*-tartaric acid may form an octahedral complex with the remaining four sites occupied by three oxo groups and a water molecule. The tartaric acid is coordinated through one carboxyl and one hydroxyl group, $2H^+[MoO_3-(H_2O)(d-C_3H_2(OH)_2(COO)_2)]^=$. Biswas suggested that with coordination the complex resulting could exist with one or more new asymmetric centers, or the tartaric acid could possess two labile forms, only one of which reacts with the molybdate radical. The result in either case would be a diminished concentration of one form of *d*-tartaric acid, and thus a change in rotation.

It is expected that the rotation of an active ligand will be changed when coordination with a metal ion occurs. Thus, when coordinated, *l*-propylenediamine (*l*pn) has quite a different specific rotation than when in the pure form. Even on dilution this ligand shows something of an anomaly. The effect of coordinating *l*-propylenediamine on its activity has been studied by a number of investigators.[20,21] Toole and O'Brien[20] have reported that molar rotation of *l*pn increases almost three-fold as the concentration is raised from 5 to 95%. In contrast, there is strict linearity (constant specific rotation) observed in the rotation of metal (as well as H^+) ionic salts of BCS^-.

The association concept[21,22] and the importance of a vicinal effect has been given credence (RCB) in a number of studies which will be later described. Refractometric, conductometric, cryoscopic data have been related to the rotation of aqueous solutions of complexes with active species. Dwyer and co-workers[23,24] have reported a number of experiments strongly favoring the concept of equilibrium shift, which according to them results in "configurational activity". In their investigations such environmental variables as ligand, active species, solvent, and oxidation state of the metal ion have been involved. The concept of configurational activity includes the idea that an asymmetric electrical field caused by the stable optical species contributes to the activity coefficients of the *d*- and *l*-metal complex antipodes, thus giving a new solution rotation. Note that the activity here means thermodynamic rather than optical activity. An experiment reported by Dwyer[23,25] proved that the solubility of both *d*- and *l*-[Ru(phen)₃]-(ClO₄)₂ was increased by either d-BCS or l-tar. However, the l-form of the complex was slightly more soluble in each case. To be exact the concentration of the d-isomer in a 3.05×10^{-2} M BCS$^-$ solution was 0.03 M lower in concentration than the l-isomer. Rates of racemization for complexes are reported as being different for complexes with and without the presence of an active species.[26,27] Specifically, it is reported that large, optically inactive ions reduce the rates of racemization of the resolved complex relative to the rates in water (with no extraneous ions), but that such inactive ions reduce the rate of racemization of each isomer *equally*. The effect of certain inactive ions is discussed later by the present authors. Optically active ions, on the other hand, are reported to reduce the rates of racemization *unequally*.

Configurational activity and equilibrium shift postulates are also supported in an experimental sequence reported by Dwyer and his co-workers. A solution of $[Ni(phen)_3]^{+2}$ when placed with such active species as l-quinine, BCS$^-$ or even the resolved complex d-$[Co(en)_3]^{+3}$ and treated with iodide ion permits a fractional precipitation of the nickel complex. It is reported that the least soluble fraction is dextro rotatory (about 0.1° as a saturated solution) and the more soluble fraction, levo. On standing this fraction becomes more levo. The separated fraction now, on fractional precipitation with iodide, showed that the least soluble fraction had *no* rotation and the most soluble fraction was levo rotatory. These observations according to Dwyer would be consistent with a d-l complex equilibrium shift and would not necessitate interaction between the complex and the resolved species.

Data derived by Dwyer on the ratio of the rate constants (k_d/k_l) has been criticized by Harris[28] since the values used by Dwyer for the rotation at the end of the racemization of d-$[Ni(phen)_3]^{++}$ were those obtained immediately after adding the optically active ion to the racemic mixture of the complex in solution. The ratio thus was not the equilibrium rotation. It was not reasonable, according to Harris, to use the data in support of an equilibrium theory for the Pfeiffer effect. Contradictory results are reported by Craddock and Jones[29] in that no difference was found for the racemization rates for either isomer (e.g., d- or l-$[Ni(phen)_3]^{2+}$) if the complex is in the presence of an optically active species. These authors point out that another environmental factor, temperature, could have accounted for unusual or anomalous rotations previously found.[26] It is evident that something more than an equilibrium shift or a configurational activity is needed to explain Pfeiffer rotations.

The importance of the solvent as an environmental factor was inferred by extraction experiments.[30] A solution of l-$[Co(acac)_3]$ (acac = acetylacetone) in CCl_3H-EtOEt mixture was extracted with 2% d-$[Co(en)_3]I_3$ saturated with ether. The portions extracted into the aqueous layer were then extracted quantitatively with $CHCl_3$ and evaporated. The green residue was levorotatory. The $CHCl_3$ extract of a solution of the acac complex and d-[Co(en)$_3$]$^{+3}$, allowed to stand for three days at room temperature, was observed to be slightly levo rotatory. Although Dwyer claims that these extraction experiments are in support of a configurational activity and equilibrium shift (an inferred independence from vicinal effects), there are other experiments which have been performed by the present authors (described later) and others[31,32] that shed light on the solvent effects that may well have been operative in Dwyer's experiments.[33,34] A very marked change in the anomalous rotation is noted when solvents with dielectric constants lower than the value for water are used.

The oxidation state of polyvalent metal ions (as well as the metal ion itself) is still another variable in an environment study. It is reported[16] that the red-ox potential of the d-[Os(dipy)$_3$]$^{2+}$/d-[Os(dipy)$_3$]$^{+3}$ and the l-[Os(dipy)$_3$]$^{+2}$/l-[Os(dipy)$_3$]$^{+3}$ couples differ slightly if the environment is changed to include an optically active species such as BCS$^-$, CS$^-$ or d- or l-[Co(en)$_3$]$^{+3}$. The data are found in Table 1. The measured oxidation

Table 1
The Differentiation of Oxidation Potentials by Optically Active Ions

Couple	Optically active ion	Slope, $mv/0.1\ (I)^{1/2}$
d[Os(bipy)$_3$]$^{+2}$-d[Os(bipy)$_3$]$^{+3}$	BCS$^-$	8.4
l[Os(bipy)$_3$]$^{+2}$-l[Os(bipy)$_3$]$^{+3}$	BCS$^-$	9.8
d[Os(bipy)$_3$]$^{+2}$-d[Os(bipy)$_3$]$^{+3}$	CS$^-$	13.6
l[Os(bipy)$_3$]$^{+2}$-l[Os(bipy)$_3$]$^{+3}$	CS$^-$	20.2
d[Os(bipy)$_3$]$^{+2}$-d[Os(bipy)$_3$]$^{+3}$	l-[Co(en)$_3$]$^{+3}$	10.8
l[Os(bipy)$_3$]$^{+2}$-l[Os(bipy)$_3$]$^{+3}$	d-[Co(en)$_3$]$^{+3}$	
d[Os(bipy)$_3$]$^{+2}$-d[Os(bipy)$_3$]$^{+3}$	l-[Co(en)$_3$]$^{+3}$	9.8
l[Os(bipy)$_3$]$^{+2}$-l[Os(bipy)$_3$]$^{+3}$	d-[Co(en)$_3$]$^{+3}$	

potentials[33] were plotted against the value of the square root of the ionic strength $\sqrt{I}$. As is evident different slopes ($mv/0.1\sqrt{I}$) were measured depending upon the couple and the active species. Pfeiffer considered this variation in slope, depending upon the system, to be evidence for the equilibrium shift and accompanying configurational activity concept. In the system of the copper(II) ion and phenanthroline there is good evidence, on the other hand, that some vicinal or bonding interaction is part of the Pfeiffer effect. The complex [Cu(phen)$_3$]$^{++}$ is known to show anomalous rotation with BCS$^-$ ion but not CS$^-$. It is known that there is a distinct change[23,35] in the blue color of the copper(II) *tris*phenanthroline complex with the BCS$^-$ but no change is observed when CS$^-$ is added to the system. Kirschner[35] reports a very slight Pfeiffer effect with Cu (EDTA) and *l*-quinine hydrobromide as the active species. No effect[36] is found with the same complex and BCS. It is difficult to believe that one can explain the Pfeiffer effect without the complex species interacting more intimately with the resolved species than is inferred by Pfeiffer.

The configurational activity concept is not in keeping with the time-rotation study of the present authors as well as Pfeiffer's[27] own studies on the system [Ni(phen)$_3$](BCS)$_2$. Since the racemization rate of the nickel(II) complex is slow, any shift in the equilibrium of the *d*- and *l*-isomers of the complex should also be slow. The Pfeiffer[3] effect for this complex would be expected to increase from a zero value to a maximum. The observed data shows an "instantaneous" increase (*vide supra*) that is about 46% of the final rotation. Although from this study an equilibrium shift may be operative it cannot be the only effect.

The Pfeiffer activation of [Co(phen)$_3$]$^{+2}$ and [Co(phen)$_3$]$^{+3}$ by BCS$^-$ as well as strychnine has been investigated by Pfeiffer as well as the present authors.[21] The complex is relatively inert to racemization if free from cobalt(II). Ellis[37] and others have studied the racemization as a function of this cobalt(II) concentration. If equilibrium shift configurational activity only were operative, it would be an extremely improbable coincidence to have the equilibrium shift exactly of the same order of magnitude to compensate for the difference in the ratio of the oxidized and the reduced complexes. If we may use Dwyer's[25] data on the osmium and ruthenium complexes (*vide supra*) in the different oxidation states and with the several resolved species, it is almost certain that the rotation of the *d*- or *l*-[Co(phen)$_3$]$^{+2}$ is different from the *d*- and *l*-[Co(phen)$_3$]$^{+3}$. Dwyer's data show *d*-[Ru(phen)$_3$]$^{+2}$ with $[M]_D = 8560°$ *vs.* *d*-[Ru(phen)$_3$]$^{+3}$ with $[M]_D = 2000°$; *d*-[Os(dipy)$_3$]$^{+2}$ with $[M]_{5461} = 21,000$ *vs.* *d*-[Os(dipy)$_3$]$^{+3}$ with $[M]_{5461} = 2500°$. These differences would have to be *exactly* compensated for by a change in the ratios of *d*- to *l*-[Co(phen)$_3$]$^{+2}$; a highly illogical expectation.

The ligands and their donor functionality present a unique situation in Pfeiffer rotations. Pfeiffer's original work showed ligands with heterocyclic nitrogen in orthophenanthroline and dipyridyl as exhibiting the effect. A limited number of other ligands have been identified by the present authors and by Kirschner.[31] One of these author's (VJL)[18] noted a small Pfeiffer rotation with the zinc(II) ion and 8-aminoquinoline. References[18,31] include a number of the ligands investigated with their rotations. It is noteworthy that the magnitude of the Pfeiffer rotation for dipridyl is only about one tenth that of orthophenanthroline. Although the 8-aminoquinoline has but one heterocyclic nitrogen and shows a very small Pfeiffer rotation, another ligand with a similar distribution of donor sites, pyridine 2-aldoxime has been shown by Kuhajek[4] to have no activity. Work is currently underway by one of the author's (LM) to investigate the very short wavelength portions of the spectrum, to see if the site(s) which might be the origin(s) of the activity is (are) simply being overlooked. Thus, the ligand, not only from the standpoint of the nature of the bonding to the metal ion but from the view of the number of ligands per metal ion, is in need of additional study.

The studies reported here by the present authors have as a common basis an investigation of environmental factors that may elucidate the source or sources of the Pfeiffer effect. These environmental factors and the variables are discussed under the following headings:

(a) Evidence for Association Between the Resolvable Complex and the Resolved Species.
(b) Temperature Dependence.
(c) Inert Ion Effects.
(d) Ligand Concentration Effects.
(e) Effect of Uncharged Optically Inert Species—Some Solvent Observations.
(f) The Resolved or Optically Active Species—Application of Optical Rotatory Dispersion Data Using Drude Plots.

EXPERIMENTAL SECTION

(a) Evidence for Association Between the Resolvable Complex and the Resolved Species

The original Pfeiffer experiments disclaimed, with very little evidence, association in solution. A variety of tools have been used by the present authors in efforts to shed light on the nature and the extent of bonding or association in the Pfeiffer active systems.

1. Refractometric and Polarimetric Observations. Standard solutions of $[Zn(phen)_3]SO_4$ and NH_4BCS were prepared and mixed to give individual samples of varying mole percent of the complex and BCS. Still other samples were prepared using strychnine sulfate in place of BCS^- as the optically active species. For each individual sample, the observed rotation and the refractive index (using the Zeiss dipping refractomer with $\pm 0.10°$ temperature control) were recorded. The n values (Fig. 1) are arbitrary scale readings from the "dipping-instrument" rather than absolute refractive indexes. It is evident that interaction between the complex and the active species is operative from the shape of the curves (deviations from linearity). The polarimetric evidence is more obvious than the refractometric though the "breaks" appear at about the same mole percent. The mole ratios for maximum interaction correspond to about one mole of complex to two moles of BCS^-, and five moles of complex to two moles of strychnine. It is interesting that

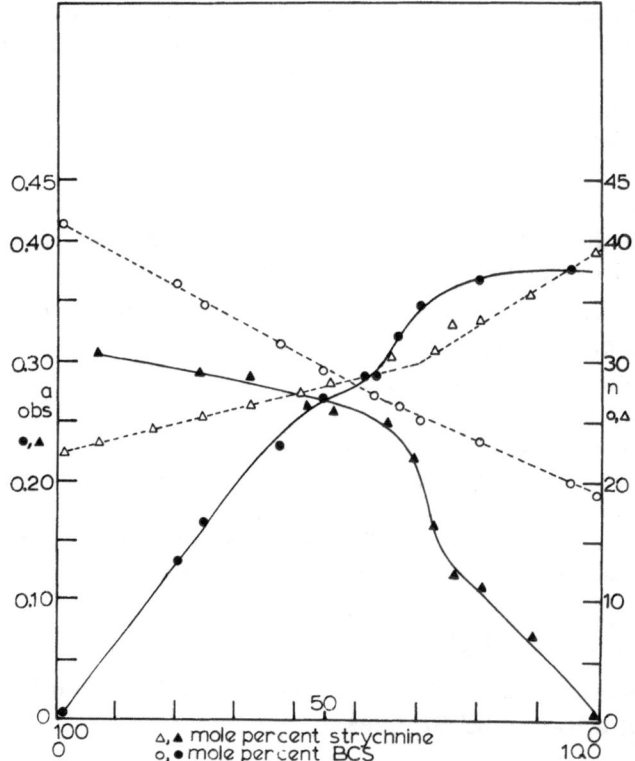

Fig. 1. Observed rotation α and refractivity n as a function of mole percent $[Zn(phen)_3]^{++}$ and the optically active species. The curves defined by circles refer to BCS^- and those with triangles refer to strychnine. Refractivity data are defined by broken lines and polarimetric data by the solid lines.

the break for the strychnine-complex system is more pronounced than that for the BCS^--complex using the polarimetric data. The sign of α obs. for the strychnine system has been reversed to facilitate intercomparison.

2. *Conductometric Observations.* Although alkali, alkaline earth, and some transition metal salts of the BCS^- ion have been suggested to be completely dissociated in solution, it does not follow that the larger transition metal coordination complexes should behave similarly. Indeed, the data of Fig. 1 do not suggest a strong electrolyte behavior. The complexes $[Zn(en)_3]$ $(BCS)_2$ and $[Zn(phen)_3](BCS)_2$ were prepared and their conductivity determined at varying concentrations. Plots were constructed for the two species according to Kohlrausch (Λ vs. $\sqrt{c}$) and according to the Arrhenius concept ($1/\Lambda$ vs. Λc). In both instances, the data fit the weak electrolyte (Arrhenius)

plot better than that for the strong electrolyte.[21] The linearity of the data using the Arrhenius plot at the low concentration permitted extrapolation to Λ_∞ giving an estimate of 50% dissociation according to the equilibrium

$$[Zn(phen)_3](BCS)_2 \rightleftarrows [Zn(phen)_3]BCS^+ + BCS^-$$

The data for the ethylenediamine complex also suggested a partial dissociation.

Using the pure complex $[Zn(phen)_3](BCS)_2 \cdot 7H_2O$, the fraction dissociation by a cryoscopic method was determined. The van't Hoff constant determined permitted the calculation of α dissociation to be 50% assuming the equilibrium above is the predominant one.

Kuhajek[4] first carried out an interesting experiment relating the metal complex concentration as a function of the active (BCS$^-$) species concentration by plotting the Pfeiffer rotation *vs.* the product of the metal complex and BCS$^-$ concentration. Kirschner and Magnell have confirmed these observations.[31] A nearly straight line function is observed (Fig. 2) at higher concentrations and deviates slightly from linearity at lower concentrations. The $[Zn(phen)_3]$ (BCS)$_2$ equilibrium just described is consistent with these data. Still another association investigation (vicinal effect) was carried out by measuring Pfeiffer activity under conditions of constant ionic

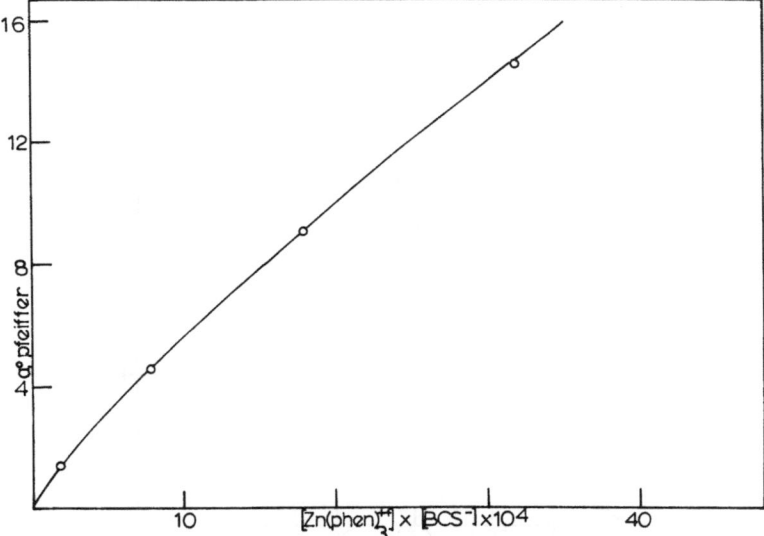

Fig. 2. Pfeiffer rotation as a function of concentration of the product of $[Zn(phen)_3]^{++}$ and $[BCS^-]$. Measurements made at 25°C, 405 mμ and using a 2 dm tube.

strength. The usual inert ion, ClO_4^-, used to maintain constant ionic strength is not suitable for use in these experiments due to solubility problems. A series of solutions was prepared by varying $[Zn(phen)_3]$-$(BCS)_2$ concentration but maintaining a constant ionic strength ($\mu = 0.30$) with the salt $(NH_4)_2SO_4$. This choice was dictated by the fact that some ammonium sulfate is already present from the *in vitro* preparation of the complex from $ZnSO_4 + phen + NH_4BCS$ in the ratio of $1:3:2$. The Pfeiffer values are plotted against the concentration of the complex (Fig. 3). A slight deviation from linearity is evident only at low concentration of the complex. Section (c) describes another experiment using sulfate ion.

(b) Temperature Dependence

It was inferred by Craddock and Jones[29] (*vide infra*) that changes in rotation attributed to "anomalous" or Pfeiffer activity might have been

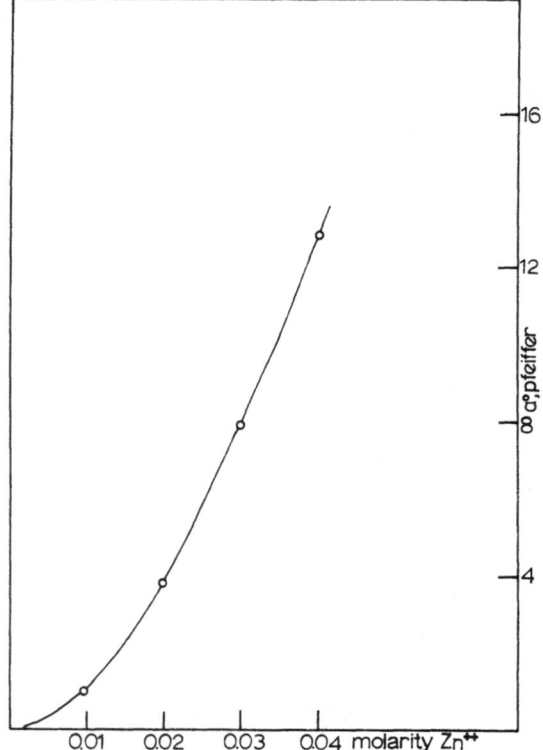

Fig. 3. Pfeiffer rotation as a function of $[Zn(phen)_3](BCS)_2$ concentration at constant ionic strength, 0.30, using $(NH_4)_2SO_4$. Temperature 25°C, 405 mμ and 2 dm tube.

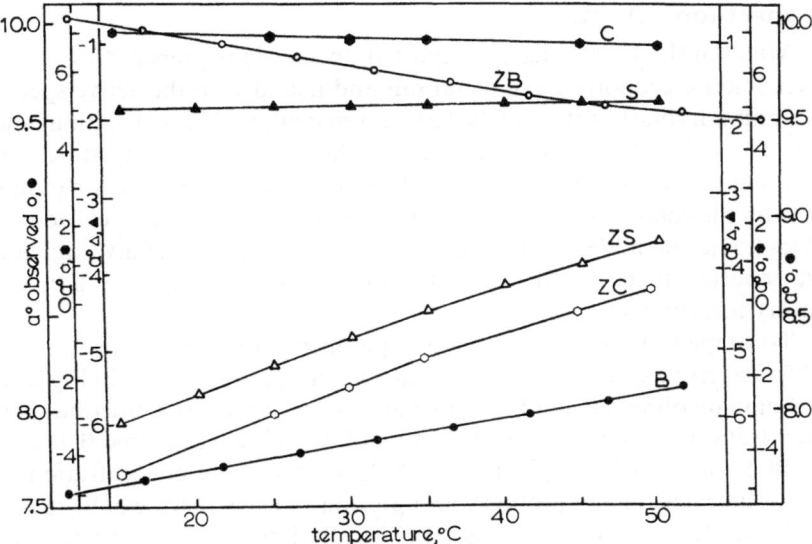

Fig. 4. Optical activity of Pfeiffer active systems as a function of temperature. Solid data points represent optically active species in solution. C, S, and B designate cinchonine hydrochloride, strychnine hydrosulfate and BCS⁻, respectively. ZC, ZS and ZB designate solutions of the active species plus the complex $[Zn(phen)_3]^{++}$. The three ordinates (α) define the rotation of the systems C + ZC, S + ZS and B + BC.

actually due to improper temperature control. Figure 4 has been constructed in such a way that three Pfeiffer active systems can be intercompared over a temperature range from approximately 15 to 50°C. Above 50°C the optics of the Rudolf polarimeter (with Faraday effect null point) used for this measurement could not be trusted to give consistent results. In every case, it is seen that whether BCS⁻ or an alkaloid is used as the active species the Pfeiffer rotation diminishes rapidly with increase in temperature. The change over this temperature range is linear or nearly so. Extrapolation to the intersection of the active species curve (C = cinchonine, S = strychnine and B = bromcamphorsulfonate ion) with the complex (Z = $[Zn(phen)_3]^{++}$) + active species could not be accomplished quantitatively, however, for two of the systems the indications are that Pfeiffer activity would disappear at about 90°C and the third (B + ZB) at about 110°C. Craddock and Jones[29] would appear to be justified in their criticism of the work of Harris[28] and others[26,27] who assumed that all Pfeiffer activity was due to configurational activity and that racemization ratios on the resolved $[Ni(phen)_2]^{++}$ were meaningful numbers.

(c) Inert Ion Effects

Much of the Pfeiffer data is reported on solutions prepared by mixing stoichiometric proportions of metal salt and ligand with the active species. The optimum concentration of the latter for maximum Pfeiffer rotation is not clear. The original work of one of the author's (RCB)[21] using strychnine infers that a 5:2 ratio of complex to alkaloid is ideal. The optical measurements on the solutions may be suspect due to the presence of so-called inert species (ionic and molecular). In the Introduction a number of investigations were reported that strongly suggested that inert ions may play a role in rotatory measurements.[38–42]

To supplement the ionic strength experiments, rotations were measured on Pfeiffer active systems using NaCl and Na_2SO_4 as inert salts. In Figures 5 and 6 data on observed and Pfeiffer rotations are plotted as a function of the concentration of the added "inert" salt. Solutions were prepared 0.02 M in NH_4BCS and 0.01 M in $[Zn(phen)_3]$ SO_4 with varying concentrations of NaCl. The extrapolated (dotted lines) portion of each curve would suggest that "pure" $[Zn(phen)_3]$ $(BCS)_2$ as a 0.01 M solution would have an observed rotation of about 10.0° or a Pfeiffer rotation of about 2.3°. It is noteworthy that extrapolation gives the same value whether chloride or sulphate is used as the inert species. When NaCl is added (Fig. 5) the rotation diminishes sharply at first then slowly. With Na_2SO_4 the same initial effect is observed,

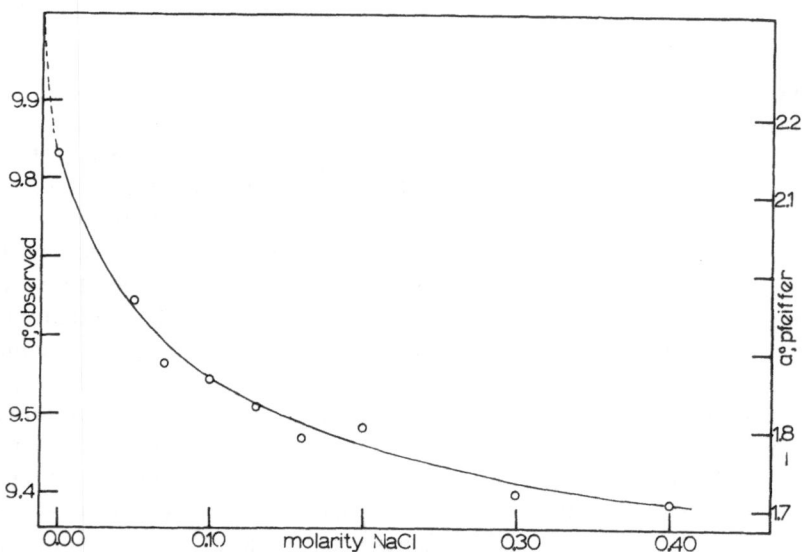

Fig. 5. Pfeiffer rotation of the system $[Zn(phen)_3](BCS)_2$(0.01 M) as a function of sodium chloride concentration. 25°C, 2 dm tube at 365 mμ.

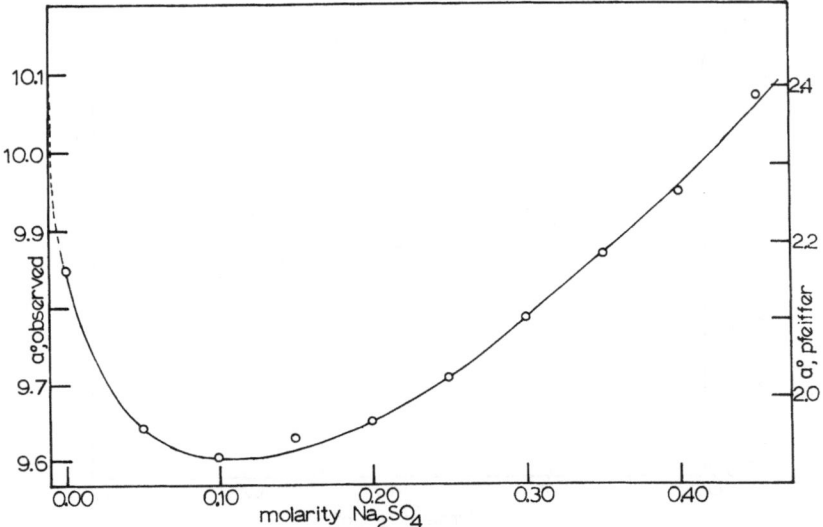

Fig. 6. Pfeiffer rotation of the system $[Zn(phen)_3](BCS)_2(0.01 M)$ as a function of sodium sulfate concentration; 25°C, 2 dm tube at 365 mμ.

however beyond 0.1 M Na_2SO_4 there is a gradual then rapid rise in the Pfeiffer rotation. No rational explanation is offered for this increase. It is possible that an equilibrium shift toward a less dissociated species is achieved with the sulfate that is not expected with the chloride. This series of experiments strengthen the belief that any information on optical activity published should be accompanied by the history and nature of the system being investigated.

(d) Ligand Concentration Effects

Among the original observations of Pfeiffer and his co-workers were data relating optical activity and the effect of varying mole ratios of phenanthroline to $Zn(BCS)_2$ solutions.[1,2] The present authors were not satisfied with Pfeiffer's explanation that the small Pfeiffer rotations at 1:1 and 1:2 ratios of $Zn(BCS)_2$: phen were due to some *tris*-complex being formed. A semiquantitative experiment was performed[18] by titrating a solution of $[Zn(phen)_3](BCS)_2$ with very strong aqueous ammonia while following the change in Pfeiffer rotation. The data are plotted (Fig. 7) as α-Pfeiffer *vs.* volume of NH_3 added. At this $NH_3 \cdot H_2O$ concentration the volume change is not an important factor in rotation change. The curve includes two independent sets of data both confirmatory to the conclusion that the α-Pfeiffer changes significantly when sufficient ammonia is present to form the complex $[Zn(phen)_2(NH_3)_2](BCS)_2$.

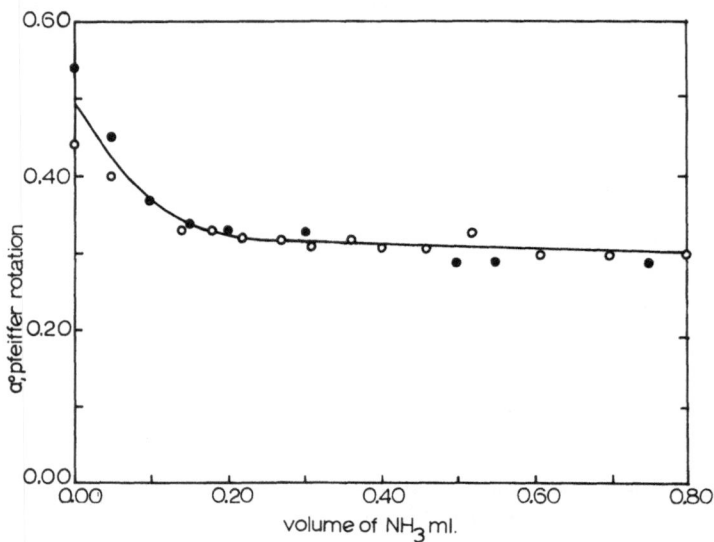

Fig. 7. Titration of $[Zn(phen)_3]^{++}$ with 15 M NH_3. Circles ∘ and solid circles • represent independent analyses.

Since a *bis*-complex in the *cis* configuration is a potentially resolvable species it could contribute to a Pfeiffer rotation either by itself or by supplementing the residual *tris*-complex. Kuhajek[4] performed a series of determinations on the effect of the *bis*-complex.

Crystalline $Zn(BCS)_2$ was used to prepare a 0.023 M solution. Two 100 ml aliquots of this solution were used, one diluted to 200 ml and the second, mixed with 0.0100 mole of phen was diluted to 200 ml. These two solutions were then mixed in various ratios totaling 20 ml volume, and then all solutions were diluted to 25 ml. Each solution had a constant (0.00982 M) zinc concentration (as $Zn(BCS)_2$). Any change in the observed rotation of these solutions is actually a Pfeiffer rotation. The observed α values for 12 such solutions were plotted in Fig. 8 with the α obs. plotted against the ratio of phen : Zn. The calculated values are incorporated in the plot as the curve C on the assumption that all of the rotation is due only to the *tris*-complex, $[Zn(phen)_3]^{++}$. The successive k dissociations for the complex[43] were used to estimate the concentration of the *tris*-complex in each of the solutions.

A similar series of solutions was prepared, using cinchonine hydrochloride (e.g., $ZnSO_4$ + cinchonine hydrochloride) instead of BCS^-. All solutions were 0.0214 M in zinc sulfate and 0.02 M in the alkaloid. The phen concentration was varied from zero to 0.072 M, a quantity slightly in excess of that needed to form the *tris*-complex. Figure 9 is a plot of the data, again, with the observed rotations plotted along with the calculated rotations on the tenuous assumption that the solutions all contained only the *tris*-complex.

As the plots are examined, it is evident that Pfeiffer rotations appear before the 3:1 complex can be present and that the discrepancy between the measured and the calculated curves is greatest at a concentration ratio of two moles of phen to one of Zn defining a complex $[Zn(phen)_2]^{++}$. That this ratio is significant is shown in Fig. 10 using the data from Fig. 9. Graphic differences between the observed and calculated curves are plotted and then compared with those curves from calculated data indicating the extent to which the *bis*-complex, $[Zn(phen)_2]^{++}$, would be formed. The two curves peak at the ratio 2 phen:1 Zn. There is no evidence for a contribution to the rotation where the ratio is 1:1. It is stressed that these experiments suggest

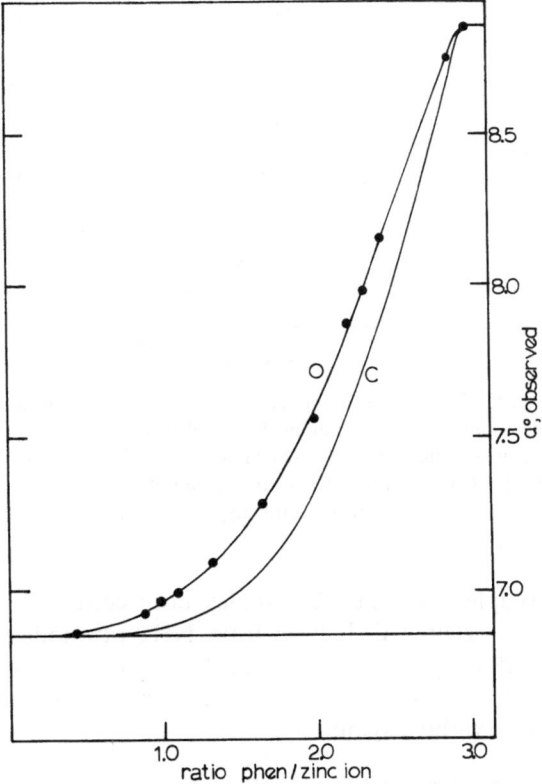

Fig. 8. Dependence of optical rotation on phen:$Zn(BCS)_2$ ratio. Curve O represents observed rotations for $Zn(BCS)_2$ with varying ratios of phenanthroline. Curve C represents the calculated values assuming the formation of $[Zn(phen)_3]^{++}$.

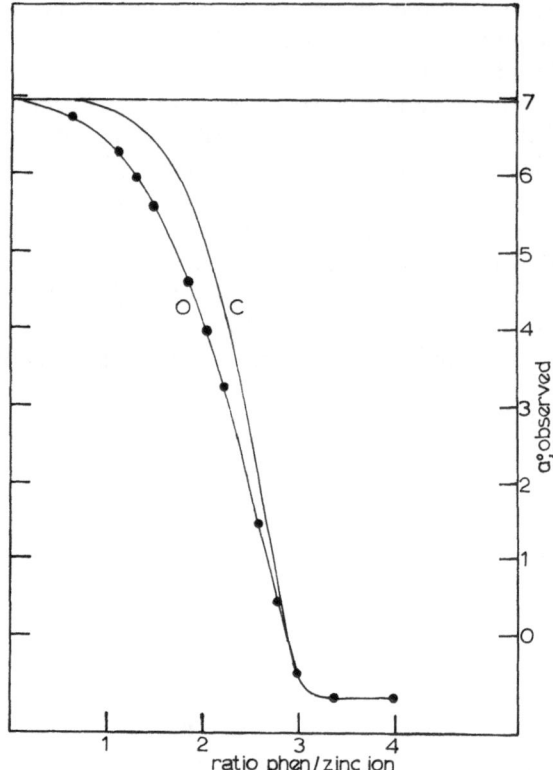

Fig. 9. Dependence of optical rotation of phenanthroline-zinc sulfate with cinchonine hydrochloride. Curve O represents rotations of $ZnSO_4$-cinchonine with varying ratios of phen. Curve C represents calculated rotations on the basis of amount of $[Zn(phen)_3]^{++}$ present.

that the *tris*-complex and the *bis*-complex both contribute to the Pfeiffer rotation. Landis' original deduction from NH_3 titrations is a reasonable one from these experiments. It is justified to assume that the zinc phenonthroline complex in solution is octahedral and not tetrahedral since the latter could not exhibit asymmetry.

(e) Effect of Uncharged Optically Inert Species—
Some Solvent Observations

A number of investigators have contributed to information on the Pfeiffer effect as it relates to solvent. Landis[18] verified data given to one of the authors (RCB) by Dwyer[33] as an informal communication, using

BCS$^-$ as the active species and $[Zn(phen)_3]^{++}$ in methanol as solvent. The loss of Pfeiffer activity in this medium has been proven. Kirschner[31] has reported data using ethanol, dimethyl formamide (DMF) and acetic acid as solvents. For all solvents (or mixtures of solvents) with lower dielectric constants than water there is a trend toward lowering the Pfeiffer rotation as the mole percent of water decreases. An exception is found with DMF where Kirschner reports an increase in α-Pfeiffer above about 40 vol.% of DMF. Nordquist[38] reports data on the $[Zn(phen)_3]^{++}$-cinchonine system in isopropanol at a number of wavelengths. The data are found in Table 2. When the active species carries a negative ionic charge (e.g., CS$^-$ or BCS$^-$) one can muster considerable support for a simple ion pair explanation for a

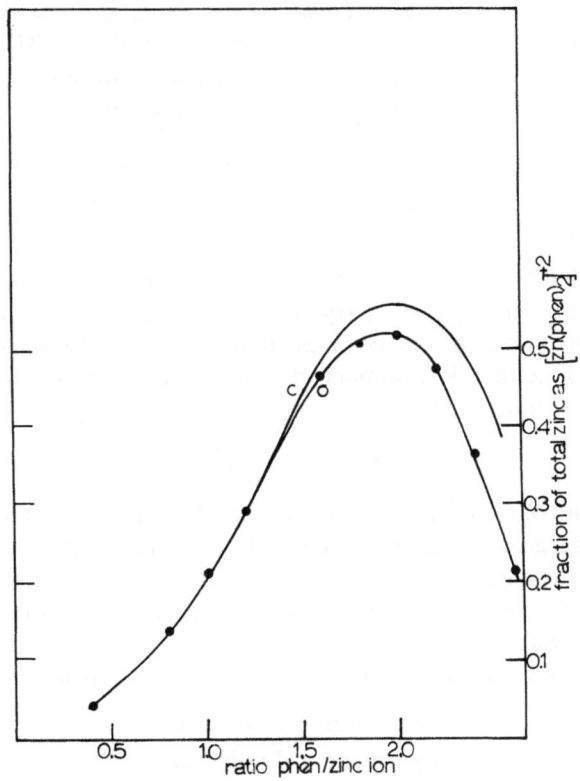

Fig. 10. Plot of graphic differences between the observed and calculated data in the ZnSO$_4$ + cinchonine system with varying phen concentrations. Observed (O) and calculated (C) curves are derived by comparing the calculated formation of $[Zn(phen)_2]^{++}$ with the (O) and (C) curves of Fig. 9.

Table 2
The Pfeiffer Effect in Solutions Containing Isopropyl Alcohol

		Wavelength, mμ			
		578	546	436	405
Water Solutions	α (solution A)	2.482*	2.861	5.198	6.544
	α (solution B)	2.268	2.604	4.654	5.784
	αp	0.214	0.357	0.554	0.760
Solutions containing 5 ml	α (solution A)	2.540	2.916		
isopropyl alcohol per	α (solution B)	2.449	2.802	+	+
20 ml A and B	αp	0.091	0.114		
Solutions containing 15 ml	α (solution A)	2.461	2.821	5.123	6.430
isopropyl alcohol per	α (solution B)	2.425	2.772	4.999	6.249
10 ml A and B	αp	0.036	0.049	0.124	0.181

A solutions: [cinchonine hydrochloride] = 0.0200 M
B solutions: [cinchonine hydrochloride] = 0.0200 M,
　　　　　　[Znphen$_3$SO$_4$] = 0.0100 M
αp = Pfeiffer rotation = α solution A − α solution B
+ excessive turbidity

*Rotation, degrees.

source of Pfeiffer activity. In the classical theory of ion association, the number of ions of opposite charge in layers of a constant thickness dr at distances r around a central ion is calculated from the Maxwell–Boltzmann distribution function. The number of ions in such a layer is found to be at a minimum at a distance

$$r = q = \frac{Z_A Z_B e^2}{2DkT}$$

where Z_A and Z_B are the charges on ions A and B, respectively, e is the electronic charge, k the Boltzmann constant, T the temperature, and D the dielectric constant of the medium. Bjerrum, in the original treatment of this problem,[38] arbitrarily assumed that ions closer than this distance, q, constituted an ion pair.

The equilibrium constant for an ion pair-forming reaction of the type

$$A^{+n} + B^{-m} \rightleftarrows A^{+n}, B^{-m}$$

is given for ions without permanent dipoles as a function of the dielectric constant[44,45] by the expression

$$K = A_0 \exp \frac{Z_A Z_B e^2}{DkTa}$$

where A_0 is a constant which includes nonelectrostatic factors and has a value on the order of unity and *a* is the ion size. It is seen that as the dielectric constant of the medium is lowered the formation of ion pairs is favored. Any experimentally measured quantities, which depend upon ion pair formation, would be expected to show a corresponding increase in media which favor ion association.

In considering the possibility that ion-pair formation is responsible for the Pfeiffer effect, we must consider as ion pairs those associated species which result from the simple electrostatic attractions between two charged particles. The ion pair hypothesis appears untenable as the only cause of Pfeiffer effect not only from the view of these observations (that the Pfeiffer effect decreases in media of low dielectric constant) but more obviously because the positively charged alkaloids (even though weak in charge) as active species exhibit strong Pfeiffer activity.

The decrease in Pfeiffer rotation observed in the alcohol solutions is not inconsistent with an associated species held together by hydrophobic forces (*vide infra*). Addition of alcohol to water changes the structural character of the water, making hydrophobic associations between nonpolar species unlikely. The decrease in observed Pfeiffer rotation, as the alcohol content of the solution increases, is here regarded as evidence of the existence of forces operating between an essentially nonpolar complex and the weakly charged or zero charged active species (large anions or alkaloids). For an extensive review of the concept of hydrophobic bonding as it relates to complex systems (such as Pfeiffer active solutions), the work of Nordquist[38] is recommended.

Although not a solvent, urea is an uncharged species that has long been known to have profound effect upon systems hydrophobically bound. If such bonds are operative in generating new asymmetric or electronic centers giving rise to Pfeiffer rotations, it is reasonable that urea would alter their rotations just as it alters the nature of and rotation of certain biologically optically active systems (e.g., proteins hydrophobically bound by solvent). Table 3 gives data for two Pfeiffer active systems in which BCS$^-$ and strychnine are the optically active species and $[Zn(phen)_3]^{++}$ the resolvable complex.[42]

It is evident that urea brings about a sharp decrease in the α-Pfeiffer. In 3.3 M urea the BCS-activated system, the Pfeiffer rotation is but 33% of the activity in absence of urea. With a strychnine-activated system the rotation in presence of urea is 55% of that in its absence.

(f) The Resolved or Optically Active Species—Application of Optical Rotatory Dispersion Data Using Drude Plots

Dwyer in his earliest work was aware that the systems now thought of as being Pfeiffer-active exhibited different activities at different wavelengths.

As early as 1942, one of the authors (RCB) suggested that dispersion data could be analyzed to advantage by expressing the observed rotation and wavelength in the form

$$\alpha = \frac{k}{\lambda^2 - \lambda_0^2} \qquad \text{or} \qquad \alpha = \sum_i \frac{A_i}{\lambda^2 - \lambda_{0_i}^2}$$

α is the measured specific or molar optical rotation. The A_i terms are constants whose values depend in part upon the units of α. The term λ is the wavelength at which the rotation is measured, and the i values of λ_0 are constants which are wavelengths of the electronic transitions producing the optical rotation. The Drude equation, of course, does not operate in the immediate region of λ_0 since this is a region of absorbance.

Table 3

The Influence of Urea on the Pfeiffer Effect—Strychnine and BCS as Active Species

Composition of solution	Rotation, α, in deg.	Composition of solution	Rotation, α, in deg.
[Stry H$^+$] = 0.0200 M	−1.203	[Stry H$^+$] = 0.0202 M	−1.205
[Stry H$^+$] = 0.0200 M		[Stry H$^+$] = 0.0202 M	
[Urea] = 3.2 M	−1.298	[Urea] = 3.2 M	−1.302
[Zn(phen)$_3$SO$_4$] =		[Zn(phen)$_3$SO$_4$] =	
0.0100 M	−2.870	0.0100 M	−2.873
[Stry H$^+$] = 0.0200 M		[Stry H$^+$] = 0.0202 M	
[Zn(phen)$_3$SO$_4$] =		[Zn(phen)$_3$SO$_4$] =	
0.0100 M	−2.247	0.0100 M	−2.247
[Stry H$^+$] = 0.0200 M		[Stry H$^+$] = 0.0202 M	
[Urea] = 3.2 M		[Urea] = 3.2 M	
αp (without urea)	−1.667		−1.668
αp (with urea)	−0.949		−0.945
αp, The Pfeiffer rotation, = $(\alpha_{\text{Complex + Stry}}) - (\alpha_{\text{Stry}})$			
[NH$_4$BCS] = 0.0199 M	3.473	[NH$_4$BCS] = 0.0200 M	3.500
[NH$_4$BCS] = 0.0199 M	3.537	[NH$_4$BCS] = 0.0200 M	3.558
[Urea] = 3.3 M		[Urea] = 3.3 M	
[Zn(phen)$_3$SO$_4$] =		[Zn(phen)$_3$SO$_4$] =	
0.0100 M	4.492	0.0100 M	4.514
[NH$_4$BCS] = 0.0199 M		[NH$_4$BCS] = 0.0200 M	
[Zn(phen)$_3$SO$_4$] =		[Zn(phen)$_3$SO$_4$] =	
0.0100 M	3.895	0.0100 M	3.918
[NH$_4$BCS] = 0.0199 M		[NH$_4$BCS] = 0.0200 M	
[Urea] = 3.3 M		[Urea] = 3.3 M	
αp (without urea)	1.019		1.014
αp (with urea)	0.358		0.360
αp, The Pfeiffer rotation, = $(\alpha_{\text{Complex + NH}_4\text{BCS}}) - (\alpha_{\text{NH}_4\text{BCS}})$			
λ = 420 mμ, T = 20.0°C			

If the observed optical rotation of a species can be satisfactorily expressed by a single term equation with only two arbitrary constants (A and λ_0), the substance is considered to show simple optical rotatory dispersion. If the observed rotation is best described by a Drude equation of more than one term, the substance is said to exhibit complex dispersion with multiple sources of electronic asymmetry. A rearrangement of the one term equation shows that if Λ^2 is plotted against $1/\alpha$ a straight line will be obtained if the substance under investigation exhibits simple dispersion.[46] The abscissa intercept of this line gives a value of λ_0 which should be the wavelength of the optically active electronic transition. The value of λ_0 so obtained should correspond rather *closely* to an absorption maximum in the absorption spectrum of the compound investigated.[47–53]

Before initiating rotatory dispersion studies on Pfeiffer active systems, it was deemed necessary to gain more of an insight on concentration effects. Due to the extremely high extinction coefficients of 1,10-phenanthroline (alone or complexed) at the low end of the spectrum (below 280 mμ), only highly diluted systems can be measured in electronic spectra studies. In Fig. 11 the data for the systems [Zn(phen)$_3$](BCS)$_2$ and Zn(BCS)$_2$ (curves labeled ZPB and ZB, respectively) are reported over the concentration range 0.04 M to 1.6×10^{-3} M at wavelengths 365 mμ and 405 mμ. The Pfeiffer rotations are directly derived by taking the differences between the two curves

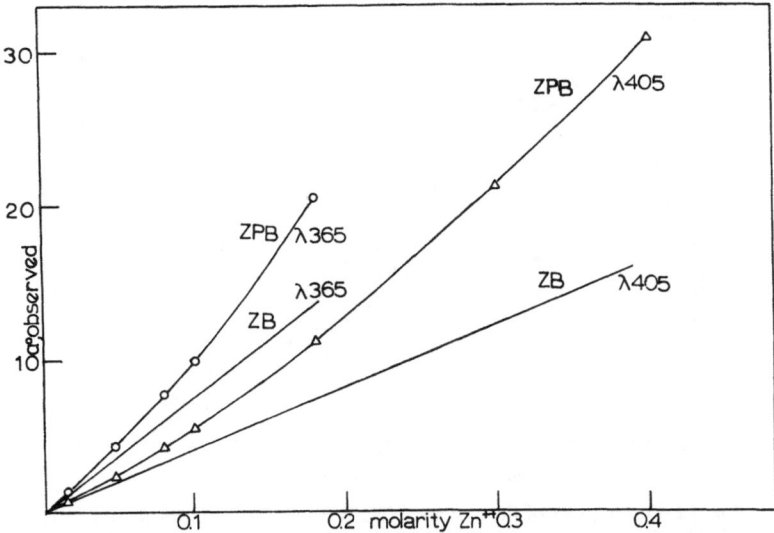

Fig. 11. Optical rotation as a function of concentration of zinc ion and wavelength. Curves ZPB represent data on [Zn(phen)$_3$]BCS)$_2$ at 365 and 405 mμ. Curves ZB represent data on Zn(BCS)$_2$ at 365 and 405 mμ. Temperature is 25°C with a 2 dm tube.

at any desired concentration. There is an increasingly high divergence in the curves (α-Pfeiffer) as the value of λ increases. At concentrations below 1.6×10^{-3}, the inference from Fig. 11 is misleading since it would appear that Pfeiffer rotations would be existent and measurable (with a sufficiently sensitive device) in virtually any concentration range. Present work by one of the authors (LM) has shown that the system just described shows *no* Pfeiffer rotation in concentration ranges of 10^{-4} to 10^{-5} M (rather than becoming asymptotic as zero concentration is approached). Real limitations are, thus, imposed on Pfeiffer studies if the active band is deep in the ultraviolet.

Landis[18] fitted Pfeiffer's original data[1-3] to the Drude equation. These data are plotted in Fig. 12, realizing that considerable error is possible since Pfeiffer only recorded dispersion data (solid lines) at three wavelengths. Landis[2] also found that solutions prepared according to Pfeiffer's directions for this study were supersaturated, the *tris*-complex separating from the solution on standing. On the same plot (Fig. 12) data from our laboratories using same general system are found. The single term Drude equation is followed for NH_4BCS and the complex $+BCS$. This straight line Drude plot for NH_4BCS intersects, at $\lambda_0 = 332$ mμ, the line (Drude plot) defined by the complex $[Zn(phen)_3](BCS)_2$. Since this wavelength does not correspond to any absorption maxima in the spectra of either BCS^- alone or the complex

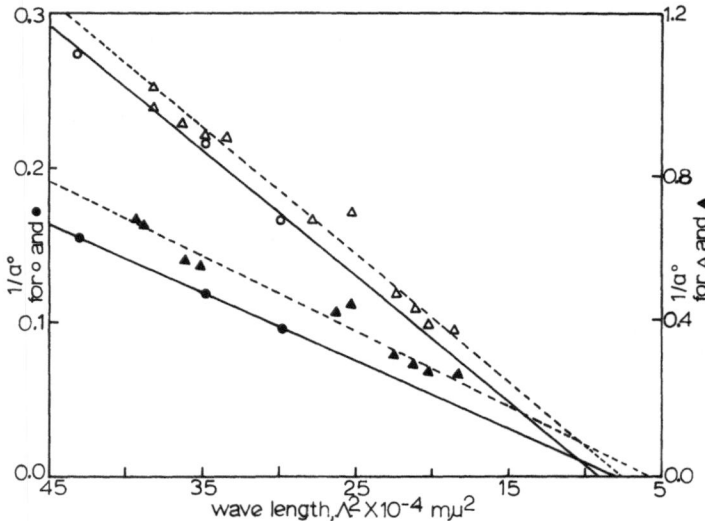

Fig. 12. Drude plots of the system: NH_4BCS and the complex $[Zn(phen)_3]$-$(BCS)_2$. The solid lines (defined by circles) represent data taken from Pfeiffer's original work. The triangles (dotted lines) are for current studies. Unfilled data points ($\bigcirc$ and $\triangle$) are for NH_4BCS while filled points ($\bullet$ and $\blacktriangle$) are for $[Zn(phen)_3](BCS)_2$. Single term Prude.

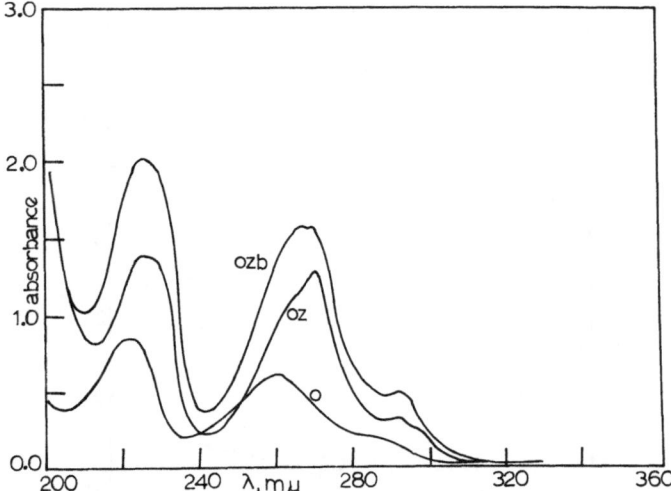

Fig. 13. Ultraviolet spectra of the systems: phen (o); phen + Zn^{++} (oz); and phen + Zn^{++} + BCS$^-$ (ozb).

+ BCS (see Fig. 13), it is concluded that the source of the Pfeiffer rotation can neither be in the metal complex alone nor in the activating agent. Although there is some scattering in the Drude plots, the intersection points for both the Landis and Pfeiffer data are concordant.

To extend the dispersion experiments to include alkaloids, Kuhajek[4] determined the ORD curves for cinchonine hydrochloride. The divergence (see Fig. 14) at shorter wavelengths between the alkaloid and the alkaloid plus the complex strongly indicates that the sources of the observed rotation is not in the alkaloid itself. It is considered a possibility that the original rotation of the cinchonine is totally unaffected by the presence of the *tris*-orthophenanthroline zinc complex. Rather, it may be simply overwhelmed by a new rotation. Kuhajek tested the validity of this assumption by computing the Pfeiffer rotation at each wavelength for the cinchonine-[Zn-(phen)$_3$]SO$_4$. If the rotation of the cinchonine is truly unaffected in the presence of the complex, and if the new rotation stems from a different asymmetric center thru the alkaloid[25] then these Pfeiffer rotation values could give straight line Drude plots. The data are found in Fig. 15. The curve labeled C is the Drude plot of cinchonine, CZ represents the Drude plot of cinchonine plus the zinc complex, and the curve P (the straight line defined by solid circles) represents the Drude treatment of the Pfeiffer rotation.

A similar Drude plot of Pfeiffer rotations in a variety of other active systems is shown in Fig. 16. The sign of rotation has been changed (those

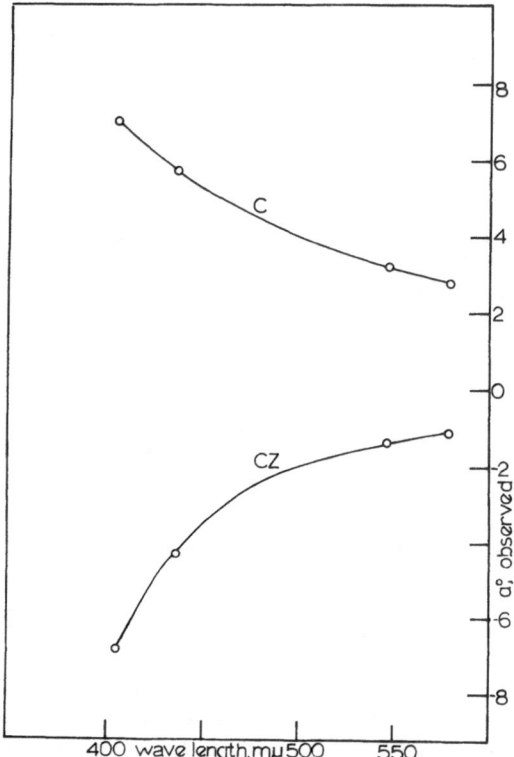

Fig. 14. Rotatory dispersion curves. Curve C represents
cinchonine hydrochloride, 0.02 M. Curve CZ represents
$[Zn(phen)_3]^{++}$ and cinchonine hydrochloride.

labeled $-$) in some of the systems, to permit comparison on the same $1/\alpha$ axis. In all cases, the relation is found to be linear (following the single term Drude equation). Of considerable interest is that all of the lines extrapolate to the same wavelength λ_0 indicating that the same absorption band must exist in each of the cases. Active species have been chosen with positive, negative, and zero charge in this study. The Drude plots for the *individual* optically active species give a wide range of λ_0 values, yet the Drude plots which are combinations of the active species and the complex all extrapolate to the same λ_0. It seems reasonable to assume that the source of the Pfeiffer rotation is some asymmetric center common to all of the systems. It is further reasonable that the complex itself fulfills this requirement, though the center may not arise in either the metal or the ligand transitions alone. A center may be created as part of the complex by the active species interacting with the complex.

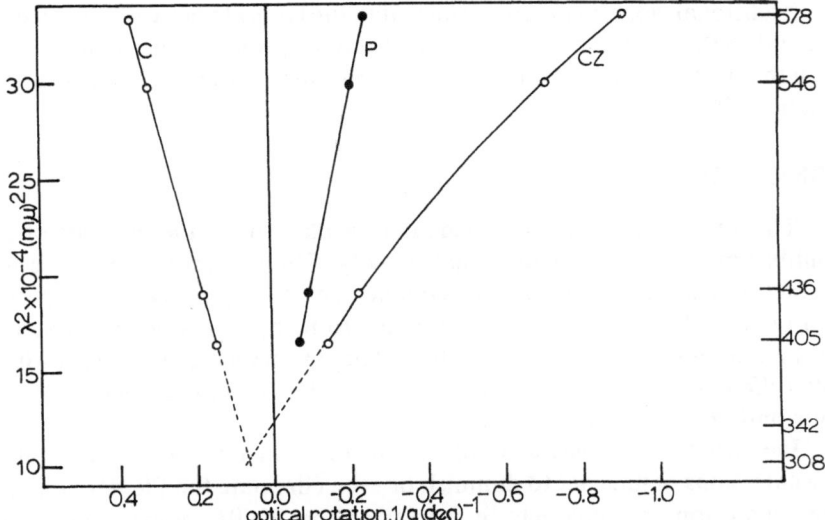

Fig. 15. Drude plots of cinchonine (C), cinchonine with $[Zn(phen)_3]^{++}$ and the resulting Pfeiffer rotation (P). Rotatory dispersion data for C and CZ taken from Fig. 14.

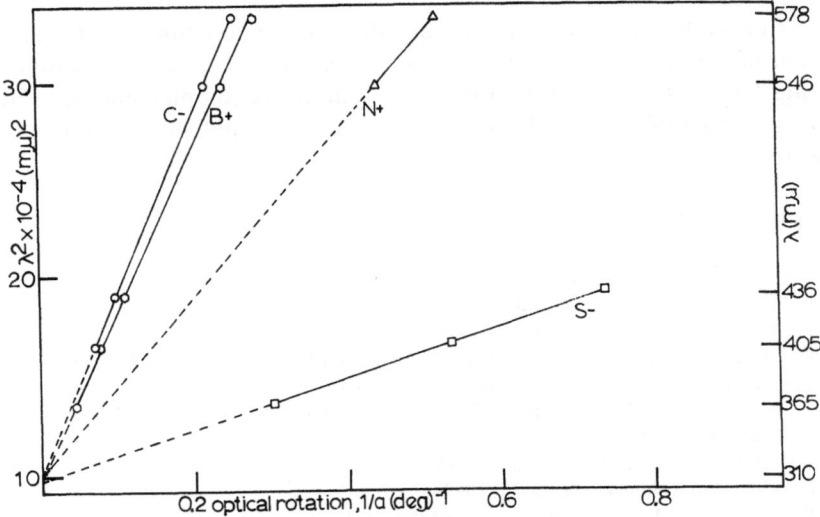

Fig. 16. Pfeiffer rotations plotted according to the Drude equation for selected Pfeiffer active systems. Data for the complex $[Zn(phen)_3]^{++}$ with cinchonine (C), the complex with BCS^- (B), the complex with nicotine (N), and the complex with strychnine (S) are recorded. The rotatory signs have been changed for C and S to allow a comparison on the same $1/\alpha$ axis.

Additional work has been completed in the laboratories with the ligand, pH, and the metal ion as variables. A computer program has been used to show that ORD data can be fitted to both one and two-term Drude rotation equations.[38]

CONCLUSIONS

The Pfeiffer effect is not adequately explained by the concept of an equilibrium shift or configurational activity. There is strong evidence for interaction between the resolvable complex and the optically active species by a bonding force that is proposed herein to be hydrophobic in character. Ion pair interaction (when oppositely charged species exist), equilibrium shift, differential association and the hydrophobic bonding are suggested as contributing to the Pfeiffer activity.

Inert (optically) ions, alcohols, an increase in temperature and high concentrations of urea act to diminish or even eliminate the Pfeiffer activity. The sulfate ion acts uniquely in increasing the Pfeiffer activity after first diminishing it at low concentrations (below 0.1 M). The mechanism by which this increase occurs is not now understood.

The *bis*- and *tris*-phenanthroline complexes have both been proven to be Pfeiffer active in the presence of such active species as the negatively charged BCS^- ion and alkaloids.

Interpretation of optical rotatory dispersion data through the mechanism of the Drude equation leads to the conclusion that the asymmetric center (or centers) created by the interaction of the complex and the active species is part of the complex rather than being located within the resolved species.

REFERENCES

1. P. Pfeiffer and K. Quehl, *Ber. Deut. Chem. Ges.* **64**: 2267 (1931).
2. P. Pfeiffer and K. Quehl, *ibid.* **65**: 560 (1932).
3. P. Pfeiffer and Y. Nakasuka, *ibid.* **66**: 410, 415 (1933).
4. E. J. Kuhajek, Ph.D. Thesis, University of Minnesota, 1962; *Diss. Abstr.* **24**: 64 (1963).
5. E. E. Turner and M. M. Harris, *Quart. Rev. (London)* **1**: 299 (1947).
6. W. W. Brandt, Dwyer, F. P., and E. C. Gyarfas, *Chem. Rev.* **54**: 959 (1954).
7. R. Kuhn and O. Albrecht, *Ann. Chem., Justus Liebig's* **455**: 272 (1927).
8. J. Read and A. McMath, *J. Chem. Soc.* 1572 (1925).
9. F. Bell and P. Robinson, *ibid.* 2234 (1927).
10. W. H. Mills and K. Elliott, *ibid.*, 1292 (1928).
11. A. Chalmers, F. Lion and A. Robson, *J. Proc. Roy. Soc. N.W. Wales* **65**: 320 (1931).
12. R. Kuhn, *Ber. Deut. Chem. Ges.* **65B**: 49 (1932).
13. P. Ritchie, *Asymmetric Synthesis and Asymmetric Inductions*, Oxford University Press, New York, 1933.
14. M. Leslie and E. E. Turner, *J. Chem. Soc.* 347 (1934).

15. J. Breckenridge and O. C. Smith, *Can. J. Research* **16B**: 109 (1938).
16. M. Jamison and E. E. Turner, *J. Chem. Soc.* 1646 (1938).
17. A. B. Biswas, *J. Ind. Chem. Soc.* **22**: 351 (1945).
18. Landis, V. L., Ph.D. Thesis, University of Minnesota, 1957; *Chemical Abstracts* **51**: 13577g (1957); see thesis refs. 25–36.
19. L. Tschagaeff and Sikoloff, *Ber. Deut. Chem. Ges.* **40**: 3461 (1907); *ibid.*, **42**: 55 (1909); see also Ref. (18), Thesis Ref. (Nos. 39–42).
20. R. Toole, Ph.D. Thesis, University of Minnesota, 1953.
21. R. C. Brasted, Ph.D. Thesis, University of Illinois, 1942; J. C. Bailar, F. Basolo, R. C. Brasted *et al.*, Chemistry of the Coordination Compounds, Reinhold Monograph No. 131, p. 581–2.
22. F. Basolo and R. Pearson, *Mechanics of Inorganic Reactions*, John Wiley and Sons, 2nd ed., 1967, p. 323.
23. N. Davies and F. Dwyer, *J. Proc. Roy. Soc. N.S. Wales* **86**: 64 (1953).
24. See Ref. 8, thesis reference (Nos. 15–22).
25. F. P. Dwyer, E. C. Gyarfas and M. F. O'Dwyer, *J. Proc. Roy. Soc. N.S. Wales* **89**: 146 (1955).
26. P. Ray and N. K. Dutt, *J. Indian Chem. Soc.* **20**: 81 (1943).
27. N. R. Davies and F. P. Dwyer, *Trans. Faraday Soc.* **48**: 244 (1952); **50**: 24 (1954).
28. Harris, M. M., *Progress in Stereochemistry* (eds. W. Klyne and P. B. de la Mare) Academic Press, Inc., New York, 1958 pp. 157–195.
29. J. H. Craddock and M. M. Jones, *J. Am. Chem. Soc.* **84**: 1098 (1962).
30. F. P. Dwyer and E. C. Gyarfas, *Nature* **168**: 28 (1951); G. Barnes *et al.*, *J. Proc. Roy. Soc. N.S. Wales* **89**: 151 (1955).
31. S. Kirschner and K. R. Magnell, Advances in Chemistry Series, No. 62, "Werner Centennial", pp. 366–377, (1966).
32. S. Kirschner, N. Ahmad, and K. Magnell, *Coordination Chemistry Reviews*, Elsevier Pub. Co., Vol. 3, 1968, p. 201–206.
33. F. P. Dwyer, personal communication to R. C. Brasted.
34. W. Schlenk, *Ann. Chem.* **565**: 204 (1949).
35. S. Kirschner, *J. Am. Chem. Soc.* **78**: 2372 (1956).
36. R. C. Brasted and M. Petrin, 1968 (in preparation).
37. P. Ellis, R. G. Wilkins, and M. J. G. Williams, *J. Chem. Soc.* **1957**: 4456.
38. P. E. R. Nordquist, Ph.D. Thesis, University of Minnesota, 1964.
39. A. Werner, *Ber.* **45**: 121 (1915).
40. M. Albinak, D. Bhatnagar, S. Kirschner, and A. Sonnessa, Proc. VI ICCC, 1961, p. 154.
41. R. Larson, *Proc.* (Abstracts) VII ICCC, 1962, p. 436.
42. A. Jensen, F. Basolo and H. M. Neuman, *J. Am. Chem. Soc.* **80**, 2354 (1958).
43. J. Brandts, Ph.D. Thesis, University of Minnesota, 1961.
44. F. Basolo and R. Pearson, *op. cit.* p. 34.
45. R. M. Fuoss and C. A. Kraus, *J. Am. Chem. Soc.* **79**: 3304 (1957).
46. T. M. Lowry, *Optical Rotatory Power*, Longmans, Green and Co., London, 1935.
47. J. A. Schellman, *Compt. Rend. Trav. Lab.* Carlsberg Ser. Chim., **30**: 363 (1958).
48. A. Moscowitz, in: *Optical Rotatory Dispersion* (by C. Djerassi) McGraw–Hill Book Co., Inc., New York, 1960, Chapter 14.
49. W. Kauzman, J. E. Walter, and H. Byring, *Chem. Rev.* **26**: 339 (1940).
50. W. Kauzman, *Quantum Chemistry*, Academic Press, Inc., New York, 1957, pp. 616–635, 703–723.
51. W. Moffitt, *J. Chem. Phys.* **25**: 1189, 467 (1956).
52. B. Bosnich, *Inorg. Chem.* **7**(11): 2379 (1968).
53. T. S. Piper and A. Karipides, *Mol Phys.* **5**: 475 (1962).

LIGAND REACTIONS WITH MINI-MOLECULES AND BIOLOGICAL MACROMOLECULES

G. L. Eichhorn

Gerontology Research Center
National Institute of Child Health and Human Development
National Institutes of Health
Public Health Service
U.S. Department of Health, Education, and Welfare
Bethesda, Maryland and the Baltimore City Hospitals
Baltimore, Maryland

INTRODUCTION

Coordination chemistry, as its name may seem to imply, furnishes a link between many fields of chemistry. I have been particularly concerned during most of my professional career with the link provided between inorganic chemistry and biological chemistry. The participation of metal ions in biochemical reactions bears many similarities to ligand reactions in inorganic coordination chemistry. Because of these similarities the biochemist can learn a great deal about metals in biology from the inorganic chemist. There are also some important differences between ligand reactions in biochemistry and those to which inorganic chemists are accustomed. One of the major reasons for these differences lies in the fact that the biochemical ligands are frequently giant macromolecules, in comparison to which the inorganic chemist's ligands could be categorized as minimolecules. The study of biochemical macromolecules, in addition to elucidating biological processes, aids the inorganic chemist in his understanding of coordination chemistry.

It will be the purpose of this presentation to compare minimolecules and macromolecules as ligands. Since I have been interested in both (although not in the sequence that would correlate with women's fashions), and in recognition of the objectives of this book, many of the illustrations will be taken from our own work. Other illustrations would obviously serve equally well.

MINIMOLECULES AS LIGANDS

The reactions of coordinated ligands have been thoroughly investigated by Busch[1] and many others. My first encounter with a ligand reaction (which had not yet been so named) was as a graduate student in collaboration with Bailar. At that time we discovered the ability of copper(II) ions to sever the double bonds of the Schiff base formed by reaction of 2-thiophenaldehyde with ethylenediamine.[2] The reaction was the following:

(I)

$$+ \quad 2H_2O \quad \longrightarrow \quad$$

The discovery was accidental; we had really anticipated that the sulfur atoms would serve as electron donors and produce Schiff base complexes somewhat analogous to the salicylal-ethylenediamine complexes discovered by Pfeiffer et al.,[3] and so widely studied before and since. To our surprise, we found that the Schiff base complex exhibited only a cursory existence and was readily dissociated as shown in the above equation. Further studies were carried out to elucidate this reaction.[4] The primary influence of the metal ion is due to its positive charge, which pulls electrons away from the ligand and thus further weakens the weakest link in the Schiff base molecule, the C=N double bond, making it more susceptible to hydrolysis.

There are numerous other instances of bond cleavage reactions initiated by metals ions, and operating through a similar mechanism. The hydrolysis of amino acid esters is catalyzed by metal ions in a fashion very similar to the Schiff base cleavage.[5] The decarboxylation of ketoacids by metal ions,[6] a reaction that serves as a model for the processes catalyzed by carboxylase enzymes, falls into the same category. In these and many other instances, the principal function of the metal ion is to polarize the electronic charges on the molecule, resulting in the cleavage of a weak bond within the molecule.

Soon after learning that the $C=N$ bond in thiophenalethylenediamine is destabilized by metal ions, we discovered that metal ions can stabilize the same kind of double bond when the structure of the Schiff base is somewhat different, as in salicylalglycine.

(II)

This reaction proceeds at pH values at which Schiff bases are ordinarily quite unstable. It is not difficult to understand why metal ions labilize one Schiff base molecule and stabilize another. In both molecules I and II the polarization effect of the metal, tending to dissociate the Schiff base, is operating. In molecule I, this polarization effect is the major factor influencing the molecule, hence cleavage of the double bond results. In molecule I, cleavage of the double bond leaves all of the bonds between the metal and the donor intact. In molecule II, cleavage of the double bond and the consequent severance of the salicylaldehyde from the complex results in the cleavage of a coordination bond between the metal and an oxygen ligand atom; consequently, Schiff base hydrolysis in II would convert a bicyclic chelate into the much less stable monocyclic chelate. This factor apparently overrides the polarization factor in molecule II, resulting in the stabilization of the double bond in this instance.

As we shall see later, there is an anology in the ability of metal ions to stabilize and destabilize minimolecules and biological macromolecules.

Among the many interesting effects of metal ions in ligand reactions is the ability of the metals to arrange reactants around themselves in such a manner as to produce one kind of molecular product, instead of a different product that is formed in the absence of the metal. This phenomenon has been termed the template effect by Busch,[1] who is also responsible for very extensive studies on the nature and mechanism of this effect. To illustrate this phenomenon, I should like to allude to some work in which we had a part, but which was brought to fruition by the work of Melson and Busch. We were interested in the reaction of metal ions and o-amino-benzaldehyde, and found rather unexpectedly that this molecule does not coordinate without a simultaneous trimerization.[8] We believed that we had isolated the metal complex of a trimer containing two C=N linkages. Melson and Busch[9] showed that all of the functional groups have produced C=N linkages to produce a macrocyclic structure, as follows

A similarly constructed tetramer also results from this reaction.[10]

These few illustrations of minimolecular ligand reactions thus have demonstrated two types of processes: bond cleavage through electron polarization and the metal template effect.

BIOLOGICAL MACROMOLECULES AS LIGANDS

These same effects are also encountered with biochemical ligands. The macromolecular nature of some of these ligands produces, nevertheless, special conditions which are not obtainable with miniligands. These special conditions may be summarized as follows[11]:

1. Whereas the metal ion in a minimolecular ligand reaction generally provides the sole impetus for the reaction, a macromolecule can furnish additional groups that act in concert with the metal.

2. A given coordination environment can be markedly influenced by the nature of the macromolecule, so that the same basic coordination features can be utilized for a number of different ligand reactions.

3. A macromolecule offers a choice of coordination sites to a metal ion; consequently the binding of the metal ion to different sites can lead to different ligand reactions.

Each of these special macromolecular effects will now be illustrated.

Functional Groups on the Ligand Acting in Concert with the Metal Ion

Perhaps the best understood metalloprotein at this time is the enzyme carboxypeptidase A. The function of this enzyme is the cleavage of the peptide bond adjacent to the carboxyl terminal of a protein molecule. It

has been known for some time that this enzyme contains zinc[12] and it has
been apparent that the zinc is at the active site of the enzyme, since the
enzymatic activity is eliminated when the zinc is removed and the activity can
be restored by recombination of the enzyme with zinc.[13] Other metal ions
also activate the enzyme when they are substituted for zinc.[14]

The recent x-ray analysis of the structure of carboxypeptidase A in the
presence and absence of a peptide substrate has given some clues concerning
the mechanism of action of this enzyme.[15] Figure 1 represents a diagrammatic
sketch of the fit of a glycyl-tyrosine substrate attached to the enzyme molecule. The zinc is bound to donor atoms on three amino acids of the enzyme.
A fourth coordination position is occupied by the carboxyl oxygen atom of
the peptide linkage. It is immediately obvious that the zinc ion can influence
the cleavage of the peptide bond in exactly the same manner as the copper
ions cause the scission of the Schiff base bond, i.e., by polarization and the
consequent weakening of the susceptible bond. In this respect, the mini-
ligand and the macroligand are affected in a similar way. However, in the
carboxypeptidase molecule the zinc does not constitute the only influence
upon the hydrolyzable bond. At the same time that zinc attaches to the
carboxyl group, a hydroxyl group of a tyrosine is in position to attack the
bond at the nitrogen atom. Meanwhile, the positive charge on an arginine
is in position to bind to the carboxyl group of the terminal amino acid, thus
holding the substrate molecule in place.

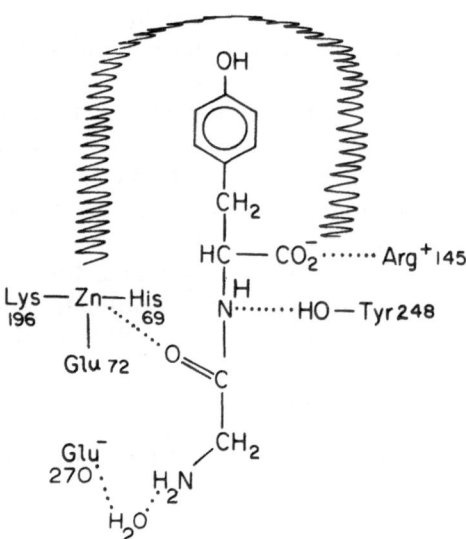

Fig. 1. Coordination of zinc in carboxypeptidase
enzyme-substrate complex (See Ref. 15).

The Variation of Ligand Properties Conferred Upon a Ligand by the Presence of Different Macromolecules

Perhaps the most versatile coordination compounds in terms of the ability to perform ligand reactions are the porphyrin complexes. The metal atom in heme, the iron(II) complex of porphyrin,

is bound to four nitrogen atoms. The heme structure is associated with many different proteins, and each association leads to a different ligand reaction.

To illustrate, let us consider the attachment of the heme to protein in hemoglobin[16] and myoglobin.[17] Figure 2[18] illustrates that the nitrogen atom of an imidazole occupies the fifth coordination position of iron. A sixth ligand is not bound in the unoxygenated form of the molecule. The heme is contained in a hydrophobic environment so that molecular oxygen can occupy a sixth coordination site without enhancing the oxidation state of the iron. The protein in hemoglobin is thus tailored to allow oxygen binding without oxidation to occur.

When the same heme is bound to the protein of cytochrome c,[19] two donor atoms from the protein are bound to the iron atom, a nitrogen from a histidine imidazole[20] and a sulfur from a methionine[21] (Fig. 3). The porphyrin is held very tightly in the protein structure by the addition of two cysteine sulfhydryl groups across the double bonds of the two vinyl side chains of the porphyrin. Since the function of cytochrome c is electron transfer, and not the attachment and detachment of additional ligands, the cytochrome c structure is tailored to produce very stable binding of the porphyrin to the protein.

A rather striking demonstration of the impact of protein structure upon the heme with which it is associated is given by rotatory dispersion

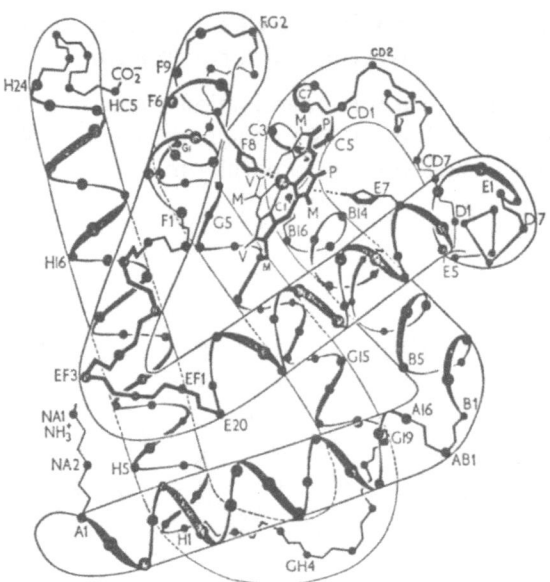

Fig. 2. Structure of myoglobin molecule, after Kendrew
et al. (from Ref. 18). Helical segments are indicated by letters
from A to H and individual amino acids are shown by number.
The heme is in the upper center of the figure, above the
C-helix. The iron atom, shown as a large black circle, is coordinated to the F8 histidine. The E7 histidine is too remote
for coordination to occur. M, V, and P on the heme indicate
methyl, vinyl and proprionic acid side chains.

studies. Some years ago we were interested in determining the relationship
between the structure of a hemoprotein and its rotatory dispersion. An
example of these studies is the comparison of the dispersion curve of the
hemoprotein peroxidase in neutral and in alkaline solution.[22] In neutral
solution peroxidase exists in its natural form, in which it retains a good deal
of ordered (helical) structure, whereas in alkaline solution the ordered
structure is destroyed. The helicity of the protein of course confers asymmetry upon the molecule. In Fig. 4 the optical rotatory dispersion curves of
peroxidase are compared in the two media. It can be seen that a Cotton
effect is observed in the region of heme absorption when the protein is
ordered, and that this Cotton effect is removed along with the asymmetry
of the protein. It is obvious that, if the same protein in different conformations
can have such a profound effect upon the heme, it is anticipated that different
proteins would drastically alter the properties of the heme and thus account
for the versatility of this structure in biochemical ligand reactions.

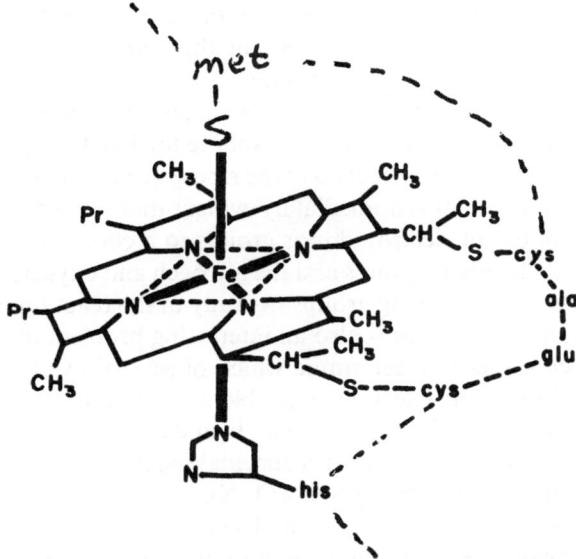

Fig. 3. Coordination of iron in cytochrome *c*. Dotted lines in the protein indicate bonds between amino acids.

Biological Macromolecules as Ligands with a Multiplicity of Sites for Metal Binding

We have been particularly interested in recent years in the interaction of metal ions with the nucleic acids. This interest has been motivated by many factors, such as the importance of the metal ions in the biological function

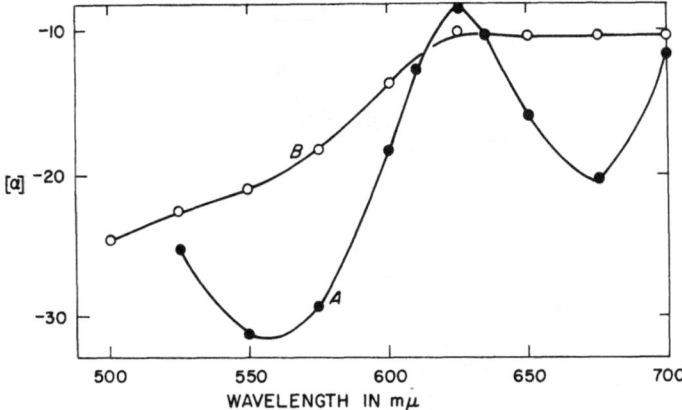

Fig. 4. Rotatory dispersion curve of peroxidase (A) at pH 7.2 and (B) at pH 11.0 (From Ref. 22.)

of these important molecules and the potential use of specific coordination reactions with the nucleic acid "bases" in the determination of the base sequence that constitutes the genetic code.

As is now generally known, the nucleic acids, DNA and RNA, contain a sugar phosphate backbone, ribose phosphate for RNA, (Fig. 5), 2'-deoxyribose phosphate for DNA. Attached to the ribose groups are the heterocyclic bases depicted in Fig. 6. It is immediately evident that these macromolecules offer a large number of electron donor groups to a coordinating metal ion. Specifically, the phosphate groups and the nitrogen and oxygen atoms on the bases are potential complexing groups. A study of the reaction of metal ions with the polynucleotides thus is also an interesting problem in coordination chemistry, since it seeks to determine which of several available coordination sites metal ions will select. (The problem is, of course, not completely unknown in reactions of minimolecular ligands, e.g., cysteine contains a sulfhydryl, an amino and a hydroxyl group, making this rather small molecule theoretically capable of forming SN, SO, NO, or SNO chelates. The possibilities for coordination in a macromolecule are, however, much greater, and produce ligand reactions that resemble those that we have discussed for the minimolecules.)

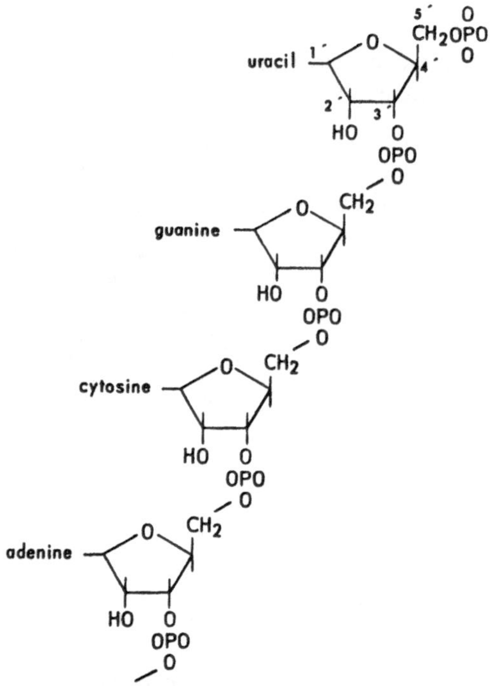

Fig. 5. The ribose phosphate backbone of RNA.

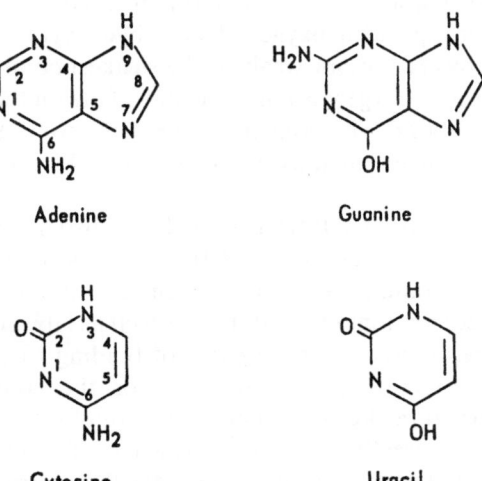

Adenine   Guanine

Cytosine   Uracil

Fig. 6. The heterocyclic "bases" of RNA. Adenine and
guanine are bound to ribose at N-9, cytosine and uracil
at N-3. DNA contains thymine (5-methyl C-uracil)
instead of uracil.

Metal ions bind to both phosphate and base "sites" on the nucleic
acid molecules, and each type of binding results in characteristic ligand
reactions. Many metals bind to both types of sites. Among these is zinc(II),
and we shall therefore concentrate our attention on the interactions of this
metal ion with nucleic acids.

The binding of zinc(II) to phosphate can bring about the cleavage of the
phosphodiester linkages and therefore the depolymerization of the polynucleotides. When high molecular weight RNA is heated with zinc(II) at
64° for 90 min, all of the polymers are degraded to small oligomers.[23] The
mechanism of this reaction can be pictured thus

The zinc ions presumably form a chelate ring by binding to an oxygen atom of the phosphate group and an oxygen of the 2'-hydroxyl group. The polarization of electrons toward the zinc weakens the weakest bond, that between the phosphorus and the 5' oxygen, leaving the phosphate on the 3'-position after hydrolysis of the 5'-bond. The presence of the 2'-hydroxyl group is a definite requirement for the reaction since DNA, which lacks this group, does not undergo this reaction.

This depolymerization of RNA by zinc thus is also quite analogous to the destabilization of thiophenal-ethylenediamine by copper ions. We shall see later that zinc ions binding to phosphate can also stabilize the nucleic acid structure, just as copper ions can stabilize, as well as labilize, a Schiff base.

Zinc ions are, however, also capable of binding to the "bases." This ability has some interesting consequences for the conformation of an ordered helical structure like DNA. Thus, we encounter an effect of coordination that depends directly upon the presence of macromolecules—the ability of coordinated metal ions to influence the three-dimensional structure of a macromolecule. We shall see, however, that this effect in some respects resembles minimolecular ligand reactions.

As is now generally known, the DNA molecule consists of two helical strands that are wound around each other in such a way that each of the bases on one strand is hydrogen-bonded to a complementary base on the other strand. (Guanine is complementary to cytosine and adenine is complementary to thymine.) The native, double helical form of DNA can be readily "denatured" to single strands by various techniques, notably by heating. These single strands lose their orderly arrangement and become "random coils" in the denatured state; it is then very difficult to rewind these two coils around each other, since the uncoiling has moved the two chains apart and the complementary base pairs are unable to find each other again. The conversion of native, double stranded DNA to single stranded DNA can be followed in the ultraviolet spectrum, since the former has a less intense absorbance than the latter, as a result of π-interactions between the stacked bases in the double helix. The irreversible denaturation is shown in Fig. 7A. The double helix is stable up to 55°; at this temperature the double helix "melts," half of the DNA having melted out at 63°, the "melting temperature," or T_m.[24] When all of the DNA has been denatured (80°) it is cooled back to room temperature. The slight decrease in absorbance in the cooling process is due to the formation of "hairpins," i.e., portions of ordered structure that are produced by complementary bases forming hydrogen bonds within the single strands.

The drastic effect of zinc(II) ions[25] on this interconversion of double and single stranded DNA is shown in Fig. 7B. Under these conditions the native DNA structure remains stable to a higher temperature, and the DNA melts out with a T_m of 68.5°. The most significant change produced by the

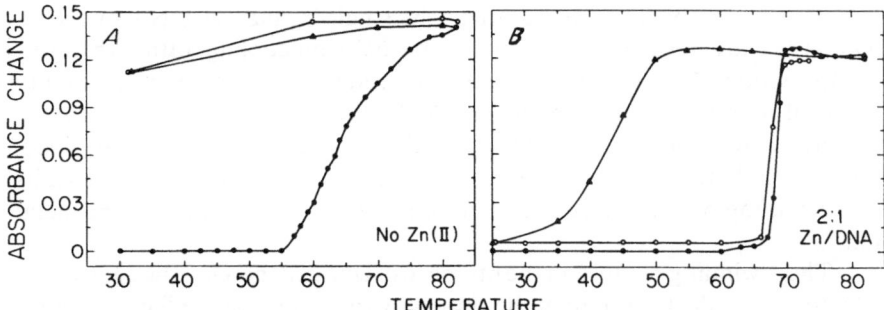

Fig. 7. Melting behavior of DNA in the absence (A) and in the presence (B) of zinc(II). (●) first heating, (▲), cooling, and (○) second heating. Solutions contained 5×10^{-5} M DNA (measured as phosphate concentration) and 5×10^{-3} M NaNO$_3$. (From Ref. 25.)

presence of zinc is that, on cooling the solution back to room temperature, the absorbance reverts to that characteristic of native DNA. Other experiments confirm that this reversion in the absorbance is indeed the result of the regeneration of native double stranded DNA. When the solution is then heated a second time, the new heating curve becomes practically identical to the original heating curve. This process can be repeated again and again.

The presence of the zinc ions thus converts the otherwise irreversible denaturation of DNA into a reversible process. The ability of zinc ions to make this reaction reversible can be understood with reference to the sketch in Fig. 8. The zinc ions can bind to the DNA bases. Thus, as the double helix unwinds at high temperature, some zinc ions bind simultaneously to a base on each of the chains that had comprised the double helix. The zinc ions form crosslinks between the two strands, which, therefore, do not completely dissociate during the denatured state. When conditions (such as lowering the temperature) again stabilize the double helix, the complementary bases are in a position to find each other and to reform the double helix.

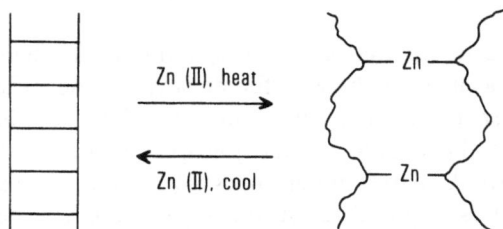

Fig. 8. Scheme for the reaction of zinc(II) with DNA, showing how zinc ions can crosslink the denatured single strands of DNA to make possible the regeneration of the double helix.

This ability of zinc ions to orient the single strands of DNA in such a way as to cause regeneration of the double helix on cooling, rather than just the formation of a few hairpins, can be considered analogous to the ability of metal ions to orient o-aminobenzaldehyde in such a way as to form a trimeric or tetrameric condensate. In both instances the metal ion, by coordination, holds several ligand molecules in a spatial orientation that permits the formation of a structure that is not attainable in the absence of the metal ion.

The rewinding of DNA in the presence of zinc can therefore be considered a "template" effect. In spite of the similarity to the template effect in a minimolecular system, the phenomenon is not quite the same in the biological macromolecule. With minimolecules the metal ions cause the formation of covalent bonds that would not otherwise be produced. In DNA covalent σ-bond formation involving the ligand is not necessarily required. The regeneration of the double helix depends upon the recognition of complementary bases which reform hydrogen bonds, and the stabilization of the double helix by π-interaction.

It is, of course, not unexpected that metal ions binding to different sites on the nucleic acid molecules produce different effects upon these molecules. Perhaps more surprising is that the metal ions binding to the *same* site can produce quite different results under different conditions.

We have seen that the phosphate interaction with RNA brings about cleavage of the phosphodiester bonds and hinted that the nucleic acid structure can also be stabilized by phosphate binding. This stabilization can be observed with DNA in Fig. 7B. The increase in T_m produced by zinc ions indicates that the double helix is stabilized relative to the single stranded DNA by the metal coordination. This stabilization is explained as follows. Native DNA contains a linear array of negative charges on the phosphate groups; these charges are close enough together to repel each other and thus destabilize the double helix.[26] The zinc ions serving as counter ions when binding to the phosphate can, therefore, preserve the double helix up to a higher temperature.

It should be noted that even though the same metal can influence nucleic acid molecules in three different ways, these different influences take place under quite different conditions. The stabilization through neutralization of the charge on the phosphate occurs at relatively low temperature, in RNA below temperatures required for cleavage and in DNA below temperatures required for strand separation. The depolymerization occurs only with RNA, because of the requirement of the 2'-hydroxyl group, and the unwinding–rewinding phenomenon can occur only with molecules like DNA that have ordered structures that can be unwound. It is not confined to DNA, but works also with polyribonucleotides with helical properties.

We have selected zinc ions for this discussion because zinc is particularly effective in all of the ligand reactions that have been discussed. The behavior of zinc is only illustrative. Other divalent metal ions generally exert similar effects. It should be obvious that the relative affinity of the metals for phosphate and base sites on nucleic acid molecules should be an important factor in determining the type of ligand reactions that occur. Such is indeed the case. Figure 9,[27] in which the T_m of DNA is plotted as a function of added metal ion concentration for a number of metal ions, gives a clue to this relative affinity. It should be remembered that an increase in T_m indicates phosphate binding, since metals binding to phosphate stabilize native DNA, whereas a decrease in T_m indicates base binding, since metals binding to bases compete with the hydrogen bonds of the double helix and therefore labilize native DNA. It is apparent that all the metals bind phosphate, since they all increase T_m at low metal concentrations. Mg(II), Ni(II), and Co(II) continue to increase T_m as the metal concentrations rise. These metals thus have a relatively strong affinity for phosphate. All the other metals, after an initial rise in T_m at low metal concentrations, produce a T_m maximum and then a decrease in T_m with further increases in the metal concentrations.

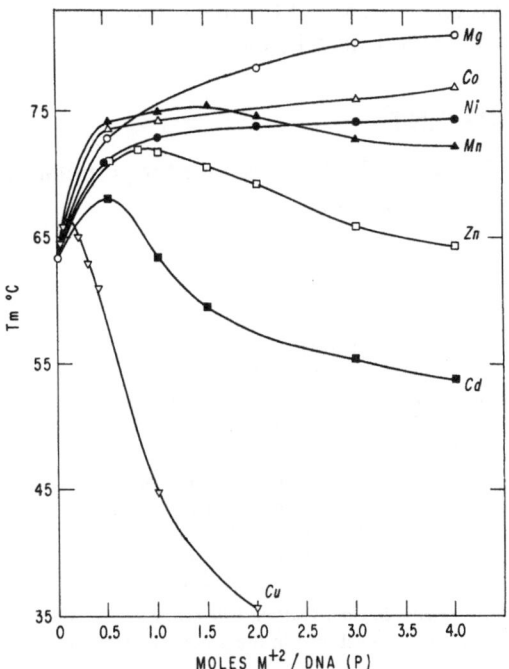

Fig. 9. The variation of DNA melting temperature as a function of metal ion concentration. (From Ref. 27.)

This phenomenon indicates that coordination to the bases is relatively more important for these metals, and it is most important for copper (II), which produces a T_m maximum at very low copper concentration.

The results of Fig. 9 correlate remarkably well with the relative effectiveness of these metal ions in producing the various ligand reactions described for zinc. Of particular interest is that the different metal ions can bring about the unwinding and rewinding of DNA under very different conditions. The differences between zinc and other metal ions in their behavior in these ligand reactions have been discussed elsewhere.[27]

CONCLUSION

We have noted that two types of ligand reactions, bond cleavage due to polarization, and template action due to the organization of reacting molecules around a metal ion, occur in biological macromolecules as well as in minimolecules. Although the same principles are involved, the macromolecular ligand reactions differ in several respects from their minimolecular counterparts. This is chiefly because of the large number of functional groups on the macromolecules that can coordinate to the metals and because of the possible existence of the macromolecules in a variety of conformations that can be influenced by metal ions.

ACKNOWLEDGMENT

I am grateful to Prof. W. N. Lipscomb for permission to use material from Ref. 15 prior to publication.

REFERENCES

1. D. H. Busch, *Adv. Chem.* **37**: 1 (1963).
2. G. L. Eichhorn and J. C. Bailar, Jr., *J. Am. Chem. Soc.* **75**: 2905 (1953).
3. P. Pfeiffer, E. Breith, E. Lübbe, and T. Tsumaki, *Ann.* **503**: 84 (1933).
4. G. L. Eichhorn and I. Trachtenberg, *J. Am. Chem. Soc.* **76**: 5183 (1954).
5. H. Kroll, *ibid.*, **74**: 2036 (1952).
6. R. Steinberger and F. H. Westheimer, *ibid.* **71**: 4158 (1949), **73**: 429 (1951).
7. G. L. Eichhorn and N. D. Marchand, *ibid.* **78**: 2688 (1956).
8. G. L. Eichhorn and R. A. Latif, *ibid.* **76**: 5180 (1954).
9. G. A. Melson and D. A. Busch, *ibid.* **87**: 1706 (1965).
10. G. A. Melson and D. A. Busch, *ibid.* **86**: 4834 (1964).
11. G. L. Eichhorn, in: *Progress in Coordination Chemistry* (ed. M. Cais) Elsevier, Amsterdam, 1968, p. 7. (Some of the ideas developed in the present manuscript were briefly discussed in this reference.)
12. B. L. Vallee and H. Neurath, *J. Biol. Chem.* **217**: 253 (1955).
13. B. L. Vallee, T. L. Coombs, and F. L. Hoch, *J. Biol. Chem.* **235**: PC 45 (1960).

14. J. E. Coleman and B. L. Vallee, *ibid.* **236**: 2244 (1961).
15. W. N. Lipscomb, J. A. Hartsuck, G. N. Reeke, Jr., F. A. Quiocho, P. H. Bethge, M. L. Ludwig, T. A. Steitz, H. Muirhead, and J. C. Coppola, *Cold Spring Harbor Symposia of Quantitative Biology* **33** 1968, (in press).
16. M. F. Perutz, M. G. Rossman, A. F. Cullis, H. Muirhead, G. Will, and A. C. T. North, *Nature* **185**: 416 (1960).
17. J. C. Kendrew, R. E. Dickerson, B. E. Strandberg, R. G. Hart, D. R. Davies, D. C. Phillips, and V. C. Shore, *Nature* **185**: 422 (1966).
18. R. E. Dickerson, in: *The Proteins* (ed. H. Neurath) Academic Press, New York, p. 603.
19. E. Margoliash and A. Schejter, *Adv. Prot. Chem.* **21**: 176 (1966).
20. M. Nakatani, *J. Biochem. Japan* **48**: 633 (1960).
21. M. W. Fanger, T. P. Hettinger, and H. A. Harbury, *Biochemistry* **6**: 713 (1967).
22. A. J. Osbahr and G. L. Eichhorn, *J. Biol. Chem.* **237**: 1820 (1962).
23. J. J. Butzow and G. L. Eichhorn, *Biopolymers* **3**: 95 (1965).
24. P. Doty, H. Boedtker, J. R. Fresco, R. Haselkorn, and M. Litt, *Proc. Natl. Acad. Sci. U.S.* **45**: 482 (1959).
25. Y. A. Shin and G. L. Eichhorn, *Biochem.* **7**: 1026 (1968).
26. J. Shack, R. T. Jenkins, and J. M. Thompsett, *J. Biol. Chem.* **203**: 373 (1953); R. Thomas, *Trans. Faraday Soc.* **50**: 324 (1954); R. F. Steiner and R. F. Beers, *Biochim. Biophys. Acta* **32**: 166 (1956); K. Fuwa, W. E. C. Wacker, R. Druyon, A. F. Bartholomay, and B. L. Vallee, *Proc. Natl. Acad. Sci. U.S.* **46**: 1298 (1960); W. F. Dove and N. Davidson, *J. Mol. Biol.* **5**: 467 (1962).
27. G. L. Eichhorn and Y. A. Shin, *J. Am. Chem. Soc.* **90**: 7323 (1968).

COORDINATION IN BIOCHEMISTRY: REACTIONS OF CHROMIUM(III)

C. L. Rollinson and E. W. Rosenbloom

University of Maryland
College Park, Maryland

INTRODUCTION

The fundamental nature of A. Werner's theoretical concepts, which have been so effectively taught, extended and applied by Bailar, is demonstrated by their great power of correlating and accounting for chemical phenomena in diverse specific areas including the biochemistry of the essential metallic elements, one of which is chromium. In the first part of this communication, we will discuss briefly the general subject and in the second we will describe in some detail our investigations of the coordination chemistry of chromium(III) in reaction mixtures at physiological pH.

COORDINATION IN THE BIOCHEMISTRY OF THE ESSENTIAL METALLIC ELEMENTS

In recent years, the inorganic constituents of biological systems have been receiving increasing attention.[1-3] As a consequence, a definable field of chemistry, encompassing all the inorganic aspects of biochemistry, is emerging. What we envision is a reorganization, from the standpoint of inorganic chemistry, of a wealth of information already made available primarily by biochemists, physiological chemists and nutritionists. Many pertinent papers are appearing in the current literature, every biochemistry text and treatise contains relevant sections, and numerous comprehensive reviews are available; some of these are devoted principally to coordination and chelation.[4-8]

This subject matter provides the basis for a graduate interdisciplinary course of exciting possibilities. It can be organized in such a way as to acquaint nonbiochemist chemistry majors with some biochemistry, and biochemistry and biology majors with some inorganic chemistry relevant to their interests. The senior author developed a course along these lines and taught it in an experimental way in the spring semester of 1968. As the course title, "Inorganic Chemistry of Biological Systems" implies, any biological phenomenon involving some aspect of inorganic chemistry is considered appropriate subject matter; this includes for example the chemistry of bones, the chemistry of respiration, biogeochemistry, the origin of life, life support systems, and, of course, the coordination chemistry of biologically important metal ions.

The majority of the essential metallic elements are transition metals whose ability to form coordination and chelate compounds is one of the most characteristic aspects of their chemistry. Of the first transition series, every element from vanadium through zinc, with the exception of nickel, is known to be essential in some biological system, and there is speculation that nickel, situated as it is in the periodic chart, will some day be found to be essential. In addition to the essential elements of the first transition series, there is one, molybdenum, in the second.

The importance of coordination in the biochemistry of essential metallic elements may be illustrated by numerous examples of metal complexes of which the following are representative: the iron complex hemoglobin and numerous enzymes containing the heme and related structures such as catalases, peroxidases and cytochromes; and the iron-containing proteins ferritin, transferrin, and hemosiderin; the zinc complexes zinc-insulin, carbonic anhydrase and the carboxypeptidases; the cobalt complex vitamin B_{12}; the copper complex, ceruloplasmin; the molybdenum-containing enzymes, xanthine oxidase, and nitrate reductase; DNA-metal ion complexes.

The question of availability of an essential metal provides further examples of coordination in biochemistry. Some foods, for example, contain considerable iron which is unavailable, generally organically-bound iron. If the food is extracted with the chelating agent, α,α-dipyridyl, whatever iron is not too tightly bound reacts to form an intensely red ferrous complex.[9] Fair agreement has been found between the availability of iron as established by this procedure and the availability established by biological evaluation with anemic rats.[10–12]

Coordination as a means of making an essential metal available is illustrated by certain organisms whose survival in a hostile environment depends on their ability to synthesize metal-binding compounds. Certain strains of soybeans, growing in alkaline soil, secrete chelating agents such as malate and citrate that are excreted into the soil and solubilize the iron required by the plants.[13] Some bacteria and fungi secrete and excrete into their alkaline environment sequestering agents that chelate with and solubilize iron that would otherwise not be available[14]; examples of such agents are the "ferrichromes," which are cyclic polypeptides, for example hydroxamic acid derivatives of ornithine, lysine, glycine and other organic acids. Some organisms can even obtain iron by etching the stainless steel of the culture tanks.

With regard to iron transport and metabolism, Saltman[15] has challenged the current textbook explanation which involves a specific transport mechanism requiring metabolic energy. In Saltman's view, primary control depends on endogenous or exogenous ligands or chelating agents capable of binding either ferrous or ferric iron. His proposed mechanism does not directly require either redox reactions or metabolic energy.

Chromium is the most recent addition to the list of essential trace metals. The first conclusive evidence demonstrating a metabolic effect of chromium was obtained by Mertz and Schwarz (National Institutes of Health) in a series of investigations of which the first report was published in 1955.[16] Much data demonstrating the essentiality of chromium has been obtained in particular by Mertz and his associates, and by Schroeder and his colleagues. Chromium, as chromium(III), acts as a glucose tolerance factor and also has an effect in fat and protein metabolism. It is generally not effective alone but acts to enhance the action of insulin. It is apparent that these findings are of great significance in the field of nutrition, and perhaps in the treatment of diabetes and related disorders. The history of this development has been briefly reviewed by the senior author.[17] An exhaustive review of the literature pertaining to the biochemistry of chromium has been compiled by Mertz.[18]

For both theoretical and practical reasons, it is important to establish the mechanism of these effects and this requires a knowledge of the reaction of chromium(III) with biological substances. This brings into the picture the fascinating and complicated chemistry of chromium(III) in basic media containing many substances in wide variety capable of reacting with chromium. The characteristic reaction of chromium(III) in solutions at physiological pH is olation, i.e., polymerization due to bridging OH^- groups.[17,19] The extent to which olation will be inhibited will depend on the complexing ability and concentration of ligands capable of competing with the OH^- ion.[17] Of course, there are substances in biological media, in addition to the OH^- ion, that may react with chromium(III) to form insoluble products or large unreactive slowly diffusing molecules; in either of these forms, the chromium could be expected to be biologically unavailable. Competing with such substances, biological coordinating and chelating agents (e.g., amino acids, keto acids, etc.) may react to form rapidly diffusing products of low ionic or molecular weight; the probability that the chromium would be biologically available in such forms is of course much higher.

In addition to ability to diffuse through biological membranes, any chromium(III) compound in which the chromium is biologically available must have at least one other characteristic. If, as seems likely, the ultimate active form of chromium(III) in the organism is produced as a result of a succession of steps, any complex in the sequence must be sufficiently labile to be converted by the appropriate ligand to the next compound. In other words, only in complexes with formation constants within a certain range will chromium(III) be biologically available (it may be that the key to the chemistry of all essential transition metals will be determination of formation constants!). Highly olated compounds are thus excluded, since reversal of olation is generally a slow process and is quite improbable under the mild reaction conditions of the biological system.

At least three factors will be operative in determining the net effect of the competitive reactions that establish the chemical form of the chromium and in consequence its biological availability: the relative coordinating tendencies of the ligands, and relative affinities of other substances, for chromium(III); the concentrations, and more specifically, the relative ratios of concentrations of the competing reactants to that of chromium(III); the relative rates of reactions. Thus, a problem in nutrition gives rise to a problem involving the basic concepts of coordination chemistry.

It may be that competitive reactions are operative even in controlling the quantity of chromium(III) absorbed through the intestine; the amount is known to be small and usually increasing the quantity in the food given to test animals does not increase appreciably the amount absorbed.[20] As chromium(III) in the ingested food passes through the stomach and the intestine, it encounters an increasingly basic environment which, in the absence of effective ligands, would cause olation resulting in formation of hydrous chromium oxide, in the limit, and as intermediate products large molecules that could not diffuse through the intestine wall; these products of course would be excreted in the feces. But as the OH^- ion concentration is increasing, amino acids and other complexing agents are being formed from the food components; thus, some chromium(III) can remain soluble and diffusible to the extent that the latter agents can prevent or minimize olation and other reactions that would lead to insoluble or large polymeric molecules or ions.[17]

REACTIONS OF CHROMIUM(III)

Principles of Dialysis Procedure for Studying Chromium(III) Reaction Mixtures

In reaction mixtures at physiological pH, whatever other properties are determined by the competitive reactions, one of the net results will be the establishing of an average molecular or ionic weight of the chromium(III) product(s) characteristic of a given reaction mixture under a given set of reaction conditions. Therefore a measure of the coordinating tendency of a ligand is its ability to compete with whatever substances are present that can react with chromium(III) to form high molecular weight products. This should be reflected in the rate of diffusion of the chromium(III), which can be established by dialysis.

In the ideal case, diffusion through a membrane in a closed-system dialyzer (equal volumes) follows the theoretical law[21]

$$\ln C_0 - \ln(C_0 - 2C_f) = \lambda t$$

in which C_0 is the initial concentration of the diffusing species (in our experiments, in the "backs" of the dialyzers), C_f is the concentration of the diffusing species in the front compartment, t is time and λ is the dialysis coefficient or specific rate constant. When different substances diffusing according to this law are dialyzed, the results are immediately comparable in terms of their individual dialysis coefficients. Chromium(III) in the reaction mixtures under discussion generally will not diffuse according to the theoretical law, because the reaction products will usually be a mixture of species of different molecular or ionic weights, and because the membrane will interfere in the diffusion of the higher molecular weight species.

In this situation a useful alternative to the dialysis coefficient is the area under the dialysis curve R vs. time (R = fractional attainment of dialysis equilibrium) (Figs. 1–4). Since these areas are proportional to the rates and completeness of diffusion, they provide a basis for comparing the results obtained with different reaction mixtures and reaction conditions. Moreover, if samples of the same reaction mixture are dialyzed at intervals, the magnitude of the change in area under the dialysis curve reflects the change in the state of aggregation of the diffusing chromium(III) species— the more stable the reaction mixture, the less will be the change in area under the curve as the reaction mixture ages. A series of ligands tested in this way thus can be placed in an order of their relative effectiveness in stabilizing the reaction mixtures, which is to say in the order of their relative coordinating tendencies.

We call this the "method of sequential dialysis." Although it is empirical the method is quantitative, in that the results are numbers which may be assumed to be related to the formation constants of the chromium(III) complexes in some definite but at the moment unspecified way.

Calculations

In the procedure, equal volumes of chemically identical chromium(III) reaction mixtures are placed in the front and back compartments of the dialysis cells; the chromium(III) in the back compartments is labeled with [51]Cr. Under these conditions, the concentrations of labeled chromium(III)

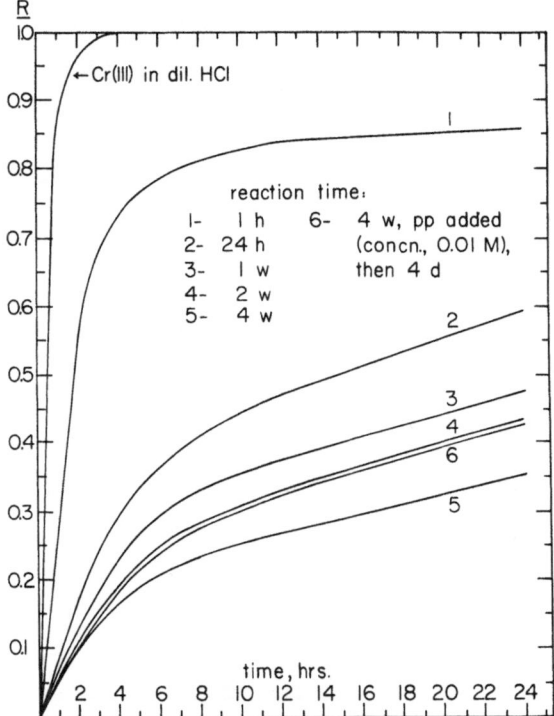

Fig. 1. Diffusion of Cr(III) in 0.02 M $PO_4^{\equiv}$ Buffer: pH 7.4.

are as follows, if all the chromium(III) is diffusible:

	initial	*at equilibrium*
back:	$C_b = C_0 = 10^{-4}$ M	$C_b = C_f = 0.5\,C_0 = 5 \times 10^{-5}$ M
front:	$C_f = 0$	$C_f = C_b = 0.5\,C_0 = 5 \times 10^{-5}$ M

If part of the chromium is not diffusible, then, of course, at equilibrium C_f will be less than C_b and $0.5\,C_0$. The fractional attainment of dialysis equilibrium in the front compartment is the ratio of C_f to $0.5\,C_0$:

$$R = C_f/0.5\,C_0 = \frac{C_f}{5 \times 10^{-5}\,\text{M}}$$

Various quantities derived from the dialysis curves (R *vs.* time, Figs. 1–4) are useful in characterizing the reaction mixtures. Some of these arise from comparison of successive dialysis curves of the same reaction mixture. The ages of the reaction mixture at the start of dialysis are generally 1 hr and

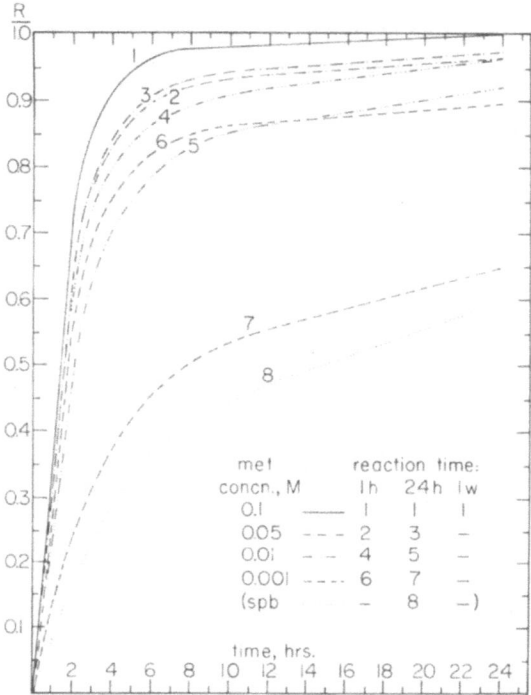

Fig. 2. Methionine: effect of concentration: pH 7.4.

24 hr for the first and second dialyses, respectively (designated by I and II). The numerical values obtained for the first dialysis of a reaction mixture are used as a basis for comparison with results of all subsequent dialyses. The various quantities are defined as follows (any letter used as a fraction is primed to indicate percent, i.e., $K' = 100\,K$).

Completeness of Diffusion (R_I, R_{II}, L)

In addition to the rate of diffusion, the completeness of diffusion, particularly after the reaction mixture has aged, is a significant indication of the effectiveness of the ligand under test. R (the ratio of concentration of tagged chromium(III) in the front compartment to the calculated equilibrium concentration) will increase from zero to 1 during the 24 hr standard dialysis period, if all the chromium(III) is diffusible and to less than 1 otherwise. Thus, a useful figure to characterize any reaction mixture is R for an arbitrarily chosen dialysis period, and another is the ratios of the R's for the fresh and aged reaction mixtures. Let R_I, R_{II}, and $R_x = R$ (24 hr dialysis) for reaction mixture aged 1 hr, 24 hr, and x hr, respectively, at the start of dialysis. Then

$$L_{(I,II)} = R_{II}/R_I \qquad L_{(I,x)} = R_x/R_I$$

Area under Dialysis Curve (A_r, F)

What we designate "area" (A_r) is actually the fraction of the total area defined by the coordinates (0–24 hr on the horizontal axis and R from 0 to 1 on the vertical axis) (Figs. 1–4). Since it differs from the actual area only by a constant factor, no inconsistency arises because of this substitution. This area is proportional to the rate of diffusion and thus generally inversely proportional to the average molecular or ionic weight of the diffusing species.

In a dilute HCl solution of chromium(III), the species present is $[Cr(H_2O)_4Cl_2]^+$ whose ionic weight is 195. In solutions of physiological pH, reactions of chromium(III) with the hydroxide ion and other substances will generally lead to formation of species of higher molecular or ionic weight.

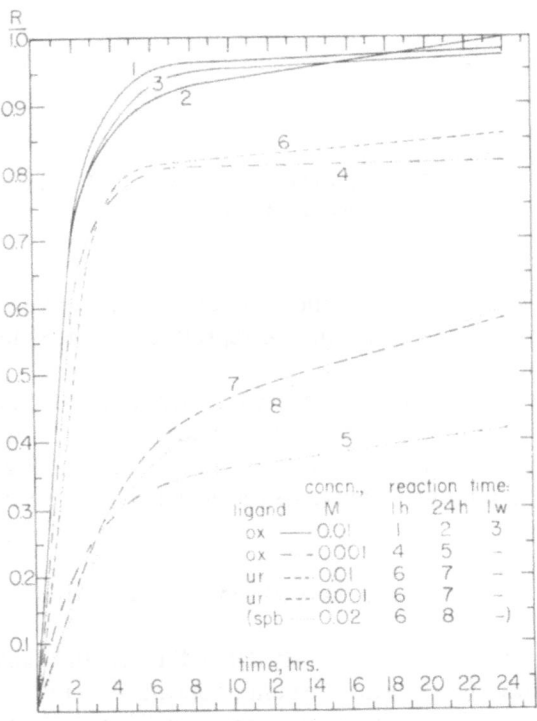

Fig. 3. Comparison of oxalate and urea: pH 7.4.

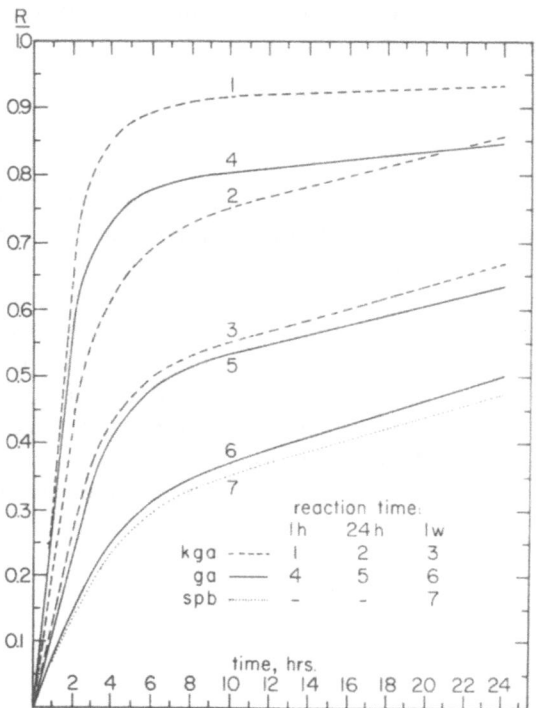

Fig. 4. Effect of chelation: comparison of glutarate and
α-ketoglutarate: pH 7.4.

Thus, 195 is about the lowest ionic weight that will ever be encountered, and this may be regarded as constant and reproducible since the ion in question is stable in dilute HCl.

The area under the dialysis curve for chromium(III) in dilute HCl may, therefore, be used as a standard of comparison; this area is found to be 95.5 % of the rectangular area defined by the coordinates (Fig. 1). A numerical criterion of the relative effectiveness of a ligand is the ratio F defined as follows:

$$F = A_r/A_{Cr} = A_r/0.955 \qquad F' = 100F$$

in which A_r is the area (as previously defined) under the dialysis curve of the reaction mixture and A_{Cr} is the area under the dialysis curve of chromium(III) in dilute HCl. Thus, F for any reaction mixture differs from A_r by the constant factor $1/0.955$ and may generally be used in its place.

Change in Area under Dialysis Curve (D, P, Q)

If the state of aggregation of the chromium(III) species changes as the reaction mixture ages, the rate of diffusion will change with resulting change in area under the dialysis curve. Any of several numerical values might be used to designate this change in area: the difference D between the area under the first and the area under a subsequent dialysis curve, the ratio P of the area under any subsequent curve to that under the first, or the fractional decrease in area Q. In our standard procedure, the first and second dialyses are started after, respectively, 1 and 24 hr aging of the reaction mixture and the dialysis period is 24 hr; with the more effective ligands it is useful to obtain dialysis curves after even longer aging periods.

The terms and relations are as follows for the first two dialyses:

A_I, A_{II} areas under first and second dialysis curves, respectively.

F_I, F_{II} ratios of the two areas to area under Cr(III)–HCl dialysis curve.

$$D = A_I - A_{II} \qquad D_{Cr} = F_I - F_{II} = D/0.955 \qquad D' = 100\,D$$

$$P = A_{II}/A_I = \frac{A_{II}/A_{Cr}}{A_I/A_{Cr}} = F_{II}/F_I \qquad\qquad P' = 100\,P$$

$$Q = \frac{A_I - A_{II}}{A_I} = 1 - A_{II}/A_I = 1 - F_{II}/F_I = 1 - P \qquad Q' = 100\,Q$$

Experimental Procedures*

For each set of dialysis curves, two chemically identical reaction mixtures are prepared, one with ^{51}Cr-labeled chromium(III), the other with inactive chromium(III). The reaction mixtures are prepared by adding a dilute HCl solution of chromium(III) (2×10^{-4} M) to an equal volume of phosphate buffer solution containing the ligand(s) and/or reactant(s) to be tested. By this procedure all competing substances (OH^-, ligand, etc.) can start reacting with monomeric chromium(III) at the same instant. The ratio of $HPO_4^=$ to $H_2PO_4^-$, established by preliminary experiments, is such that the desired pH of the reaction mixture is automatically attained when the solutions are mixed. With only a few exceptions, the characteristics of the reaction mixtures we have studied so far, including those discussed in this paper, have been as follows: Cr(III), 10^{-4} M; pH, 7.0 or 7.4; $PO_4^=$, 0.02 M.

Enough reaction mixture is prepared for at least two dialysis experiments, generally with reaction mixture 1 hr and 24 hr old; if one of the more effective ligands is being tested, the rate of change is low and it is useful to

---

*The authors' experimental work has been facilitated by the assistance of a succession of competent and enthusiastic undergraduate chemistry and biology majors.

obtain dialysis curves after even longer periods. Reaction mixtures are aged 1 and 24 hr in a shaker at 37°, longer periods without shaking at the laboratory temperature.

Into each of the two identical compartments (each 7 ml capacity) of a dialysis cell[22] is charged 5 ml of reaction mixture—the ^{51}Cr-labeled chromium(III) in the back, the unlabeled chromium(III) in the front. The membrane separating the compartments is "Visking," a regenerated cellulose whose average pore diameter is 4.8 mμ: its diameter is 38 mm.

The dialyzers are shaken in a constant temperature shaking machine at 37°C. At the desired intervals two dialyzers are removed and 0.5 ml counting samples are taken from front and back compartments; thus, each point on the dialysis curve is established by means of individual dialyzers in duplicate. No dialyzer is returned to the shaker and sampled again after a further interval since errors are introduced if several samples are taken from the same dialyzer because of the change in the ratio of solution volume to membrane area. The samples are counted with a gamma scintillation spectrometer. Although we base our calculations on the counts of the samples taken from the front compartments (in which the concentration of labeled chromium(III) is increasing), counts of the samples from the backs are obtained as a check.

With the compartments charged with chemically identical reaction mixtures, dialysis equilibrium exists even at time zero, with chromium(III) moving both ways at the same rate through the membrane with redistribution of the [51]Cr(III). If chromium(III) were placed initially in only one compartment, it would be progressively diluted during the dialysis (ultimately 1:1) with possible shifting of all hydrolytic equilibria. This variable is eliminated since the chromium(III) concentration remains constant in both compartments. Another variable is also eliminated, namely osmotic pressure, which, although it would be small, would continuously change during the dialysis if initially only the back compartment contained chromium(III). It would appear that the only driving force under these conditions is diffusion which depends on, among other factors, the relative weights or perhaps diameters of the diffusing species. Since the concentration of each species is the same on both sides of the membrane in our procedure, the driving force is evidently not osmotic pressure as usually stated in discussions of dialysis.

Furthermore, since vigorous agitation continually renews the solutions in contact with the membrane surfaces, the rate is determined for transport of chromium(III) across the membrane only. Thus, for any given reaction mixture containing some specific distribution of molecular weights of chromium(III) species, the measured rate will be a function of the ratio of membrane area to volume of solution if agitation is vigorous enough to eliminate any lag in renewal of solution in contact with the membrane. On this basis, it appears possible to magnify the difference between two effective ligands, which would otherwise appear equally effective, by making this ratio very small.

The areas under the dialysis curves are determined by weighing. For each individual dialysis, a curve is drawn, R vs. t, on a separate sheet of graph paper. The paper used has been tested and found to meet the requirement that the weight per unit area of an individual sheet is constant over its entire area, i.e., the area will be proportional to weight for that sheet. The rectangular area, designated A_t, defined by the coordinates (0 to 24 hr and R from 0 to 1; Figs. 1–4) is carefully cut out and weighed on an analytical balance. This piece is then cut along the curve and the part under the curve A_r is weighed. The area under the curve is thus

$$A_r = A_t(w_r/w_t)$$

in which w_t and w_r are the weights of the rectangular area and the area under the curve, respectively. The ratio w_r/w_t is used instead of the actual area and is designated "area"; it differs from the actual area only by the constant factor A_t, which is 60 sq in. with the graph paper we use. With the procedure described, it is not essential that the sheets of graph paper have the same weight since each sheet serves as its own standard; nevertheless, we always use the same kind of paper for all the curves.

Results

By sequential dialysis, we have determined the relative coordinating tendencies of many biological substances as measured by their relative abilities to stabilize chromium(III) reaction mixtures. There is wide variation in the effectiveness of amino acids which depends on their molecular weight and structure. On the basis of limited data, we have tentatively concluded that sugars such as glucose and fructose are quite ineffective. The effectiveness of a ligand is markedly dependent on its concentration or perhaps more specifically the ratio of its concentration to that of chromium-(III). The dialysis method is applicable to systems containing metals other than chromium(III) and preliminary work indicates that different metal ions can compete for the same ligand, a phenomenon that we will investigate further.

The most potent ligands investigated so far are citrate, pyrophosphate and triphosphate which, in appropriate concentrations, stabilize the reaction mixtures for weeks; moreover, they are capable of reversing (at least partially) the polymerization caused by the hydroxide ion, and reactions of chromium(III) with such substances as fatty acid ions.[23] Citrate, of course, is abundant in biological systems, e.g., in the Krebs cycle. The concentration of pyrophosphate and triphosphate, as inorganic ions, is zero or at least very small in biological systems. There are however organic derivatives of pyrophosphate and triphosphate (e.g., ADP, ATP) in significant concentration in biological systems; since some of these may have considerable ability to coordinate with chromium(III), we have started to investigate them.

We plan to make a systematic survey of all substances in the biological cycles (such as the Krebs and glycolytic cycles) whose structures indicate that they have coordinating ability. Almost every compound in the Krebs cycle is a possibility. The objectives of further investigation will be to ascertain the composition of the more promising complexes, to determine their relative biological effects, and to relate biological activity to structure.

From a large number of results, we have selected typical data to illustrate the utility and versatility of the method of sequential dialysis; these results are summarized in Figs. 1–4 and the corresponding Tables 1–4. In comparing the results of many experiments, the actual curves are cumbersome to use; the tables show how the salient features of the dialysis curves can be displayed as numerical values.

All the reaction mixtures had the following characteristics:

$$\text{Cr(III)}, \ 10^{-4} \text{ M} \qquad \text{PO}_4^{\equiv}, \ 0.02 \text{ M} \qquad \text{pH}, \ 7.4$$

The reaction mixtures were aged at 37° in the shaking machine for the 1 and 24-hr periods and at laboratory temperature without agitation for the longer periods.

Diffusion of Chromium(III) in Phosphate Buffer, pH 7.4

Figure 1 and Table 1 show how the rate of diffusion of chromium changes on aging of the standard 0.02 M $\text{PO}_4^{\equiv}$ buffer solution to which no ligand has been added; similar results are obtained at pH 7.0. Since there is nothing to compete with the OH^- ion, the rate of diffusion continually decreases

Table 1
Diffusion of Cr(III) in 0.02 M $\text{PO}_4^{\equiv}$ Buffer; pH 7.4

Curve Fig. 1	t^*	R'	L'	F'	Q'
1	1 hr	86.0		80.9	
2	24 hr	59.5	69.2	45.1	44.3
3	1 week	47.8	55.6	36.8	54.5
4	2 weeks	43.4	50.5	31.6	60.9
5	4 weeks	35.3	41.0	27.5	66.0
6	4 weeks, 4 days†	42.9	49.9	30.7	62.2

*t = aging time.
†4 days after pyrophosphate (0.01 M) was added to 4–week-old reaction mixture (curve 5).

and this is reflected in the changing area under the dialysis curve, F' decreasing from 80.9 to 27.5, and Q' increasing from 44.3 to 66.0 in the four-week period.

If the reaction mixture had contained an effective ligand such as pyrophosphate, triphosphate, or citrate (concentration $\sim 5 \times 10^{-3}$ M), F' would have been over 90 and would have decreased very little in several weeks. These substances can compete with OH^- for chromium(III) and thus prevent the polymerization.

They can also slowly reverse the olation. Pyrophosphate (0.01 M), added to the four-week old reaction mixture, in four days increased F' from 27.5 to 30.7, and displaced the curve back to about where it was at two weeks (curves 6,4).

Methionine: Effect of Concentration

Of several amino acids tested, methionine is one of the more effective in stabilizing chromium(III) reaction mixtures, but its action is strongly dependent on concentration, the effect of which is shown in Fig. 2 and Table 2.

Table 2
Methionine: Effect of Concentration: pH 7.4

Curve Fig. 2	c^*	t	R'	L'	F'	Q'
1	10^{-1}	1 hr	101.0		96.1	
1		24 hr	100.6	99.6	97.4	-1.4
1		1 week	99.6	98.6	96.3	-0.2
2	5×10^{-2}	1 hr	96.2		91.0	
3		24 hr	97.2	101.0	91.8	-0.9
4	10^{-2}	1 hr	96.1		89.3	
5		24 hr	92.1	95.8	82.1	8.1
6	10^{-3}	1 hr	89.8		83.9	
7		24 hr	64.3	71.6	51.4	38.7
8	(spb)†	24 hr	59.5	69.2	45.1	44.3

$^*c$, concentration, M.
†Standard phosphate buffer only, 0.02 M $PO_4^=$, pH 7.4.

At 10^{-1} M (ratio met/Cr(III), 10^3) the mixture is so stable over a one week period that one curve fits all the data quite well (curve 1). At half this concentration (curves 2 and 3) there is a little change in 24 hr, in the direction of an increased area under the curve (very slight) which may reflect experimental error or the possibility that equilibrium is not reached in the first hour. However, even though the reaction mixture is essentially stable over the 24-hr period, the rate of diffusion of the chromium(III) is less than it is at the higher methionine concentration, an indication of higher molecular weight of the diffusing species. At 10^{-2} M methionine (ratio met/Cr(III), 10^2; curves 4 and 5) there is a significant decrease in rate of diffusion as well as decreased stability, and at the next lower concentration (ratio met/Cr(III), 10) the stability is little greater than it is in the absence of methionine (curves 6, 7, and 8).

Comparison of Oxalate and Urea: Effect of Concentration

Oxalate ion is a powerful chelating agent whose stable chromium(III) complexes are well known. Urea on the other hand coordinates but does not chelate and its complexes with chromium(III) generally are not formed by the action of urea on chromium(III) in dilute aqueous solution. The difference in coordinating tendency is strikingly demonstrated by the marked difference in the ability of these two ligands to compete with OH^- for chromium(III) (Fig. 3, Table 3).

Table 3
Comparison of Oxalate and Urea: pH 7.4

Curve Fig. 3	r^*	c	t	R'	L'	F'	Q'
1	ox†	10^{-2}	1 hr	97.3		94.4	
2			24 hr	100.0	102.8	92.4	2.1
3			1 week	98.1	100.9	92.8	1.7
4	ox	10^{-3}	1 hr	81.2		78.9	
5			24 hr	41.8	51.5	35.9	54.5
6	ur†	10^{-2}	1 hr	85.9		81.6	
7			24 hr	58.1	67.6	45.6	44.1
6	ur	10^{-3}	1 hr	84.8		79.2	
7			24 hr	58.5	69.0	45.4	42.7
6	(spb)		1 hr	86.0		80.9	
8			24 hr	59.5	69.2	45.1	44.3

*reagent.
†ox, oxalate; ur, urea.

The curves (6 and 7) for the two concentrations of urea are identical, and not greatly different from those for the phosphate buffer (6 and 8). It is evident that urea does not coordinate with chromium(III) under the reaction conditions of the experiment. Thus urea, which is abundant in animal organisms, is ruled out as an effective competitor for chromium(III); far more powerful ligands are available.

On the other hand, oxalate (10^{-2} M, curves 1, 2, and 3) allows only minor changes over a period of a week; oxalate, which is quite abundant in some foods, is thus more likely to combine with chromium(III) than is urea. At 10^{-3} M, however, it is no more effective than urea. Curve 5 for oxalate at the lower concentration is below that for the buffer (8) for the 24-hr old reaction mixtures; this apparently indicates combination of oxalate with chromium-(III) in the polymeric complex without causing de-olation.

Glutarate and α-Ketoglutarate: Effect of Chelation

The utility of the sequential dialysis procedure is demonstrated in a particularly striking fashion by its ability to reflect differences between related ligands, e.g., α-ketoglutarate, a chelating agent, and glutarate, a coordinating agent which probably does not chelate; these are important metabolic intermediates. Figure 4 and Table 4 show the dramatic effect of this difference.

Table 4
Effect of Chelation: Comparison of Glutarate and α-Ketoglutarate, pH 7.4

Curve Fig. 4	r	t	R'	L'	F'	Q'
1	kga*	1 hr	93.1		90.3	
2		24 hr	85.8	92.2	75.0	16.9
3		1 week	67.3	72.3	55.9	38.1
4	ga*	1 hr	84.5		78.9	
5		24 hr	63.8	75.5	53.4	32.3
6		1 week	50.4	59.6	38.4	51.3
7	(spb)	1 week	47.8	55.6	36.8	54.5

*kga, α-ketoglutarate, 10^{-2} M; ga, glutarate, 10^{-2} M.

The difference in coordinating ability of these two ligands is shown very clearly by the following arrangement of some of the data:

	1 hr	24 hr	1 week
F', α-keteroglutarate	90.3	75.0	55.9
F', glutarate	78.9	53.4	38.4
increase due to keto group	11.4	21.6	17.5

This is an example of the modification of a basic structure by a functional group, with a change in coordinating tendency. To extend this series, related substances such as glutamate are under investigation. In progress also is a study of the analogous biological sequence based on succinate.

ACKNOWLEDGMENTS

We thank Dr. Walter Mertz (Chief, Dept. of Biological Chemistry, Walter Reed Army Institute of Research) for making available to us his extensive knowledge of the biochemistry of metals. We are indebted to Col. Edward C. Knoblock (Director, Div. of Biochemistry, WRAIR) for continued encouragement and support. Financial support provided by the U.S. Army Medical Research and Development Command is gratefully acknowledged (Contract No. DA-49-193-MD-2444).

REFERENCES

1. E. J. Underwood, *Trace Elements in Human and Animal Nutrition*, Academic Press, Inc., New York, 1962.
2. C. L. Comar and F. Bronner (eds.), *Mineral Metabolism*, Academic Press, Inc., New York, 1962–1964.
3. T. Bersin, *Biochemie der Mineral- und Spurenelemente*, Akademische Verlagsgesellschaft, Frankfurt am Main, 1963.
4. S. Chaberek and A. E. Martell, *Organic Sequestering Agents*, Chapter 8, John Wiley and Sons, Inc., New York, 1959.
5. M. J. Seven (ed.), *Metal Binding in Medicine*, J. B. Lippincott Co., Philadelphia, 1960.
6. G. L. Eichhorn, in: *Reactions of Coordinated Ligands* (Advances in Chemistry Series 37), Chapter 3, American Chemical Society, Washington, D.C., 1963.
7. F. P. Dwyer, Chapter 8; A. Shulman and F. P. Dwyer, Chapter 9; J. E. Falk and J. N. Phillips, Chapter 10: in: *Chelating Agents and Metal Chelates* (eds. F. P. Dwyer and D. P. Mellor), Academic Press, Inc., New York, 1964.
8. B. L. Vallee and J. E. Coleman, in: *Comprehensive Biochemistry* (eds. M. Florkin and E. H. Stotz), Vol. 12, Chapter VI, Elsevier Publishing Company, Amsterdam, 1964.
9. R. Hill, *Proc. Roy. Soc. (London)*, **B 107**: 205 (1930).
10. W. C. Sherman, C. A. Elvehjem and E. B. Hart, *J. Biol. Chem.* **107**: 383 (1934).
11. D. R. Bergen and C. A. Elvehjem, *ibid.* **119**: 725 (1937).
12. M. C. Smith and L. Otis, *J. Nutrition* **14**: 365 (1937).
13. J. C. Brown and L. D. Tiffin, *Plant Physiol.* **40**: 395 (1965).
14. J. B. Neilands, *Bact. Rev.* **21**: 101 (1957).
15. P. Saltman, *J. Chem. Educ.* **42**: 682 (1965).
16. W. Mertz and K. Schwarz, *Arch. Biochem. Biophys.* **58**: 504 (1955).
17. C. L. Rollinson, in: *Radioactive Pharmaceuticals* (eds. G. A. Andrews, R. M. Kniseley, and H. N. Wagner, Jr.), Chapter 24, U.S. Atomic Energy Commission, CONF-651111, Oak Ridge, Tenn., 1966.
18. W. Mertz, *Physiol. Rev.* (in press).
19. J. C. Bailar, Jr., in: *Preparative Inorganic Reactions* (ed. W. L. Jolly) Vol. 1, Interscience Publishers, New York, 1964, p. 3.
20. H. A. Schroeder, J. J. Balassa, and I. H. Tipton, *J. Chron. Dis.* **15**: 941 (1962).

21. M. Kirchgessner and U. Weser, *Z. Tierphysiol. Tierernähr. Futtermittelk.* **20**: 34 (1965).
22. B. G. Malström, *Arch. Biochem. Biophys.* **46**: 349 (1953).
23. C. L. Rollinson, E. Rosenbloom, and J. Lindsay, in: *Proceedings of the VIIth International Congress of Nutrition* (ed. J. Kühnau), Vol. 5, Friedr. Vieweg und Sohn GMBH, Braunschweig, W. Germany, 1967, p. 692.

MANGANESE PORPHYRIN COMPLEXES II—ELECTRONIC SPECTROSCOPY AND STRUCTURE

L. J. Boucher

Carnegie–Mellon University
Pittsburgh, Pennsylvania

THE ROLE OF MANGANESE IN PHOTOSYNTHESIS

In green plant photosynthesis several transition elements are needed for normal function of the chloroplast. One of these, manganese, is required for the oxygen evolution that occurs in photosystem II.[1] Several investigations have suggested, on the basis of photochemical experiments, that the manganese is involved in the breakdown of a primary oxidant which is formed in the photolytic splitting of water.[2] The nature of the primary oxidant and the detailed description of the action of manganese on it are still to be determined. It appears that there is one manganese atom present per 50 chlorophylls in the photosynthetic unit.[3] The metal atoms are physically located in the protein portion of photosystem II.[4] The manganese is difficult to remove from the chloroplast and no physical property (namely, ESR spectra) of the metal can be observed *in vivo*. The ligand atoms binding the metal are not known. The manganese complex does not appear to be a special chlorophyll molecule associated with chlorophyll in the lipid phase.[4] The evidence, however, does not rule out a similar complex contained in the protein phase. Recent evidence[4] indicates that the manganese is in a high oxidation state, $+3$ or $+4$. Calvin has suggested that the manganese atoms in the chloroplast are bound to porphyrin-like ligands.[5] In fact, it has been demonstrated that porphyrin complexes can be generated in which the manganese atom is in the $+3$ and $+4$ oxidation states.[6] Although the initial work supports the relevance of manganese porphyrins as photosynthetic model compounds, no detailed study of the chemical and physical properties of these materials has been carried out.

126

CHARACTERIZATION OF MANGANESE(III) PORPHYRIN COMPLEXES

A manganese porphyrin complex was first prepared by Zaleski[7] and later more fully investigated by Taylor.[8] When a manganese(II) salt and the free porphyrin are heated in glacial acetic acid, octahedral complexes of manganese(III) appear to form exclusively. In fact, the manganese(II) complexes can only be prepared by the reduction of the +3 complex in the absence of air. The manganese(III) porphyrin complexes can be precipitated from aqueous solution by the addition of suitable coordinating anion. A structural representation of the complexes formed in that way is given in Fig. 1. Their elemental analyses, physical and spectral properties are consistent with the formulation given there.[9] Magnetic susceptibility measurements in the solid state and solution and at a number of temperatures yield effective magnetic moments of 4.8 to 5.0 B.M. The expected value for a spin-free d^4 manganese-(III) complex is 4.9 B.M. There appears to be no spin pairing even for the strongest field anion. Infrared spectra indicate that the porphyrin macrocycle remains intact in the formation of the metalloporphyrin.[10] In addition to porphyrin absorption, the far infrared spectra show metal–anion stretching absorptions in the 160 to 500 cm^{-1} range. This is consistent with a strong binding of the anion in the axial coordination position of the metal. In the

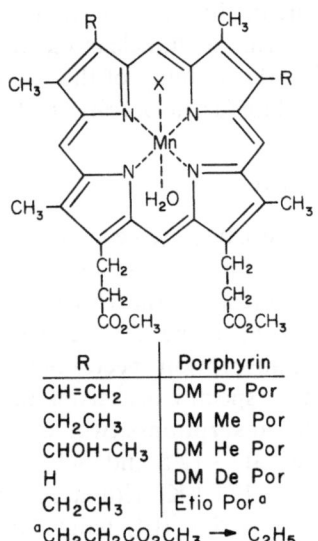

R	Porphyrin
CH=CH$_2$	DM Pr Por
CH$_2$CH$_3$	DM Me Por
CHOH-CH$_3$	DM He Por
H	DM De Por
CH$_2$CH$_3$	Etio Por^a

^{a}CH$_2$CH$_2$CO$_2$CH$_3$ → C$_2$H$_5$

Fig. 1. Structural representation of manganese(III) porphyrin complexes X = F$^-$, Cl$^-$, Br$^-$, I$^-$, CN$^-$, SCN$^-$, N$_3^-$, CNO$^-$.

rock salt region the CN^-, SCN^-, CNO^-, and N_3^- complexes also show anion stretching absorptions in the 2150 to 2000 cm^{-1} region.[11] The frequency of the bands indicates that CN^- and CNO^- ligands are bound to metal via the carbon atom, while the SCN^- is bound through the nitrogen atom.

In solution, the water ligand and then the coordinated anion are displaced by good coordinating solvents. Preliminary spectrophotometric determinations[12] yield values for the equilibrium constant for water replacement in pyridine-chloroform of about 5. The equilibrium constant is about 10^{-2} M^{-1} for anion replacement and is larger for the iodide than for the fluoride complex. The low value for the displacement of the anion may indicate relatively strong binding of the anion to the metal. In mixed solvents, such as dimethylsulfoxide-pyridine, the spectra show two isobestic points which indicates that the two disolvent complexes are in equilibrium with the mixed solvent species. In this system, the dimethylsulfoxide complexes are favored. However, preparation of solid solvent complexes has not been successful. From the anion binding and solvation equilibria, the manganese-(III) appears to retain its hard acid behavior even when bound to the strong field porphyrin ligand. Manganese(III) complexes of porphyrin-like ligands have also been studied. A chlorophyll complex has been prepared with methyl pheophorbide a[13] and several phthalocyanine complexes have been reported.[14,15] In a general way, their chemistry appears to parallel that of the porphyrins. A major point of difference is that while the manganese(III) phthalocyanine complex forms a dinuclear oxygen bridged complex, the porphyrins do not.[16] Also, the stability of the +2 oxidation state is much greater for phthalocyanine than for the porphyrins.[15]

ABSORPTION SPECTRA

One of the most characteristic physical properties of a metalloporphyrin is its visible absorption spectrum. A typical spectrum would consist of two medium-intensity bands, α and β, in the 16 to 20 kK ($\varepsilon \sim 1 \times 10^4$) region and a very intense Soret band in the 22 to 25 kK region ($\varepsilon \sim 2 \times 10^5$).[17] From this point of view the absorption spectra of manganese(III) porphyrins are unusual. Representative spectra are given in Fig. 2. In general, the spectra show six prominent absorptions: two, low-intensity bands ($\varepsilon \sim 1$–2×10^3) in the near infrared at ~ 12.5 and ~ 14.6 kK; a medium-intensity band ($\varepsilon \sim 5$–10×10^3) at ~ 17.8 kK with a red shoulder at ~ 16.8 kK; and two, high-intensity ($\varepsilon \sim 4$–8×10^4) bands at ~ 21.0 and ~ 27.0 kK. In some cases the infrared bands appear to be split into two closely spaced components. There is also a medium-intensity shoulder at ~ 23.5 kK. The extreme broadness of the highest energy band (half width at half height ~ 1800 cm^{-1}) indicates that more than one absorption is located under its

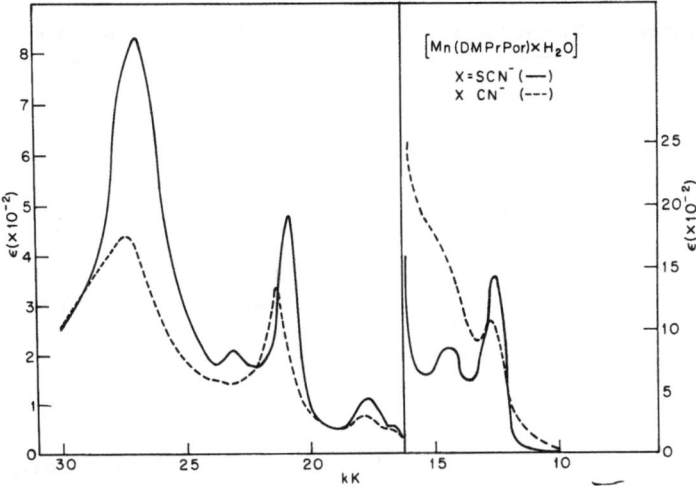

Fig. 2. Electronic absorption spectra of manganese(III) protoporphyrin IXDimethyl ester complexes in chloroform.

envelope. In fact, a shoulder on the red side of this band at ~ 29 is almost discernible and the ultraviolet region ($< 40\,kK$) shows an absorption at $\sim 33\,kK$. The near infrared region from 6.0 to 10.0 kK does not show any additional low-energy bands.

The unusual nature of the electronic spectra of the manganese(III) porphyrins might be diagnostic of a unique electronic structure. This could give rise to interesting physical–chemical properties. A study of the spectral properties of a number of complexes was initiated to learn something about their electronic structure. The work includes the effect of coordinated anion, porphyrin substituent, and solvent on the visible spectra of manganese(III) porphyrins. The absorption maxima, labeled I–VI, of some protoporphyrin IX-dimethylester complexes, in chloroform, are given in Table 1. Similar data are gotten from nitromethane or benzene solutions. Because of broadness of the bands there is some uncertainty in the position of the maxima. From multiple readings and duplicate samples the uncertainty is estimated to be less than $\pm 0.1\,kK$. All the solutions obey Beer's law in the concentration range, 10^{-3} to 10^{-6} M. Previous evidence also shows that the complexes are monomeric in this concentration range.[6] The band maxima vary measurably with the anion coordinated to the metal. Band I and band II vary in the range 12.8 to 12.3 and 14.7 to 14.1 kK. The ranges for bands III and IV are about the same : 16.9 to 16.7 and 17.8 to 17.3 kK. Conversely, the range of variation for bands V and VI is by far the greatest : 21.3 to 20.0 and 27.4 to

Table 1
Absorption Maxima (in kK[a]) of [Mn(DMPrPor)XH$_2$O] in Chloroform

Band	X$^-$							
	CN	F	CNO	Cl	N$_3$	Br	SCN	I
I	12.8	12.7	12.7	12.7	12.7	12.6	12.4	12.3
II	∼14.7	∼14.6	∼14.7	∼14.6	∼14.7	∼14.4	∼14.5	∼14.1
III	∼16.9	∼16.8	∼16.8	∼16.7	∼16.7	∼16.7	∼16.8	—
IV	17.8	17.7	17.6	17.6	17.6	17.5	17.6	17.3
V	21.3	21.0	21.1	20.9	20.8	20.7	20.8	20.0
VI	27.4	27.3	27.4	27.2	27.1	27.0	26.8	26.5

[a] 1 kK = 1000 cm^{-1}.

26.5 kK. A general anion order can be easily extracted from the frequencies of all the maxima

$$CN^- > CNO^- > F^- > Cl^- > N_3^- > Br^- > SCN^- > I^-$$

The effect of solvent on the visible spectra of the complexes can arise from two sources. In good coordinating solvents, the donor molecule displaces the coordinated anion to give disolvent complex in solution. The spectra for all the anion complexes would then be the same for a particular solvent. The data for three such solvents, pyridine, dimethylsulfoxide, and methanol, are given in the right side of Table 2. The two oxygen donors give similar spectra which are different from that of the nitrogen donor, pyridine. For example, the former show absorptions at higher energy than the latter. The largest differences appear for bands I, V, and VI. It is interesting to note that bands V and VI are significantly blue-shifted from the values seen for all of the anion complexes. A second solvent effect arises from the

Table 2
Absorption Maxima (in kK) of [Mn(DMPrPor)IH$_2$O] in Various Solvents

Band	Benzene	Chloro-form	Nitro-methane	Acetone	Aceto-nitrile	Pyridine	Dimethyl-sulfoxide	Methanol
I	12.5	12.3	12.4	∼12.4	12.5	12.4	12.8	12.6
II	∼14.5	∼14.1	∼14.2	∼14.2	∼14.5	∼14.1	∼14.6	∼14.6
III	—	—	∼16.8	∼17.2	∼16.8	∼17.1	∼17.1	∼17.2
IV	17.4	17.3	17.8	18.0	18.0	18.0	18.1	18.2
V	19.8	20.0	21.1	21.3	21.4	21.0	21.6	21.6
VI	16.5	26.5	26.7	26.7	27.2	26.5	27.0	27.0

change in dielectric constant and refractive index (polarizability) of the solvent. Typical data for a variety of solvents is given in the left-hand side of Table 2. In these solvents the anion is not displaced by a donor molecule, i.e., the spectra are different for each anion complex. Bands I and II do not appear to be much affected by the solvent. Conversely, bands III, IV, V, and VI vary significantly. The solvent order is: acetonitrile > acetone > nitromethane > chloroform > benzene. The ranges are: III, 16.8 to 17.2; IV, 17.3 to 18.0; V, 19.8 to 21.4; and VI, 26.5 to 27.2. In most cases, the variation in the frequency of the maxima is comparable or greater than for the various anions. In general, the anion order observed in chloroform is maintained in the other solvents.

By varying the substituents on the β positions of the porphyrin, the basicity of the pyrrole nitrogens can be varied. The spectral data for a number of manganese(III) porphyrin complexes in pyridine are given in Table 3. The frequency of the absorption maxima vary with porphyrin substituent. The frequency range of the maxima is not large for bands I to V, > 300 cm^{-1}; the range for band VI is somewhat larger, 26.5 to 27.1. In general, the energy order of the maxima is

$$\text{DMDePor} > \text{EtioPor} > \text{DMMePor} > \text{DMPrPor}$$

Table 3
Absorption Maxima (in kK) of [Mn(Porphyrin)XH$_2$O] in Pyridine

Band	Porphyrin			
	DMPrPor	DMDePor	DMMePor	EtioPor[15]
I	12.4, 12.8	12.6, 12.9	12.5	12.5, 13.5
II	14.2, 14.5	14.3, 14.7	14.2, 14.6	14.8
III	~17.1	~17.2	~17.2	17.2
IV	18.0	18.2	18.3	18.3
V	21.0	21.3	21.2	21.2
VI	26.5	27.1	26.9	27.0

ASSIGNMENT OF ELECTRONIC SPECTRAL TRANSITIONS

A striking feature of the electronic spectra of manganese(III) porphyrin complexes is the occurrence of two low-energy bands of modest intensity in the near infrared. Divalent and trivalent metalloporphyrins, including manganese(II), do not show near infrared absorptions.[18] Most octahedral manganese(III) complexes show ligand field absorption in the 6 to 20 kK region.[19] In an octahedral ligand field only one spin allowed d–d electronic transition is expected, $^5Eg \rightarrow {}^5T_{2g}$, for the high-spin d^4 case. In a lower symmetry ligand field D_{4h} the degenerate d levels are split and three spin allowed d–d transitions are expected.[20] Considering only the donor atoms, the complexes of manganese(III) porphyrins belong to the D_{4h} point group. An appropriate d-level ordering, originally derived from molecular orbital calculation on the (chloro)iron(III) porphyrin is given in Fig. 3.[21] The four unpaired electrons singly occupy the four bottom levels. The three expected ligand field transitions, in order of increasing energy, are: $d_{z^2} \rightarrow d_{x^2-y^2}(\Delta_1)$, $d\pi \rightarrow d_{x^2-y^2}(\Delta_2)$, $d_{xy} \rightarrow d_{x^2-y^2}(\Delta_3)$. The possibility exists that the near infrared bands observed here have a ligand field origin. Several pieces of information are not consistent with this proposition. First, although the anion order roughly parallels the spectrochemical series, the variation in band maxima is small, <500 cm^{-1} for a wide range of anionic ligands. The anions include the strong field CN$^-$ and the weak field I$^-$. The change in ligand field bands with anion for the manganese(III) complexes is usually 1.0 to 3.0 kK.[19] Second, the positions of the maxima are at too low energy for the strong field porphyrin donor. Transition Δ_3 should give the ligand field parameter for these complexes. Even if the highest energy case is taken, 10 Dq would be $\sim 15,000$ cm^{-1}. This value is lower than for any other manganese(III) complex.[20] If band III or IV were taken to be d–d bands, the ligand field parameters would still be too low in energy. Third, the intensity of the near infrared bands is between 1 to 2 orders of magnitude larger than the intensities for the bands for manganese(III) complexes. In fact, the observed intensities are more in the range expected for normal charge transfer bands.

A molecular orbital model of metalloporphyrins predicts that a number of low energy charge transfer bands are possible. The porphyrin–metal transitions involve the highest filled porphyrin π M.O. and the partially filled d levels on the metal. The highest filled and lowest empty π levels of the porphyrin are shown in Fig. 3. The lowest energy charge transfer transitions are from the filled porphyrin levels a_{1u}, a_{2u}, b_{1u}, a_{2u}, to the metal $d\pi$, d_{z^2} levels. It is not unreasonable to assign transition I, II and perhaps III to these charge transfer bands. The visible absorption bands of metalloporphyrins have been assigned to $\pi \rightarrow \pi^*$ transitions of the porphyrin macrocycle. The $a_{2u}(\pi) \rightarrow e_g^*(\pi)$ and $a_{1u}(\pi) \rightarrow e_g^*(\pi)$ transitions are near degenerate in energy and mixed because of electronic configuration interaction.[20] The resulting excited states are a low-energy state (α, β) in which the transition dipoles nearly cancel (low intensity) and higher energy state (Soret) in which the transition dipoles add (high intensity). Lower intensity N and L bands appear in the 30 to 37 kK range for most metalloporphyrins. These bands are assigned to $a_{2u}(\pi)$ and $b_{2u}(\pi)$ to e_g^* transitions.[22] For the manganese(III) complexes, the Soret band appears to be split into two near equal intensity

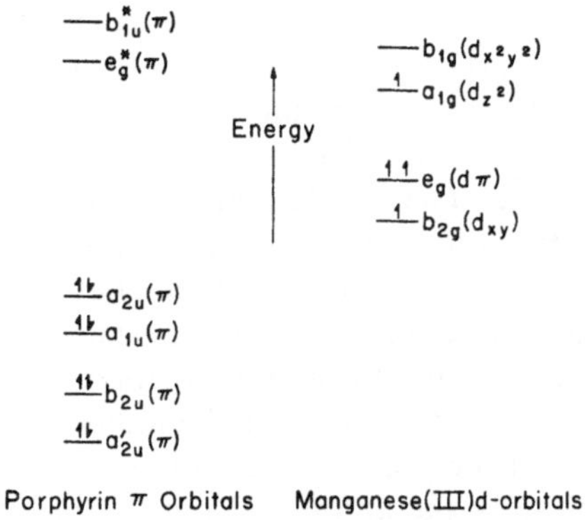

Fig. 3. Molecular orbital scheme for manganese(III) porphyrin complexes.

bands at ~ 27.0 and ~ 21.0 kK. The higher energy band is quite broad and probably contains a contribution from both the N and L bands under its envelope to the blue side. The ultraviolet bands at ~ 29, ~ 33 kK could be assigned to the N and L bands of the manganese(III) porphyrins. Although somewhat broad, band IV appears in the region expected for the α, β bands. The intensity of band IV, V, and VI is about one-half that of the corresponding α, β and Soret bands of a normal metalloporphyrin.

The assignment of the split Soret-like bands V and VI, to porphyrin $\pi \rightarrow \pi^*$ transitions is straightforward. The energies and intensity are in the region expected for such allowed bands. The detailed assignment of the bands to transitions between specific energy levels is, of course, difficult. In the absence of configuration interaction, the two low-energy $a_{1u}(\pi) \rightarrow e_g^*(\pi)$ and $a_{2u}(\pi) \rightarrow e_g^*(\pi)$ should give rise to maxima of different energy and equal intensity with the former at higher energy than the latter. Assignment of the near equal intensity bands VI and V to these transitions would be a reasonable, although naive proposal. Band IV would then be assigned to a charge transfer band.

The anion dependence of the electronic spectra can arise in a number of ways. One mechanism would depend on the specific bonding properties of the anion. For example, sigma bonding of the anion to the metal would be antibonding with respect to the metal d_{z^2} orbital. If the near infrared bands of the manganese(III) porphyrin originated in transition from the filled porphyrin π level to the half-empty metal d_{z^2}, then the energy order of the anion should be directly related to the relative sigma bonding ability of the anion. In fact, the anion order ($CN^- > I^-$) does correspond to the predicted one. Unfortunately all of the bands, including the $\pi \rightarrow \pi^*$ transitions, are blue shifted and not just the near infrared bands. The d_{z^2} metal orbital can mix into the filled $a_{2u}(\pi)$ porphyrin orbital if the metal is out of the plane of the porphyrin. In this way, some anion character can be mixed into the π system of the porphyrin. The axial anions can also interact with the $d\pi$ orbital of the manganese. The π interaction would be antibonding with respect to the metal orbitals for the halide ions (π donor), and bonding for the pseudohalides (π acceptors). If the near infrared charge transfer bands of the manganese(III) porphyrin complexes are assigned porphyrin $(\pi) \rightarrow d\pi$ transitions, then its energy should be highest for I^- and lowest for CN^-. This is just the reverse of what is observed. An indirect effect of the anion on the $\pi \rightarrow \pi^*$ transitions of the porphyrin could arise because of competition for the metal $d\pi$ orbital by the anion and porphyrin ligand. To the extent that the axial anion is involved in the π bonding with the metal then the in-plane porphyrin π interaction would be less.

An alternate rationalization of the spectral data can be obtained by considering a polarization of the porphyrin charge cloud by the axial anion. The π charge cloud extends above and below the plane of the porphyrin for several angstroms. Binding an anion to the metal in the axial coordination position could place the anion close to the π electron cloud. A distortion of the cloud would occur and the extent of this distortion would depend on the polarizability of the anion. This general anion interaction could affect all the π levels of the porphyrin in the same way (slightly stabilizing them or de-stabilizing them). In fact, the anion order follows closely the order of polariza-bility of the anion.[23] The most polarizable anion, iodide, shows the maximum red shift of the bands. This is consistent with a destabilization of the filled π porphyrin orbitals. The explanation is also supported by the fact that solvent effects on the spectra are large (as large or larger than the anion dependence). The solvent order corresponds to the polarizability of the solvent molecule.[24] The most polarizable solvent (benzene) gives the largest red shift. The solvent effect can be larger than anion effect, since the solvation of the porphyrin involves many more than the two molecules that can axially coordinate to the metal. The magnitude of the shift of band V with solvent, ~ 1.5 kK is comparable to the shift of the Soret peak with donor solvent for metalloporphyrins.[25]

The dependence of the spectra on porphyrin substituents is typical of all metalloporphyrins.[25] Electron withdrawing substituents which raise the pK of the porphyrin red shift $\pi \rightarrow \pi^*$ transitions. Increasing the electron density in the vicinity of the nitrogen atom tends to stabilize the porphyrin π orbitals. In the case of the manganese(III) porphyrins similar substituent effects are noted. The magnitude of the shifts is about the same as for the normal metalloporphyrins. Finally, the same anion order appears to also hold for the various porphyrins.

ELECTRONIC STRUCTURE OF MANGANESE(III) PORPHYRIN COMPLEXES

It has been concluded, from the observed spectral patterns and molecular orbital calculations, that the metal and porphyrin π systems are only weakly interacting.[20] Since there appears to be no appreciable ligand–metal π bonding for the divalent metalloporphyrins, the degeneracy of the filled porphyrin π levels, $a_{1u}(\pi)$ and $a_{2u}(\pi)$, is not removed. Of these orbitals only the $a_{2u}(\pi)$ can effectively overlap with the metal $d\pi$ orbitals. Significant π mixing of the metal ion and porphyrin levels could result in the partial lifting of the accidental degeneracy of the $a_{1u}(\pi)$ and $a_{2u}(\pi)$ levels. This would destroy the normal Soret and α, β peak spectrum, since configuration interaction would not effectively take place. The $\pi \rightarrow \pi^*$ spectrum would now resemble more closely a polyene type spectrum with two low-energy bands of equal intensity, $a_{1u} \rightarrow e_g^*$ and $a_{2u} \rightarrow e_g^*$.[26] The manganese(III) porphyrin spectra are not at all like those of other metalloporphyrins. In fact, they appear to resemble spectra expected for the case of strong metal porphyrin π mixing. The anion and solvent dependence is only a perturbation on the more important manganese(III)–porphyrin interaction. If the above interpretation is correct, then manganese(III) is the first metalloporphyrin in which there appears to be significant metal–porphyrin π bonding. The exact nature of the porphyrin–metal d-orbital π interaction is difficult to describe since the molecular structure of the complexes is unknown. One possibility is a tetragonal geometry with the metal ion in the plane of the four porphyrin nitrogen donors, with the axial anion and solvent molecules at a somewhat larger distance from the metal. The $d\pi$ orbitals would be of proper symmetry for overlap with the filled $a_{2u}(\pi)$ porphyrin orbital. The interaction would be antibonding with respect to metal levels. Another likely choice for the geometry of the complexes would be one in which the metal atom is above the plane of the porphyrin (~ 0.5 Å) closer to the axial anion. This situation is, in fact, observed for the iron(III) porphyrins.[27] In this case, separation of the d orbitals into sigma and π bonding sets is not straightforward. Not only the $d\pi$ but the d_{z^2} can mix with the $a_{2u}(\pi)$ orbital of the porphyrins. Mixing of the $d\pi$ metal orbitals with the porphyrin π levels would, of course, measurably increase the intensity of charge-transfer transitions from the porphyrin π levels to the $d\pi$ orbital. The near infrared bands may correspond to this transition.

It is unusual that only manganese(III) shows the π-bonding effect. None of the divalent metalloporphyrins, including manganese(II), show any evidence of π bonding. The observation seems to be general for manganese(III) when bound to porphyrin-like macrocycles. For example, both the phthalocyanines[15] and chlorophyll[13] complexes of manganese(III) have very unusual visible spectra, indicative of some special bonding by the metal. The metal porphyrin π interaction can occur if the orbitals are of comparable energy. Crossing the periodic chart from Sc to Zn the metal d orbitals are stabilized while the porphyrin levels, of course, remain at constant energy.[28] At some point in the transition series the two systems have the appropriate energy for π interaction. Apparently this occurs at manganese(III). The energy of the manganese(II) d levels should be of higher energy than for manganese-(III) and apparently there is no appreciable π mixing and the visible spectra are normal. The energy requirement for mixing of π levels is not an exact one in that the levels can still mix over a range of energies for the metal orbitals. Therefore, π mixing should occur to some extent for the manganese neighbors, either chromium or iron. Chromium(II) shows a normal spectrum,[29] while iron(III) spectra are somewhat anomalous. This may be indicative of a π interaction for iron(III) porphyrins.[21] Of course the π bonding of the metal and porphyrins is not only dependent on the energy of the metal orbitals but also on the geometry of the complexes. For instance, the mixing of the $d\pi$ orbitals with the porphyrin π orbital would be more efficient if the metal were exactly in the plane of the porphyrin. An x-ray crystallographic structure would be a valuable aid in any extensive discussion of the manganese(III) porphyrin. Molecular orbital calculations similar to those carried out for the iron(III) porphyrins would also be necessary to make a detailed assignment of the electronic spectra.

PHOTOSYNTHETIC IMPLICATIONS

Calvin has shown that the manganese porphyrins can exist in a number of oxidation states, e.g., $+2$, $+3$ and $+4$. While the $+3$ oxidation state is stable at physiological pH in air, the $+2$ state is rapidly oxidized and the $+4$ state is a powerful oxidant. The high ligand field strength of the porphyrins permits even the transient formation of a high oxidation state. In the chloroplast compounds like these could then further oxidize a primary oxidant to form oxygen while the metal atom is reduced to the $+2$ oxidation state. Although photoreduction of manganese(III) has been observed,[15] no photo-oxidation to manganese(IV) appears to occur in the absence of oxygen. Perhaps a specific electron acceptor (like the chloroplast component plastoquinone) is needed for this to occur. The short induction period in oxygen evolution in illuminated chloroplasts might be evidence for a photochemical activation of the manganese complex from the stable $+3$ state to the reactive $+4$ state.[2] The photochemical utility of the manganese complex in photosynthesis would depend on its ability to obtain radiant energy while in the chloroplast where the competition for light is great. One advantage that manganese(III) porphyrins have is that they absorb light at different wavelengths from the normal metallochlorophylls, chlorophylls, or carotenoids. For example, the complexes show appreciable absorption of far-red light because of the absorption maxima near 12.5 kK. The presence of the near infrared charge transfer bands could provide a convenient route for photoactivity of these complexes. Many simple manganese(III) complexes show low-energy charge-transfer bands[19] and are also photoactive,[30] presumably for the same reason. The basic pattern of the absorption spectra of manganese(III) porphyrins is not disturbed by its environment (no spin pairing). It is true, however, that the anion and other ligands, the solvent, and porphyrin ring substituents will affect the detailed electronic structure of the complex. There is evidence that the manganese complex is associated with a protein phase in the chloroplast. An intriguing possibility is that the manganese is bound by a porphyrin which is bonded to an apoprotein. This would give a material similar to the cytochromes[31] which are active in photosystem I in the chloroplast.

REFERENCES

1. E. Kessler, *Arch. Biochem. Biophys.* **59**: 527 (1955).
2. B. Kok and G. M. Cheniae, *Current Topics in Bioenergetics* **1**: 1 (1966).
3. G. M. Cheniae and I. F. Martin, *Biochem. Biophys. Acta* **153**: 819 (1968).
4. G. M. Cheniae and I. F. Martin, *Energy Conversion by the Photosynthetic Apparatus*, Brookhaven Symposia in Biology **19**: 406 (1967).
5. M. Calvin, *Rev. Pure Appl. Chem.* **15**: 1 (1965).
6. P. A. Loach and M. Calvin, *Biochemistry* **2**: 361 (1963).
7. J. Zaleski, *J. Physiol. Chem.* **43**: 11 (1904).
8. J. F. Taylor, *J. Biol. Chem.* **135**: 569 (1940).
9. L. J. Boucher, *J. Am. Chem. Soc.* **90**: 6640 (1968).
10. L. J. Boucher and J. J. Katz, *J. Am. Chem. Soc.* **89**: 1340 (1967).
11. L. J. Boucher, to be published.
12. Sr. M. C. LeMaire and L. J. Boucher, unpublished work.
13. P. A. Loach and M. Calvin, *Nature* **202**: 343 (1964).
14. A. Yamamoto, L. K. Phillips, and M. Calvin, *Inorg. Chem.* **7**: 847 (1968).
15. G. Engelsma, A. Yamamoto, E. Markham, and M. Calvin, *J. Phys. Chem.* **66**: 2517 (1962).
16. L. H. Vogt, Jr., A. Zalkin, and D. H. Templeton, *Science* **151**: 569 (1966).
17. C. W. Weiss, H. Kobayashi, and M. Gouterman, *J. Mol. Spect.* **16**: 415 (1965).
18. M. Zerner and M. Gouterman, *Theoret. Chim. Acta* (*Ber.*) **4**: 44 (1966).
19. R. Dingle, *Acta Chem. Scand.* **20**: 33 (1966).
20. T. S. Davis, J. P. Fackler, and M. J. Weeks, *Inorg. Chem.* **7**: 1994 (1968).
21. M. Zerner, M. Gouterman, and H. Kobayashi, *Theoret. Chim. Acta* (*Ber.*) **6**: 363 (1966).
22. W. S. Caughey, R. M. Deal, C. Weiss, and M. Gouterman, *J. Mol. Spect.* **16**: 451 (1965).
23. K. M. Ibne–Raza, *J. Chem. Educ.* **44**: 89 (1967).
24. G. R. Seely, *Spectrochimica Acta* **21**: 1835 (1965).
25. W. S. Caughey, W. Y. Fujimoto and B. P. Johnson, *Biochemistry* **5**: 3830 (1966).
26. M. Gouterman, *J. Mol. Spect.* **6**: 138 (1961).
27. D. F. Koenig, *Acta Cryst.* **18**: 663 (1965).

28. M. Gouterman, private communication.
29. M. Tsutsui, M. Ichikawa, F. Vohwinkel, and K. Suzuki, *J. Am. Chem. Soc.* **88**: 853 (1966).
30. P. H. Homan, *Biochemistry* **4**: 1902 (1965).
31. T. L. Fabry, C. Simo and K. Javaherian, *Biochem. Biophys. Acta* **160**: 118 (1968).

COORDINATION CHEMISTRY OF CYCLIC AMIDES

S. K. Madan

State University of New York at Binghamton
Binghamton, New York

INTRODUCTION

The donor characteristics of amides of the general formula $RC(O)NR_2$ have been extensively investigated.[1-4] The sulfur analog $RC(S)NR_2$ (where $R = CH_3$) has also been studied.[5] In most cases, it has been found that the oxygen of the amide is the principal coordination site, and the carbonyl oxygen of the amide is a better donor than is the carbonyl oxygen of acetone. This has been attributed to delocalization of the nonbonding electron pair on the nitrogen into a π-molecular orbital involving oxygen, carbon, and nitrogen.

The synthesis and spectral properties of a series of nickel(II) perchlorate and chromium(III) perchlorate complexes of the substituted amides have been reported.[2,3] Based on measured electronic parameters of these amide complexes, a series of amides have been arranged in the following order in the spectrochemical series:

$HC(O)N(CH_3)_2$ (850 cm^{-1}) > $HC(O)NHCH_3$ (838 cm^{-1})

$\sim HC(O)N(C_2H_5)_2$ (840 cm^{-1}) > $CH_3C(O)NH_2$ (824 cm^{-1})

$\gg CH_3C(O)N(CH_3)_2$ (769 cm^{-1}) > $CH_3(O)NHCH_3$ (752 cm^{-1})

It has been shown that when both the carbon and the nitrogen contain an alkyl substituent, a steric effect exists between the coordinated amides, causing Dq to decrease, thereby dividing the above substituted amides into two groups.

This article is intended to summarize recent additional work on cyclic amides of the general formula I (where $n = 3, 4, 5$ and $R = H$ and CH_3).

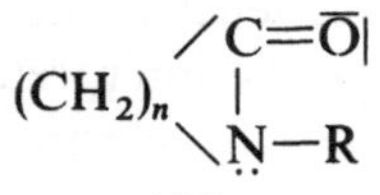

$$(CH_2)_n \begin{array}{c} \diagup C{=}\overline{O}| \\ | \\ \diagdown \underset{\cdot\cdot}{N}{-}R \end{array}$$

139

The following abbreviations have been employed for the various lactams: BuL for γ-butyrolactam ($n = 3$), NMBuL for N-methyl-γ-butyrolactam, Val for δ-valerolactam ($n = 4$) NMVaL for N-methyl-δ-valerolactam, CaL for ε-caprolactam ($n = 5$), and NMCaL for N-methyl-ε-caprolactam.

TRANSITION METAL COMPLEXES OF LACTAMS

The reaction between BuL, VaL, or CaL and the perchlorates of transition metals, yields well-defined crystallized products which on the basis of their analyses and conductivity measurements can be formulated as the perchlorates of the hexakis-(lactam) metal cation, $[M(BuL, VaL \text{ or } CaL)_6]^{n+}$ ($M = Mg^{2+}$, Ca^{2+}, Cr^{3+}, Mn^{2+}, Fe^{2+}, Fe^{3+}, Co^{2+}, Ni^{2+}, Cu^{2+}, Zn^{2+}, Cd^{2+}). These compounds are remarkably stable to air and light, and moderately stable to heat. In aqueous solution, they are less stable and tend to decompose quickly with separation of hydrolysis products.[7] Transition metal chlorides, bromides, iodides, and thiocyanates, in most cases, likewise yield hexacoordinated cations $[M(BuL \text{ or } VaL \text{ or } CaL)_6]^{n+}$, but the halide salts are extremely soluble and, therefore, difficult to isolate. On prolonged standing, the reaction mixtures yield sparingly soluble tetrahalometallates of the complex cations $[M(BuL \text{ or } CaL)_6]^{n+}$.[8] These tetrahalometallates are analogous in their general properties to the corresponding perchlorates, except that they decompose at lower temperatures than the perchlorates. They are also hygroscopic. Unlike the other halides, nickel(II), copper(II), manganese(II), and iron(III) chlorides do not yield hexakis cations with CaL[7] and BuL.[8] Nickel(II) chloride and copper(II) chloride with CaL give instead a yellowish material of undetermined structure.[7] Compounds of the type $MLCl_2$ and ML_2Cl_2 have been isolated for BuL and NMBuL, but similar CaL-complexes have not been isolated.[8] Iron(III) gives the unipositive cations $[Fe(CaL)_4Cl_2]^+$, isolated as the tetrachloroferrate.[7] With the exception of nickel(II), copper(II), and manganese(II) chloride-lactam complexes, all the investigated lactam complexes behave in solution as strong electrolytes. Their molar conductivities in nitromethane λ_m are in good agreement with those reported for complexes of hexamethylphosphoramide and other complexes.[9,10]

Anhydrous nickel(II) chloride in nitromethane, containing a stoichiometric amount of CaL, yields a light green solution.[7] When the temperature is raised, an intense blue color develops; this suggests the existence of an equilibrium similar to the one observed for nickel(II) chloride solution in dimethylsulfoxide.[11]

The carbonyl stretching frequency of the lactam is lower in the complex than in the free ligand. This can be interpreted to indicate oxygen coordination. Elemental analyses and the electronic spectra indicate that the

Table 1
Spectral Bands for Ni(II) and Cr(III) Complexes of the Lactams

Compound	Solvent	λ_{max} cm^{-1}	ε_{max}[a]	Band assignment[b]
[Ni(BuL)$_6$](ClO$_4$)$_2$	BuL	8,104	7.25	$^3A_{2g}(F) \rightarrow {}^3T_{2g}(F)$
		13,350	6.69	$\rightarrow {}^3T_{1g}(F)$
		24,630	16.3	$\rightarrow {}^3T_{1g}(P)$
[Ni(NMBuL)$_6$](ClO$_4$)$_2$	NMBuL	7,800	5.61	$^3A_{2g}(F) \rightarrow {}^3T_{2g}(F)$
		12,930	5.00	$\rightarrow {}^3T_{1g}(F)$
		24,150	14.5	$\rightarrow {}^3T_{1g}(P)$
[Ni(VaL)$_6$](ClO$_4$)$_2$	3.0 M VaL in CH$_2$Cl$_2$	8,327	8.60	$^3A_{2g}(F) \rightarrow {}^3T_{2g}(F)$
		13,350	5.94	$\rightarrow {}^3T_{1g}(F)$
		24,600	17.7	$\rightarrow {}^3T_{1g}(P)$
[Ni(NMVaL)$_6$](ClO$_4$)$_2$	NMVaL	7,587	6.81	$^3A_{2g}(F) \rightarrow {}^3T_{2g}(F)$
		12,660	5.62	$\rightarrow {}^3T_{1g}(F)$
		23,750	23.0	$\rightarrow {}^3T_{1g}(P)$
[Ni(CaL)$_6$](ClO$_4$)$_2$	4.3 M CaL in CH$_2$Cl$_2$	8,337	9.25	$^3A_{2g}(F) \rightarrow {}^3T_{2g}(F)$
		13,430	5.95	$\rightarrow {}^3T_{1g}(F)$
		24,680	18.9	$\rightarrow {}^3T_{1g}(P)$
[Ni(NMCaL)$_6$](ClO$_4$)$_2$	NMCaL	7,494	7.66	$^3A_{2g}(F) \rightarrow {}^3T_{2g}(F)$
		13,710	5.25	$\rightarrow {}^3T_{1g}(F)$
		23,980	19.8	$\rightarrow {}^3T_{1g}(P)$
[Cr(BuL)$_6$](ClO$_4$)$_3$	BuL	(14,430)		
		16,230	40.0	$^4A_{2g}(F) \rightarrow {}^4T_{2g}(F)$
		22,940	42.8	$\rightarrow {}^4T_{1g}(F)$
[Cr(NMBuL)$_6$](ClO$_4$)$_3$	NMBuL	(14,370)	—	
		15,750	52.0	$^4A_{2g}(F) \rightarrow {}^4T_{2g}(F)$
		22,470	48.7	$\rightarrow {}^4T_{1g}(F)$
[Cr(VaL)$_6$](ClO$_4$)$_3$	2.5 M VaL in CH$_2$Cl$_2$	(14,290)	—	
		16,210	54.1	$^4A_{2g}(F) \rightarrow {}^4T_{2g}(F)$
		22,620	49.3	$\rightarrow {}^4T_{1g}(F)$
[Cr(CaL)$_6$](ClO$_4$)$_3$	4.4 M CaL in CH$_2$Cl$_2$	(14,250)	—	
		16,180	58.1	$^4A_{2g}(F) \rightarrow {}^4T_{2g}(F)$
		22,620	47.7	$\rightarrow {}^4T_{1g}(F)$

[a] ε is given in 1. mole^{-1} cm^{-1} from the equation $A = \varepsilon cl$.
[b] The spectra of Ni(II) and Cr(III) complexes of these ligands are very similar to the spectra of the corresponding hexaaquo complexes except for the wave lengths of λ_{max}.
[c] The parentheses designate a low-intensity shoulder on the main peak of the spectrum for the Cr(III) complexes. Shoulders appeared for the Ni(II) complexes of the unsubstituted lactams at a higher frequency than the second band, but were too low in intensity and too broad to be measured accurately. For the Ni(II) complexes the calculated values for λ_{max} of the second band agree very well with the experimental values.

perchlorate-complexes are six-coordinate, octahedral species. This conclusion is substantiated by the constancy of the intensity ratio[2,12,13] of the $^3A_2(F) \rightarrow {}^3T_1(P)$ to $^3A_2(F) \rightarrow {}^3T_1(F)$ absorption bands in the case of nickel(II) complexes in solution, and by the good agreement between the calculated and experimental values for λ_{max} of the second band (Table 1 and 2).

Table 2

Summary of the Calculated Ligand Field Parameters

Compound	Solvent	$E(P)-$ $E(F)$	$Dq,^a$ cm^{-1}	β^b
[Ni(BuL)$_6$](ClO$_4$)$_2$	BuL	13,860	810	0.88
[Ni(NMBuL)$_6$](ClO$_4$)$_2$	NMBuL	13,830	780	0.87
[Ni(VaL)$_6$](ClO$_4$)$_2$	3.0 M VaL in CH$_2$Cl$_2$	13,460	833	0.85
[Ni(NMVaL)$_6$](ClO$_4$)$_2$	NMVaL	13,730	759	0.87
[Ni(CaL)$_6$](ClO$_4$)$_2$	4.3 M CaL in CH$_2$Cl$_2$	13,540	834	0.85
[Ni(NMCaL)$_6$](ClO$_4$)$_2$	NMCaL	14,140	749	0.89
[Cr(BuL)$_6$](ClO$_4$)$_3$	BuL	10,170	1623	0.74
[Cr(NMBuL)$_6$](ClO$_4$)$_3$	NMBuL	10,320	1575	0.75
[Cr(VaL)$_6$](ClO$_4$)$_3$	2.5 M VaL in CH$_2$Cl$_2$	9,579	1621	0.70
[Cr(CaL)$_6$](ClO$_4$)$_3$	4.4 M CaL in CH$_2$Cl$_2$	9,484	1618	0.69

[a]The error limits of Dq are ± 5 cm^{-1} for Ni(II) and Cr(III) complexes; this error arises in determining λ_{max}.
[b]The error limits of β are $\pm 1.2\%$ for Ni(II) and Cr(III) complexes. The quantity β is defined as the ratio B'/B where B' is the $P-F$ term splitting for the complex and B the value for the gaseous ion with no crystalline field.

The spectrochemical series for these lactam complexes of nickel(II) and chromium(III) shows the predominance of the same type of steric effect between ligands that existed in the amides.[2] The Dq values for all the N-methyl lactams are significantly lower than those for the unsubstituted compounds. It is interesting that the Dq values for the nickel complexes of the N-methyl lactams (780, 759, 749 cm^{-1}) are close to those in the completely alkyl-substituted amide, N,N-dimethylacetamide (767 cm^{-1}),[2] while those for the unsubstituted lactams (810, 833, 834 cm^{-1})[14] are close to those in the disubstituted amide N,N-dimethylformamide (850 cm^{-1}).[2] Drago and co-workers proposed that the N-methyl group in CH$_3$C(O)NHCH$_3$ is cis to the carbonyl; this amide, therefore, encounters the same steric effect leading to a lower Dq as was observed for N,N-dimethylacetamide in an octahedral nickel(II) complex. In the unsubstituted lactams the methylene group bonded to nitrogen is trans to the carbonyl and much higher Dq

values result. This result confirms the earlier proposal of the existence of a steric effect in the N-methylacetamide complex.[2]

The effect of ring size on the donor strengths of the lactams is not easily assessed. The Dq values for the nickel(II) complexes of the unsubstituted lactams suggest a donor order based on ring size 5- < 6- ~ 7-membered. In contrast, in the chromium(III) complexes of the same lactams, no variation of donor strength with ring size is found. A monotonic decrease in donor strength with increasing ring size is indicated by the Dq values for the N-methyl lactam complexes of nickel(II). Here, however, steric factors may be of considerable importance in determining the order.

The phenol frequency shifts for the methyl lactams give a donor order of 5- < 6- ~ 7-membered ring.[2] This is the same order given by the Dq values for the unsubstituted lactam complexes of nickel(II).[14]

NONTRANSITION METAL COMPLEXES OF LACTAMS

Few stable complexes of alkali metals have been reported. Complexes of the alkali metals with β-diketone derivaties, o-nitrophenol, and salicylaldehyde have been described in the literature.[15,16] The sodium compounds ranged from hydrated ionic salts such as Na (acac)·2 H_2O to adducts like Na(Sal) HSal (HSal = Salicylaldehyde). Pfeiffer[17] and others[18] have reported the isolation of a complex of sodium perchlorate with 1,10-phenanthroline. Brady and Badger[19] have presented evidence for chelation in sodium salts of o-hydroxybenzaldehydes. More recently, Popp and Joesten[20] have isolated and characterized a number of alkali metal and alkaline earth metal salts with the ligand octamethylpyrophosphoramide (OMPA). Gentile and co-workers[21] have also reported complexes of diacetamide with alkali and alkaline earth metal salts.

Pursuant to our interest in substituted amides as ligands, we studied reactions of NMBuL with nontransition metal ions.[22]

The reaction between NMBuL and most of the nontransition metal perchlorates, nitrates and halides yields well-crystallized products on the basis of the analyses and infrared measurements. The perchlorates salts may be formulated as the perchlorates of hexakis-(N-methyl-γ-butyrolactam) metal cation, $[M(NMBuL)_6]^{n+}$ (M = Ca^{2+}, Mg^{2+}, Sr^{2+}, Cd^{2+}, Al^{3+}, and In^{3+}), and tetrakis (N-methyl-γ-butyrolactam) metal cation, $[M(NMBuL)_4]^{n+}$ (M = Li^+ and Zn^{2+}).

The infrared spectra of $Zn(NMBuL)_2(NO_3)_2$, $Cd(NMBuL)_3(NO_3)_2$, $Mg(NMBuL)_3(NO_3)_2·2H_2O$, and $Li(NMBuL)_2NO_3·H_2O$ show evidence of ionic and coordinated nitrate ions. The infrared spectrum of coordinated nitrate ion has previously been described.[23-27] More recently, Curtis[28] has reported that the nitrate ion in ionic compounds exhibits three bands: v_2

(out-of-plane deformation), v_3 (doubly degenerate stretch), and v_4 (doubly degenerate in-plane bending). When the nitrate-ion is coordinated as a monodentate or bidentate ligand, all bands become active, shifts in band positions and intensities are observed, and the degeneracy of v_3 and v_4 is lifted. For a coordinated monodentate nitrate ion, v_3 is weakened, and there is splitting of v_3 and v_4, and a small shift of v_2 to lower frequency. However, for both a coordinated bidentate and a monodentate nitrate ion, an enhancement and further splitting of the above modes is observed. The v_3 and v_4 split into four bands, while the v_2 split into two bands. Table 3 shows a comparison of the infrared spectra of the NMBuL-nitrate complexes, along with some model compounds in which a monodentate, monodentate and bidentate, or ionic and bidentate nitrate ions are present.[28]

The infrared spectrum of $Zn(NMBuL)_2(NO_3)_2$ provides evidence that this complex contains both an ionic and a bidentate nitrate ion. The spectrum of $Cd(NMBuL)_3(NO_3)_2$ seems to indicate the presence of a monodentate nitrate ion and also a bidentate one, whereas the infrared spectra of Mg-$(NMBuL)_3(NO_3)_2 \cdot 2H_2O$ and $Li(NMBuL)_2(NO_3) \cdot H_2O$ point to the fact that both these compounds contain monodentate nitrate ions.

The positions and relative intensities of d_{hkl} for the 6:1 complexes are given in Table 4. The d_{hkl} values of $Ni(NMBuL)_6(ClO_4)_2$ are included for comparison. The x-ray data indicate that 6:1 complexes are isomorphous. Earlier work with the spectra of $Ni(NMBuL)_6(ClO_4)_2$ dissolved in nitromethane and NMBuL supported the assignment of an octahedral configuration to this complex in solution.[14] The similarity of x-ray data for the 6:1 complexes in this study confirms the assignment of an octahedral configuration to the metal ions in these complexes.

LANTHANIDE METAL COMPLEXES OF LACTAMS

It is well known that the lanthanides form complexes most easily with oxygen donor ligands, so one would expect lanthanide amide complexes to form easily. However, upon examining the literature, this author found only a few examples of such complexes. Several years ago Moeller and co-workers[29,30] prepared some lanthanide perchlorate complexes with N,N-dimethylformamide (DMF). More recently, Moeller and Vicentini[31] have reported a series of crystalline lanthanide perchlorate complexes of N,N-dimethylacetamide (DMA) which were found to be either eight, seven, or six-coordinated, depending upon the size of the central metal ion. Krishnamurthy and Soundararajan,[32] on the other hand, have reported a series of DMF complexes with rare earth nitrates which were found to be nine-coordinated for all the metals used. In another study of lanthanide nitrate complexes of monodentate ligands, using dimethyl sulfoxide, Ramalingam

Table 3

Infrared Bands Pertinent to the State of Nitrate Ion in the NMBuL-Nitrate Complexes

Observed bands,[a] cm⁻¹					Accepted bands, cm⁻¹ [28]		
$Zn(NMBuL)_2(NO_3)_2$	$Cd(NMBuL)_3(NO_3)_2$	$Li(NMBuL)_2NO_3 \cdot H_2O$	$Mg(NMBuL)_3(NO_3)_2 \cdot 2H_2O$	Assignment	$Ni(en)_2(NO_3)_2$ monodentate nitrate	$Ni(dien)(NO_3)_2$ monodentate and bidentate nitrate	$[Ni(tetb)NO_3]NO_2$ ionic and bidentate nitrate
840 m,sp	838 m,sp	827 m,sp	810 m,sp	v_2 (out-of-plane def)	818 m,sp	816 m,sp	825 m,sp
815 m,sp	818 m,sp					808 m,sp	806 m,sp
	1330 s			v_3 (doubly degenerate str)		1315 s	
1370 s,br	1470 s					1440 s	1375 s,br
1290 s	1300 s	1300 s	1305 s		1307 s	1300 s	1280 s
1475 s	1490 s	1430 s	1420 s	v_4	1420 s	1480 s	1490 s
720 vw	725 w,br	675 w,sp	685 w,sp	(doubly degenerate in-plane bending)	708 w,sp	714 vw,sp	695 vw
680 vw	740 w,sp	722 w,br	725 w,br		728 w,sp	736 vw,sp	745 w,sp
750 w,br	675 vw					698 vw,sp	
	752 w,sp					849 w,sp	

[a] s, strong; m, medium; w, weak; v, very; br, broad; sp, sharp.

Table 4
X-ray Powder Pattern Data

$Mg(NMBuL)_6(ClO_4)_2$		$Ca(NMBuL)_6(ClO_4)_2$		$Sr(NMBuL)_6(ClO_4)_2$	
d_{hkl}	I/I_{100}	d_{hkl}	I/I_{100}	d_{hkl}	I/I_{100}
10.52	18	11.04	4	10.77	29
6.10	20	6.32	14	6.27	56
5.30	42	5.47	9	5.21	12
4.79	15	4.92	26	4.87	44
4.37	100	4.48	63	4.43	9
3.72	28	3.93	4	3.76	24
3.53	18	3.78	9	3.63	50
3.26	15	3.63	14	3.37	6
3.11	8	3.14	5	3.29	5
2.74	8	2.88	2	2.81	15
2.31	8	2.83	4	2.32	13

$Cd(NMBuL)_6(ClO_4)_2$		$In(NMBuL)_6(ClO_4)_2$		$Ni(NMBuL)_6(ClO_4)_2$	
d_{hkl}	I/I_{100}	d_{hkl}	I/I_{100}	d_{hkl}	I/I_{100}
10.91	75	10.27	75	11.18	100
6.27	47	6.50	11	6.96	10
5.43	4	5.30	5	5.63	10
4.87	50	4.69	13	4.97	40
4.43	95	4.15	64	4.76	100
3.62	35	4.00	45	4.31	50
3.27	23	3.91	13	4.05	30
2.90	5	2.90	5	3.00	10
2.71	18	2.84	5	2.85	5
2.37	17	—	—	2.50	8

and Soundararajan[33] obtained a series of compounds which they believe to contain two monodentate groups and one bidentate nitrate group. A large number of lanthanide nitrate complexes was also obtained by Cousins and Hart,[34] using triphenylphosphine oxide as the monodentate ligand. These authors suggest that in many of these complexes, all three of the nitrate groups are present as bidentates.

As a continuation of our work with amides, we have now prepared lanthanide perchlorate and nitrate complexes of BuL and perchlorate complexes of NMBuL, and have examined the effects of unsubstituted and substituted lactams on the coordination number.[35] Lanthanide perchlorate complexes of BuL were found to be eight-coordinated, $[Ln(BuL)_8](ClO_4)_3$ (Ln = La, Pr, Nd, Sm, Gd; Dy, Er, Yb, Y), whereas two different types of compounds are present in the NMBuL series $[Ln(NMBuL)_8](ClO_4)_3$ (Ln = La, Pr, Nd, Sm, Gd) and $[Ln(NMBuL)_7](ClO_4)_3$ (Ln = Dy, Er, Yb, Y). The infrared spectra of all of the BuL complexes are very similar, as are the spectra of all the NMBuL complexes. No change is apparent in the spectra of the NMBuL complexes when the coordination number of the metal ion decreases from eight to seven. The spectrum of each complex in the BuL series showed that the frequency of the v CO band of the coordinated amide was lowered by approximately 33 cm^{-1} relative to the free ligand. The NMBuL series showed a slightly larger decrease in the v CO frequency of about 38 cm^{-1}. These results are in agreement with those reported by Madan and Sturr[8] for BuL and NMBuL complexes of both transition and nontransition metals,[22] all of which involve coordination of the amide through the oxygen.

A comparison of the visible spectra of each BuL complex in a BuL solution with the corresponding NMBuL complex in NMBuL showed that, in general, the spectra were very similar. A few differences were noted, however, in the absorption bands, due to the hypersensitive $^4I_{9/2} \rightarrow {}^4G_{5/2}$, $^2G_{7/2}$ transitions in Nd and the hypersensitive $^4I_{15/2} \rightarrow {}^2H_{11/2}$ transition in Er.[36] In both of the Nd complexes, the hypersensitive transition band occurs as a doublet at 581 and 578 mμ. In the NMBuL complex the molar absorptivities of the two bands are 10.6 and 11.2 M^{-1} cm^{-1}, respectively, while in the BuL complex both bands have a molar absorptivity of 12.4 M^{-1} cm^{-1}. In the case of Er complexes the hypersensitive transition band appeared as a doublet at 524 and 520 mμ. The molar absorptivities of the bands in the BuL complex are 16.3 and 21.6 M^{-1} cm^{-1}, respectively, and are 5.48 and 9.47 M^{-1} cm^{-1} in the corresponding NMBuL complex. The molar absorptivities of the bands in the two Nd complexes are only slightly different because in both complexes the Nd has a coordination number of eight. However, in the Er complexes the intensity of the bands in the BuL complex is much higher than that of those in the NMBuL complex. This is due to the decrease in the coordination number of the erbium from eight to seven when NMBuL is substituted for BuL. The decrease in intensity of the hypersensitive transition bands resulting from a decrease in coordination number has also been observed in a series of rare earth complexes reported by Karraker.[36]

It appears as though no six-coordinated complexes of NMBuL are formed, as in the case of the series of DMA complexes reported by Moeller and Vicentini.[31] This is probably due to the greater chance for steric interactions between the coordinated ligands in the DMA complexes. As a consequence of the presence of substituent groups on the amide nitrogen, a highly substituted amide will interact more strongly with neighboring ligands than would an unsubstituted amide as the central metal ion in a complex becomes smaller. This situation can be relieved either by decreasing the coordination number of the metal ion, or by replacing the substituted amides with smaller ligands. In the case of the NMBuL complexes, both a decrease in the coordination number and replacement of the NMBuL ligands by a smaller ligand, i.e., water, are observed. The increasingly hygroscopic nature of the complexes as the radii of the rare earth ions decreases is probably attributable to a ligand exchange reaction in which the ligands are partially replaced by smaller water molecules.

In the case of the BuL complexes, the unsubstituted amide ligands apparently interact less strongly with one another than in the NMBuL complexes. This is evidenced by the fact that all of the BuL complexes are eight-coordinated and are nonhygroscopic. It appears from steric considerations that the BuL complexes are generally more stable than the NMBuL ones, even though the inductive effect of the methyl group in NMBuL would make the oxygen a slightly better donor than the oxygen in BuL.

The analytical data in the case of lanthanide nitrate-BuL complexes clearly indicate that two different types of compounds are obtained: those of the lanthanides with the larger ionic radii having the general formula $[Ln(BuL)_8](NO_3)_3$ (where Ln = La, Nd, Gd), and those of the lanthanides with smaller radii having the general formula $[Ln(BuL)_3(NO_3)_3]$. A comparison of the infrared spectra of these two types of compounds in Nujol mull (Figs. 1 and 2) indicates that definite structural differences exist between them.

The ionic nitrate group, which has D_{3h} symmetry, has four fundamental vibrations, three of which are active in the infrared.[26] Upon coordination as either a monodentate or a bidentate ligand, the symmetry becomes C_{2v}, which has six, infrared active normal modes of vibration. As a result of coordination, the doubly degenerate v_4 band of the ionic nitrate group in the 700 cm^{-1} region[28] splits into the v_3 and v_5 bands of the coordinated nitrato group.

In a study of a series of uranyl complexes containing bidentate nitrato groups, Bullock[37] has found that there are two infrared bands between 700 and 750 cm^{-1}, one in the range of 710 to 725 cm^{-1} which was assigned to the v_5 vibration, and a stronger band in the 735 to 750 cm^{-1} region which was assigned to the v_3 vibration of the nitrato group. Furthermore, in a series of lanthanide nitrate complexes with triphenylphosphine oxide reported by Cousins and Hart,[34] the authors found that those complexes containing ionic nitrate had an infrared band at 830 cm^{-1}, which was assigned to the v_2 normal mode of vibration for the nitrate ion. On the other hand, those complexes which had coordinated nitrato groups present had an infrared band between 813 and 819 cm^{-1}, which the authors assigned to the v_6 vibration of the nitrato group. In agreement with this work, Krishnamurthy and Soundararajan[32] have found that in a series of complexes of

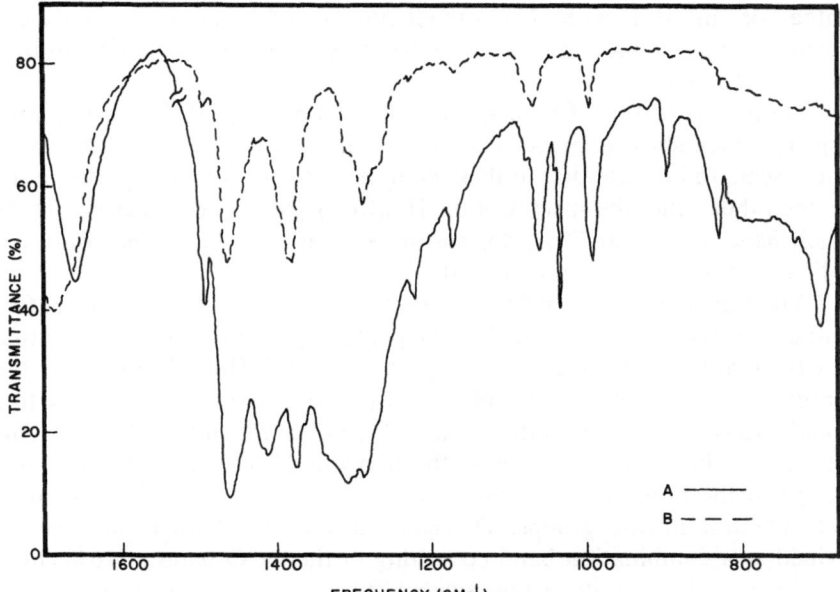

Fig. 1. Infrared spectra of [La(BuL)$_8$](NO$_3$)$_3$ in Nujol A (——) and BuL, γ-butyrolactam in Nujol B (– – – – –).

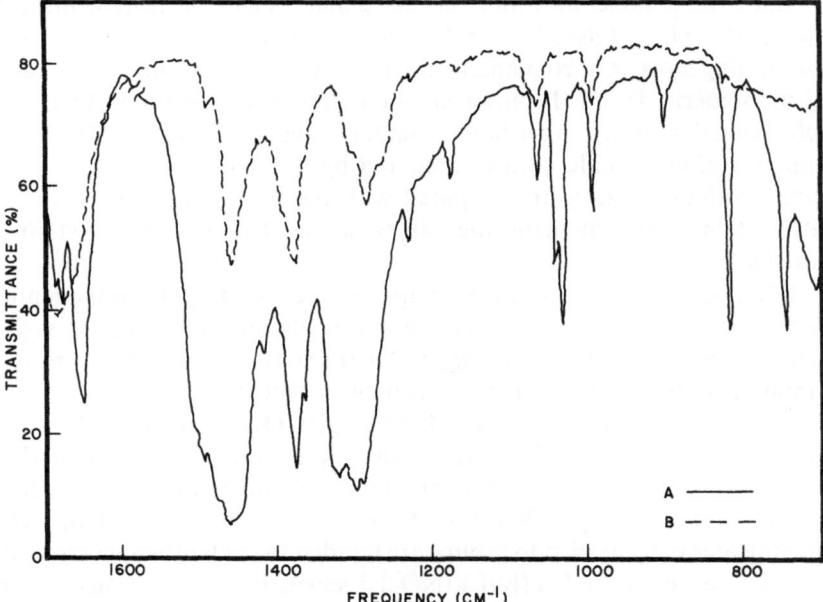

Fig. 2. Infrared spectra of [Dy(BuL)$_3$(NO$_3$)$_3$] in Nujol A (——) and BuL, γ-butyrolactam in Nujol B (– – – – –).

lanthanide nitrates with N,N-dimethylformamide containing covalent nitrato groups, a medium intensity band was obtained in the infrared spectra at 819 cm^{-1}.

On this basis, the infrared spectra of the [Ln(BuL)$_8$](NO$_3$)$_3$ complexes (Fig. 1), which show a single band at 701 cm^{-1} and a band at 829 cm^{-1}, would seem to indicate that in these complexes the nitrate groups are ionic. On the other hand, the spectra of the [Ln(BuL)$_3$(NO$_3$)$_3$] complexes (Fig. 2), which have a band at 705, 746, and at 815 cm^{-1}, indicate that only co-ordinated nitrato groups are present.

The frequency of the ν CO vibrational band at 1691 cm^{-1} in the free amide is decreased to 1648 cm^{-1} in the [Ln(BuL)$_8$](NO$_3$)$_3$ series, indicating that the amide is coordinated through the oxygen.[8] The ν CO band in the [Ln(BuL)$_3$(NO$_3$)$_3$] series is also of lower frequency than the band in the free ligand. However, it is split into a triplet at 1649, 1675, and 1648 cm^{-1}. This splitting could possibly be due to the nonequivalence of the three amide groups in the molecule, as a result of the presence of both monodentate and bidentate nitrato groups. The band at 1684 cm^{-1} might also be explained as a combination band consisting of the ν CO band at 1648 cm^{-1} and the ν_3 and ν_5 nitrato bands, since ν CO + ν_3 − ν_5 = 1689 cm^{-1}. Both Curtis[28] and Bullock[37] have shown that nitrate groups can combine to form a large number of overtone and combination bands.

Both the conductance and the molecular weight measurements substantiate the infrared data for the [Ln(BuL)$_3$(NO$_3$)$_3$] complexes. In the poor coordinating solvent nitromethane, the complexes were found to be neutral and monomeric. In DMF, however, the complexes were found to be 1:2 electrolytes, demonstrating behavior similar to that of the dimethyl sulfoxide complexes of lanthanide nitrates prepared by Ramalingam and Soundararajan.[33] Values for λ_m were compared with those of Ames and Sears[38] in DMF, while in nitromethane the values obtained by Gill and Nyholm[10] were used as a reference.

Since only two of the nitrato groups were replaced by the monodentate DMF molecules, it appears as though one of the nitrato groups is bound differently from the other. This suggests that two of the nitrato groups may be monodentate, which the third is probably a bidentate ligand.

The solution behavior of the [Ln(BuL)$_8$](NO$_3$)$_3$ complexes, however, is not in agreement with what one would expect for an ionic compound. In nitromethane the complexes are nonconducting, indicating covalent, rather than ionic, nitrate groups, while in DMF the compounds are 1:2 electrolytes. This indicates that the nitrato groups are bonded to the lanthanide ion in the same manner as in the [Ln(BuL)$_3$(NO$_3$)$_3$] series. Further evidence for the covalency of the nitrate ion in solution is obtained from the solution spectra of these complexes represented in Fig. 3.

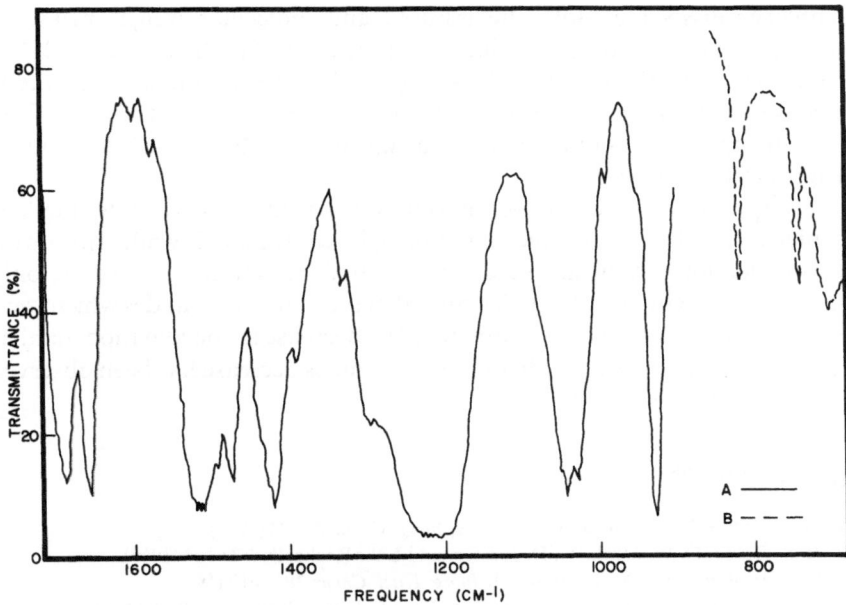

Fig. 3. Infrared spectra of $[Gd(BuL)_8](NO_3)_3$ in nonaqueous solvents A (———) represents spectra in chloroform from 1700 cm^{-1} to 900 cm^{-1} and B (–––––) represents spectra in nitromethane from 850 to 675 cm^{-1}.

The band at 829 cm^{-1} in the Nujol mull spectra has shifted to 817 cm^{-1} in nitromethane; and the band at 701 cm^{-1} in Nujol has split into two bands, one at 701 cm^{-1} and the other at 739 cm^{-1}. A comparison of the 700 to 900 cm^{-1} region in Figs. 2 and 3 indicates that the complexes are very similar.

The infrared spectra of these complexes in both nitromethane and chloroform also showed a strong band at 1690 cm^{-1}, which is characteristic of the free ligand. After the absorbance of v CO band for several concentrations of BuL in nitromethane had been measured, a Beer's law plot was made. A measured amount of each complex was then dissolved in nitromethane and the absorbance of the v CO band at 1690 cm^{-1} was measured. A comparison of the absorbance of the v CO band of the complex in solution at 1690 cm^{-1} with the Beers' law plot was then used to obtain a quantitative estimate of the amount of free ligand in solution. The analyses showed that the lanthanum complex dissociated in solution to give four BuL molecules and probably $[La(BuL)_4(NO_3)_3]$, while the neodymium and gadolium complexes dissociated to give five BuL molecules and $[Ln(BuL)_3(NO_3)_3]$. These values were in agreement with the molecular weight measurements[39] which indicates that there are five particles in solution when the lanthanum complex dissociates, and six particles in solution when the neodymium and gadolinium complexes dissociate. Both the infrared and molecular weight measurements were performed with similar concentrations of each complex (0.01M). The presence of the $[Ln(BuL)_x(NO_3)_3]$ ($x = 3,4$) in solution is not unlikely, since we have obtained both $[La(BuL)_4(NO_3)_3]$ and $[Gd(BuL)_3(NO_3)_3]$ by crystallizing solutions of the lanthanide nitrates and BuL in 2,2-dimethoxy-propane-ether mixtures.

It appears from the above information that in the solid state, the lanthanides with the larger ionic radii are eight-coordinated, while those with the smaller ionic radii are seven-coordinated. In solution, however, only lanthanum is eight-coordinated, while all of the other lanthanides which were used appeared to be seven-coordinated. The decrease in coordination number from eight to seven as the radii of the lanthanides decrease has been observed in several systems.[29,30,33,35,40]

REFERENCES

1. W. E. Bull, S. K. Madan, and J. E. Willis, *Inorg. Chem.* **2**: 303 (1963).
2. R. S. Drago, D. W. Meek, M. D. Joesten, and L. LaRoche, *ibid.* **2**: 124 (1963).
3. S. K. Madan and A. M. Donohue, *J. Inorg. Nucl. Chem.* **28**: 1617 (1966).
4. B. B. Wayland, R. J. Fitzgerald, and R. S. Drago, *J. Am. Chem. Soc.* **88**: 4600 (1966).
5. S. K. Madan and D. Mueller, *J. Inorg. Nucl. Chem.* **28**: 177 (1966).
6. For a summary of the literature see R. L. Middaugh, R. S. Drago, and R. J. Niedzielski, *J. Am. Chem. Soc.* **86**: 388 (1964).
7. S. K. Madan and H. H. Denk, *J. Inorg. Nucl. Chem.* **27**: 1049 (1965).
8. S. K. Madan and J. A. Sturr, *ibid.* **29**: 1669 (1967); R. J. Niedzielski and G. Zinder, *Can. J. Chem.* **43**: 2618 (1965).
9. J. T. Donoghue and R. S. Drago, *Inorg. Chem.* **1**: 866 (1962).
10. N. S. Gill and R. S. Nyholm, *J. Chem. Soc.* 3997 (1959).
11. D. W. Meek, D. K. Straub, and R. S. Drago, *J. Am. Chem. Soc.* **82**: 6013 (1960).
12. C. K. Jørgensen, *Acta Chem. Scand.* **9**: 1362 (1955).
13. W. Low, *Phys. Rev.* **109**: 247 (1958).
14. J. H. Bright, R. S. Drago, D. M. Hart, and S. K. Madan, *Inorg. Chem.* **4**: 18 (1965).
15. N. W. Sidgewick and F. M. Brewer, *J. Chem. Soc.* **127**: 2379 (1925).
16. F. M. Brewer, *ibid.* **361** (1931).
17. P. Pfeiffer and W. Christeleit, *Z. anorg. allgem. Chem.* **239**: 133 (1938).
18. A. A. Schilt and R. O. Taylor, *J. Inorg. Nucl. Chem.* **9**: 211 (1959).
19. O. L. Brady and W. H. Badger, *J. Chem. Soc.* 952 (1932).
20. C. J. Popp and M. D. Joesten, *Inorg. Chem.* **4**: 1418 (1965).
21. P. S. Gentile and T. A. Shankoff, *J. Inorg. Nucl. Chem.* **27**: 2301 (1965).
22. S. K. Madan, *Inorg. Chem.* **6**: 421 (1967).
23. S. Buffagni, L. M. Vallarino, and J. V. Quagliano, *ibid.* **3**: 480 (1964).
24. M. B. Gatehouse, S. E. Livingstone, and R. S. Nyholm, *J. Chem. Soc.* 4222 (1957); *J. Inorg. Nucl. Chem.* **8**: 75 (1958).
25. C. C. Addison and G. M. Gatehouse, *J. Chem. Soc.* 613 (1960); B. M. Gatehouse and A. E. Comyns, *ibid.* 3965 (1958).
26. K. Nakamoto, *Infrared Spectra of Inorganic and Coordination Compounds*, John Wiley and Sons, Inc., New York, 1963; E. Bannister and F. A. Cotton, *J. Chem. Soc.* 2276 (1960).

27. E. P. Bertin, R. B. Penland, S. Mizushima, C. Curran, and J. V. Quagliano, *J. Am. Chem. Soc.* **81**: 3818 (1959).
28. N. F. Curtis and Y. M. Curtis, *Inorg. Chem.* **4**: 804 (1965).
29. T. Moeller and V. Galasyn, *J. Inorg. Nucl. Chem.* **12**: 259 (1960).
30. T. Moeller, V. Galasyn, and J. Xavier, *J. Inorg. Nucl. Chem.* **15**: 259 (1960).
31. T. Moeller and G. Vicentini, *J. Inorg. Nucl. Chem.* **27**: 1477 (1965).
32. S. S. Krishnamurthy and S. Soundararajan, *J. Inorg. Nucl. Chem.* **28**: 1689 (1966).
33. S. K. Ramalingam and S. Soundararajan, *ibid.* **29**: 1763 (1967).
34. D. R. Cousin and R. A. Hart, *ibid.* **29**: 1745 (1967).
35. W. V. Miller and S. K. Madan, *ibid.* **30**: 3287 (1968).
36. D. G. Karraker, *Inorg. Chem.* **6**: 1863 (1967).
37. J. I. Bullock, *J. Inorg. Nucl. Chem.* **29**: 2257 (1967).
38. D. P. Ames and P. G. Sears, *J. Phys. Chem.* **59**: 16 (1955).
39. W. V. Miller and S. K. Madan, *J. Inorg. Nucl. Chem.* **30**: 2785 (1968).
40. V. N. Krishnamurthy and S. Soundararajan, *J. Inorg. Nucl. Chem.* **29**: 517 (1967).

SYNTHESES AND REACTIONS OF SOME DIACIDOBIS(ETHYLENEDIAMINE)-IRIDIUM(III) COMPLEXES

R. A. Bauer and F. Basolo

Northwestern University
Evanston, Illinois

INTRODUCTION

Cobalt(III) complexes of the type $[Co(en)_2X_2]^+$ have long been known, and extensively investigated with regard to mechanisms of reactions in these systems.[1] Some years ago it was decided to extend these studies to include analogous compounds of the other members of this triad, rhodium(III) and iridium(III). It was then learned that very few such compounds had been prepared and it was necessary to start the laborious task of the syntheses and characterizations of these desired complexes. Several salts of $[Rh(en)_2X_2]^+$ were prepared[2] and the rates of substitution of some of these have been investigated.[3]

This paper reports the syntheses and properties of salts of $[Ir(en)_2X_2]^+$, where the X^- groups are the same or different anionic ligands. When this work was started the only complex of this type that had been reported[4] was *cis*-$[Ir(en)_2(NO_2)_2]^+$. Kida[5] has recently described the preparation of *cis* and *trans*-$[Ir(en)_2Cl_2]Cl$. We have repeated his syntheses and also designed new methods to prepare the same complexes as well as other compounds of iridium(III).

The method of approach to the syntheses of these iridium(III) complexes is not new. Delepine[6] noted that it was difficult to prepare *trans*-$[Rh(py)_4Cl_2]Cl$, but that the addition of primary alcohols to the same reaction mixtures readily resulted in the formation of the desired compound in good yield. This reaction has been studied and it was found that only catalytic amounts of alcohol are required and that other reducing agents such as N_2H_4, BH_4^- and H_3PO_2 are also effective.[7] Although there are differences of opinion as to the mechanism of this catalytic effect,[7,8] it is agreed that reducing agents can be effective catalysts for reactions in these systems. Hypophosphorous acid was the reducing agent used in the synthesis[9] of trans-$[Ir(en)_2Cl_2]^+$.

The photochemical approach used to prepare the mixed chloroacido complexes of the type trans-$[Ir(en)_2XCl]^+$, starting with trans-$[Ir(en)_2Cl_2]^+$, is a new method.[10] The success of this method is in accord with the general observation made by Adamson[11] that photochemically induced substitution reactions of an octahedral complex of less than O_h symmetry takes place on the low-field axis and involves the replacement of the group of higher ligand field on that axis.

EXPERIMENTAL

K_2IrCl_6

This compound was the starting materials for the other syntheses and it was readily prepared from iridium metal. The metal was initially obtained from Engelhard Industries, Inc. and later recovered from iridium residues produced during this investigation. The method used was similar to that described by Gire.[12]

Trans-$[Ir(en)_2Cl_2]ClO_4$

This compound was prepared by two methods, and only the better one is described here. Twenty grams of K_2IrCl_6 was placed in 100 ml of hot (80°) water and reduced by slowly adding 3.82 g of $K_2C_2O_4 \cdot H_2O$. After effervescence had subsided, the solution was boiled to remove the remaining CO_2 and 20 g of ethylenediamine dihydrochloride were added. A green-brown precipitate was produced by the addition of 0.1 g of KH_2PO_2, and 20 ml of concentrated HCl was added to complete precipitation. The solution was cooled to room temperature, filtered, and the green-brown precipitate washed with 20 ml of cold water. The precipitate was transferred to 100 ml water and dissolved by the addition of 8 ml of neat ethylenediamine. The solution was boiled for 5 min and an additional 8 ml of ethylenediamine was added. After boiling for 1 hr and concentrating to 50 ml, the solution was neutralized with concentrated HCl, boiled for 10 min, cooled, and treated with 10 ml of 70% $HClO_4$. The immediate precipitate was collected on a filter and recrystallized from 70 ml of hot water by adding 10 ml of 70% $HClO_4$ and allowing it to stand overnight. The yellow crystals were collected on a filter, washed with 10 ml portions of cold water, ethanol, ether, and air-dried. The yield was 4.3 g (47%). (For analyses, see Table 1.)

Cis-$[Ir(en)_2Cl_2]Cl$

Potassium hexachloroiridate(IV) (20.0 g) was stoichiometrically reduced to $IrCl_6^{3-}$ in 100 ml of hot (80°) water by the slow addition of 3.81 g of $K_2C_2O_4 \cdot H_2O$. After the vigorous evolution of CO_2 had subsided, the

Table 1

Elemental Analyses of Diacidobis(ethylenediamine)iridium(III)

Compound	%C	%H	%N
trans-[Ir(en)$_2$Cl$_2$]ClO$_4$	9.71 (9.95)[a]	3.23 (3.34)[a]	11.42 (11.61)[a]
trans-[Ir(en)$_2$Br$_2$]ClO$_4$	8.42 (8.40)	2.85 (2.82)	9.82 (9.80)
trans-[Ir(en)$_2$I$_2$]I	7.10 (6.93)	2.26 (2.33)	8.25 (8.09)
trans-[Ir(en)$_2$(NO$_2$)$_2$]ClO$_4$	9.64 (9.54)	2.95 (3.20)	17.93 (16.7)
trans-[Ir(en)$_2$(N$_3$)$_2$]PF$_6$	8.83 (8.87)	3.11 (2.98)	25.32 (25.87)
trans-[Ir(en)$_2$BrCl]ClO$_4$	9.12 (9.11)	3.02 (3.06)	10.70 (10.63)
trans-[Ir(en)$_2$ICl]ClO$_4$	8.54 (8.36)	2.89 (2.81)	9.94 (9.76)
trans-[Ir(en)$_2$NO$_2$Cl]ClO$_4$	9.75 (9.74)	3.32 (3.27)	14.18 (14.20)
trans-[Ir(en)$_2$NCSCl]ClO$_4$	10.64 (11.84)	2.49 (3.19)	13.88 (13.86)
trans-[Ir(en)$_2$H$_2$OCl](ClO$_4$)$_2$	8.66 (8.51)	3.24 (3.21)	9.99 (9.92)
cis-[Ir(en)$_2$Br$_2$]Br·H$_2$O	8.47 (8.43)	2.99 (3.18)	9.88 (9.83)
cis-[Ir(en)$_2$I$_2$]Cl·H$_2$O	7.73 (7.75)	2.88 (2.93)	9.08 (9.04)

[a] The numbers in parentheses are the theoretically calculated values.

solution was boiled for 2 min to complete the reduction and remove CO_2. A solution of 8 ml of 90% aqueous ethylenediamine in 100 ml of H_2O was neutralized and buffered at pH 7 with glacial acetic acid. The two solutions were combined and heated to boiling. Solid NaCl and water were added until the boiling solution was saturated in NaCl and had a total volume of 600 ml. The solution was refluxed for 4 hr. The volume of the solution was reduced to 50 ml in several operations of successively boiling, quickly cooling to room temperature, and removing the precipitated NaCl. Upon standing at room temperature for 2 hr, large light yellow crystals of cis-[Ir(en)$_2$Cl$_2$]Cl were collected on a filter, washed with 5 ml cold water and redissolved in 20 ml of hot (80°) water and recrystallized by adding 10 ml of concentrated HCl and allowing it to stand at room temperature overnight. The crystals were collected on a filter, washed with 5 ml portions of cold H_2O, ethanol and ether, and allowed to air dry. The yield was 2.3 g.

Upon standing 24 hr at room temperature 2.5 g of orange cis-[Ir(en)$_2$Cl$_2$][Ir(en)Cl$_4$] precipitated from the reaction solution. The orange crystals were dissolved in 50 ml of hot water and passed through an anion exchange column, Cl$^-$ form. The resulting solution (eluate) was evaporated to 10 ml. Upon the addition of 10 ml of concentrated HCl, cis-[Ir(en)$_2$Cl]Cl precipitated, was collected on a filter after 3 hr, and washed and dried as before. The yield was 1.0 g.

Resolution of cis-[Ir(en)$_2$Cl$_2$]ClO$_4$

The resolution of cis-[Ir(en)$_2$Cl$_2$]$^+$ was successfully accomplished with the ammonium salt of (+)-α-bromocamphor-π-sulfonic acid. Cis-[Ir(en)$_2$Cl$_2$]Cl (4.0 g) was dissolved in 75 ml of H_2O at room temperature and 5.87 g of ammonium $(+)$-α-bromocamphor-π-sulfonate $((+)$-BCS) was added and quickly filtered, after it had dissolved. The filtrate began to crystallize on standing for 2 min and was allowed to crystallize for 30 min with occasional stirring. The solution was then cooled to $10°$ for 10 min, filtered, and the light yellow fluffy precipitate washed with H_2O (5 ml), ethanol (20 ml) and ether (20 ml) and air dried. The yield was 2.5 g. The precipitate was converted to the chloride by grinding in an ice cold mortar for 10 min with 12 ml of cold $1:1:1$ HCl:ethanol:ether. The precipitate was collected, washed with ethanol (5 ml) ether (10 ml), and air dried. The yield was 1.1 g. The chloride was converted to the perchlorate salt by dissolving it in 20 ml of H_2O and adding 5 ml of 70% $HClO_4$. The fluffy needles were collected, washed as above and air dried. The yield was 1.0 g. $[M]_{350}^{250} = +107$.

To the initial filtrate above was added 3 ml of concentrated HCl and the solution heated at $80°$ for 5 min, it then was evaporated on a steam bath $(55°)$ in a stream of air to 30 ml when 0.4 g of crystals separated on cooling to room temperature (fraction 2). The filtrate was evaporated to 10 ml at $55°$ in a stream of air, cooled to room temperature, and fraction 3 removed by filtration. Fraction 3 was dissolved in 10 ml of H_2O and crystallized by adding 2 g of NH_4I and cooling at $5°$ for 30 min. To the remaining filtrate was added 2 g of NH_4I and it was allowed to crystallize at room temperature for 1 hr. Fraction 4 was collected, washed with H_2O, ethanol, ether, and air dried, with a yield of 0.8 g.

Fractions 2 and 3 were found to be $(+)$-cis-$[Ir(en)_2Cl_2]^+$ as the iodide and chloride salts, and contaminated with a large amount of resolving agent. Fraction 4, $(+)$-$[Ir(en)_2Cl_2]I$, was converted to the perchlorate by dissolving it in 30 ml of H_2O at $80°$ and adding 5 ml of 70% $HClO_4$, cooling to room temperature for 2 hr, filtering, washing with water (5 ml), ethanol (10 ml), ether (10 ml), and air drying. $[M]_{350}^{250} = -97$.

Cis and Trans-$[Ir(en)_2X_2]ClO_4$

The syntheses of the diacido complexes where both of the acido ligands are the same were accomplished by the reaction of the dichloro complex with an excess of NaX at an elevated temperature. The salts were often isolated as the perchlorates by the addition of $HClO_4$ to the cold reaction mixture and overall yields of about 70% were obtained. The method of preparation of trans-$[Ir(en)_2Br_2]ClO_4$ described here was the procedure generally used for the syntheses of salts of other cis and trans-diacido complexes.

To a solution of 0.25 g (0.5 mmol) of trans-$[Ir(en)_2Cl_2]ClO_4$ in 30 ml of water was added 0.51 g (5.0 mmol) of NaBr. The reaction mixture was placed in a polymer pressure bottle and kept at $140°$ for 4 hr. At the end of this time, the solution was concentrated on a steam bath to give a volume of approximately 15 ml. After cooling to room temperature, 2 ml of 70% $HClO_4$ was added and yellow orange crystals separated. The crystalline product was collected on a filter, washed, in turn, with 5 ml portions of cold water, ethanol and ethyl ether and then allowed to air-dry. The yield was 80% trans-[Ir(en)$_2$Br$_2$]ClO$_4$. (See Table 1 for analyses, and Table 2 for spectra of this and other compounds prepared by the same method.)

Table 2
Visible-Ultraviolet Spectra of Ir(III) Complexes

Complex	λ, mμ	ε, 1-mole^{-1}, cm^{-1}
trans-[Ir(en)$_2$Cl$_2$]Cl	425, 345, 273(sh)	12.7, 51.5, 58.4
trans-[Ir(en)$_2$(Br$_2$)]ClO$_4$	450, 364, 285(sh), 230(sh)	12.5, 73.5, 46.5, 3010
trans-[Ir(en)$_2$I$_2$]ClO$_4$	490(sh), 398, 282.5, 230	18, 183.3, 1.37 × 10^4, 4.72 × 10^4
trans-[Ir(en)$_2$(N$_3$)$_2$]PF$_6$	412, 341, 255(sh)	78, 511, 3490
trans-[Ir(en)$_2$(NO)$_2$)$_2$]ClO$_4$	250	3790
trans-[Ir(en)$_2$BrCl]ClO$_4$	358, 275(sh)	57.4, 37.4
trans-[Ir(en)$_2$ICl]ClO$_4$	393, 356, 280(sh)	65.5, 58.6, 314
trans-[Ir(en)$_2$NCSCl]ClO$_4$	295	336
trans-[Ir(en)$_2$(H$_2$O)Cl](ClO$_4$)$_2$, pH = 2	∼350(sh), ∼285(sh)	20, 52
trans-[Ir(en)$_2$(OH)Cl]$^+$, pH = 11[a]	∼360(sh), ∼290	30, 90
trans-[Ir(en)$_2$(NO$_2$)Cl]ClO$_4$	280, 258(sh)	1.04 × 10^3, 963
trans-[Ir(en)$_2$(ONO)Cl]ClO$_4$	288	1.11 × 10^3
trans-[Ir(en)$_2$IBr]ClO$_4$	400, 258	267, 6810
cis-[Ir(en)$_2$Cl$_2$]Cl	380, 315, 293, 255	17, 92, 120, 127
cis-[Ir(en)$_2$Br$_2$]Br	310, 265(sh), 230(sh)	205, 210, 1730
cis-[Ir(en)$_2$I$_2$]Cl	346, 283, 226	393, 2750, 2.9 × 10^4

[a] This compound has not been isolated.

Trans-[Ir(en)$_2$XCl]ClO$_4$

The trans-acidochloro complexes were prepared from the aquochloro complex, which was generated in solution by the photochemical aquation of the dichloro complex. Typical of this method of preparation is the procedure described here for the synthesis of the chlorobromo salt.

A solution of 0.5 g of trans-[Ir(en)$_2$Cl$_2$]ClO$_4$ in 50 ml of H$_2$O was irradiated for 2 hr in a water-cooled (25°) reaction dish 25 cm from a GE UA11 1200-W vicor filtered mercury arc lamp. The depth of the solution was 1 cm. The free chloride ion released during this time was precipitated by adding 0.177 g of AgNO$_3$ dissolved in 10 ml of H$_2$O (Ag:Ir = 1:1). The solution was centrifuged and filtered and 0.1070 g of NaBr was added and dissolved. A drop of 70% HClO$_4$ was added and the solution was boiled for 15 min and evaporated on a steam bath to 25 ml. The solution deposited large yellow feathery crystals upon cooling to room temperature. These were collected after 3 hr, washed with H_2O (5 ml), ethanol (10 ml), ether (20 ml), and air-dried. The yield was 60% trans-$[Ir(en)_2BrCl]ClO_4$. (See Table 1 for analyses, and Table 2 for spectra of this and other compounds prepared by this method.)

Trans-[Ir(en)$_2$H$_2$OCl] (ClO$_4$)$_2$

A 130-ml solution containing 1.0 g of trans-$[Ir(en)_2Cl_2]ClO_4$ and 0.02 M in NaOH was irradiated for 1.5 hr by a method similar to that described above. The solution was reduced in volume on a steam bath to 7 ml, and a drop of $HClO_4$ (70%) was added to precipitate unreacted trans-$[Ir(en)_2Cl_2]ClO_4$. After filtration, the solution was cooled at room temperature for 24 hr, and white crystals of the desired compound separated. These were collected on a filter, washed with ice cold water (10 ml), ethanol (10 ml), and ether (10 ml), and air-dried. The yield was 0.30 g (26%). (For analyses, see Table 1.)

Trans-[Ir(en)$_2$(ONO)Cl]ClO$_4$

Five-tenths grams of trans-$[Ir(en)_2Cl_2]$-ClO_4 was dissolved in 75 ml of water, and the solution was irradiated 1 hr as mentioned above. The solution was treated with 0.200 g of $AgClO_4$ and 1 ml of 70% $HClO_4$, reduced in volume to 20 ml on a steam bath, filtered, cooled to $5°$ for 30 min, and filtered again. The solution was cooled in an ice bath, 0.9 g of $NaNO_2$ was added, and the mixture was stirred until dissolved. The reaction mixture was maintained at this low temperature for 20 min and then filtered to remove the light yellow product that formed. The crystals of the nitrito isomer were washed with 5-ml cold water and separate 10 ml portions of ethanol and ether and then air dried. The yield was 0.25 g (51%). (For analyses, see Table 1.)

Trans-[Ir(en)$_2$(NO)$_2$Cl]ClO$_4$

The nitro isomer was prepared from the nitrito isomer *in situ* either in solution by heating at $100°$ for a minute (the nitrito isomer rearranges in aqueous solution to the nitro isomer with a half-life of less than 80 sec at $80°$) or in the solid state by heating at $100°$ for three days.

RESULTS AND DISCUSSION

The elemental analyses of the compounds prepared are given in Table 1 and their absorption spectra are recorded in Table 2.

The *cis* and *trans* structures of the complexes were identified by the method of Baldwin.[13] The *cis* isomers have two infrared bands in each of the

CH_2-rocking (800 to 900 cm^{-1}) and NH_2-asymmetric deformation (1600 cm^{-1}) regions, whereas the *trans* isomers exhibit only one band in each of these regions. Furthermore, positive proof of the structure of *cis*-[Ir(en)$_2$Cl$_2$]$^+$ was obtained by resolving it to give the optical isomers.

A general method for the preparation of *trans*-diacidotetraammine-iridium(III) complexes has been developed which depends on the use of reducing agents such as KH_2PO_2 and KBH_4. The nature of the catalysis is uncertain, but complexes of iridium(III) containing amine ligands seem to be stabilized towards reduction. Therefore, catalysis seems to depend upon the reduction of ICl_6^{3-}, which then promotes substitution by amines. Since this method gives exclusively *trans*-[Ir(en)$_2$Cl$_2$]$^+$, it is tempting to suggest that the catalyst is iridium(I) and that a two-electron redox process is involved.[7]

The *cis* isomer is obtained in low yield from reaction mixtures, to which no reducing agent has been added. The secret of improved yields seems to be the presence of a high concentration of sodium chloride. The reason for this is not known, but it may be because the *cis*-dichloro complex forms more stable ion pairs than does the *trans* isomer.[14] Thus, at higher chloride ion concentration there is a greater tendency to generate the more stable ion pair of the *cis* isomer.

Photoinduced substitution reactions of the *trans*-dichloro complex provides a convenient method for the synthesis of *trans*-[Ir(en)$_2$XCl]$^+$. No evidence has been found for photoreduction or photoisomerization in these systems. The method of synthesis is that represented by reaction sequence (1):

$$\text{\textit{trans}-[Ir(en)}_2\text{Cl}_2]^+ \xrightarrow[\text{H}_2\text{O}]{h\nu} \text{\textit{trans}-[Ir(en)}_2\text{H}_2\text{OCl}]^{2+} + \text{Cl}^-$$

$$\xrightarrow{\text{Ag}^+} \text{AgCl} + \text{\textit{trans}-[Ir(en)}_2\text{H}_2\text{OCl}]^{2+} \xrightarrow{\text{X}^-} \text{\textit{trans}-[Ir(en)}_2\text{XCl}]^+ \qquad (1)$$

The photoaquation reaction at the experimental conditions used (see the Experimental Section) has a half-life of 13 min at 25°C. This is five times faster than the thermal aquation at 140°C and 10^5 times faster than the thermal reaction is estimated to be at 25°C.

The photochemical reactions of *trans*-[Ir(en)$_2$Cl$_2$]$^+$ in aqueous solutions containing added nucleophiles were also examined by monitoring spectral changes during the reactions. The results of these observations and other similar studies are shown in Table 3. These are all photochemical reactions because they are too fast to involve thermal substitutions. Thus, for example, the reactions of *trans*-[Ir(en)$_2$Cl$_2$]$^+$ in the presence of bromide ion to form *trans*-[Ir(en)$_2$Br$_2$]$^+$ must be a primary substitution process and not a photo-induced aquation with subsequent thermal replacements.

The results obtained (Table 3) are in good agreement with Adamson's[11] empirical rule mentioned in the introduction. For example, the photoaquation of *trans*-[Ir(en)$_2$Cl$_2$]$^+$ does not give the diaquo product but generates in high

Table 3
Photoinduced Substitution Reactions of Trans-[Ir(en)$_2$X$_2$]$^+$

Reactant	Nucleophile	Major product
trans-[Ir(en)$_2$Cl$_2$]$^+$	H$_2$O	trans-[Ir(en)$_2$(OH$_2$)Cl]$^{2+}$
	OH$^-$	trans-[Ir(en)$_2$(OH)$_2$]$^{+}$ [a]
	I$^-$	trans-[Ir(en)$_2$I$_2$]$^+$
	Cl$^-$	trans-[Ir(en)$_2$(OH$_2$)Cl]$^{2+}$
	Br$^-$	trans-[Ir(en)$_2$Br$_2$]$^+$
trans-[Ir(en)$_2$Br$_2$]$^+$	Cl$^-$	trans-[Ir(en)$_2$BrCl]$^+$
trans-[Ir(en)$_2$I$_2$]$^+$	Br$^-$	trans-[Ir(en)$_2$IBr]$^+$
	Cl$^-$	trans-[Ir(en)$_2$ICl]$^+$

[a] The salt of this complex has not been isolated.

yield trans-[Ir(en)$_2$H$_2$OCl]$^{2+}$. According to the rule, photochemical substitution of the chloroaquo complex would involve water exchange rather than the replacement of the chloro group because the relative ligand field strengths are H$_2$O > Cl$^-$. However, in the presence of added iodide ion, the major product is trans-[Ir(en)$_2$I$_2$]$^+$. Thus, the photoinduced reaction of trans-[Ir(en)$_2$ICl]$^+$ results in the replacement of the chloro group in accord with the rule and the ligand field strengths of Cl$^-$ > I$^-$. The only exception to this rule was the observation that trans-[Ir(en)$_2$OHCl]$^+$ forms trans-[Ir(en)$_2$(OH)$_2$]$^+$ although the ligand field strengths are OH$^-$ > Cl$^-$. However, this may be due to the hydroxide ion generally being a very strongly bonded ligand in most metal complexes. It may also be pointed out that the results obtained are in accord with the trans-effect behavior in these systems.[3b]

The photochemistry of cis-[Ir(en)$_2$Cl$_2$]$^+$ is complicated and it has not been extensively studied. An examination of spectral changes accompanying the photoaquation of the cis isomer suggests that this may involve chelate ring opening rather than the release of the chloro group. For example, after irradiation for what was estimated to be five half-lives of reaction, the solution contained little or no free chloride ion. The rule for photoinduced substitutions in these systems would predict an initial opening of the chelate ring.

Some studies have been made of the rates and stereochemistry of thermal substitution reactions. The stereochemistry is readily summarized by indicating that in none of the many reactions investigated was there any evidence for a change in stereochemistry. For example, (-)-cis-[Ir(en)$_2$Cl$_2$]$^+$ was allowed to react with excess alkali to generate the dihydroxo complex and this to react with excess hydrochloric acid to regenerate the dichloro complex. Within the limit of experimental error there was no loss in optical activity of the complex, which suggests that all of these substitutions take place with retention of configuration.

The rate of release of chloride ion from aqueous solutions of the complex *trans*-$[Ir(en)_2Cl_2]^+$ does not depend on the nature of the nucleophile (OH^-, Br^-, I^-, N_3^- or NCS^-) present nor on its concentration. Thus, there is no direct displacement of chloride ion by these reagents and the rate determining step must be equation (2) followed by the rapid replacement of water (3).

$$trans\text{-}[Ir(en)_2Cl_2]^+ + H_2O \longrightarrow trans\text{-}[Ir(en)_2H_2OCl]^{2+} + Cl \qquad (2)$$

$$trans\text{-}[Ir(en)_2H_2OCl]^{2+}X^- \longrightarrow trans\text{-}[Ir(en)_2XCl]^+ + H_2O \qquad (3)$$

If the reagent has a high *trans* effect, such as I^-, then the initial slow aquation is readily followed with the formation of the disubstituted product *trans*-$[Ir(en)_2X_2]^+$. These results do not permit a further assignment of reaction mechanism. The rate constant for reaction (2) at 105°C is $k = 5.9 \times 10^{-6} \, sec^{-1}$ and the activation parameters are $\Delta H^\ddagger = 29.2 \, kcal$, $\Delta S^\ddagger = -5.7 \, e.u.$

Finally, the rates of aquation of several acidochloro complexes have been determined in (4):

$$trans\text{-}[Ir(en)_2XCl]^+ + H_2O \rightarrow trans\text{-}[Ir(en)_2H_2OX]^{2+} + Cl^- \qquad (4)$$

At 105°C, the rates of reaction increase in the order

$$NCS^- < Cl^- < Br^- < NO_2^- < I^-$$

This is the same order of *trans*-effect that was found for the corresponding rhodium(III) complexes.

ACKNOWLEDGMENT

This work was supported in part by the U.S. Atomic Energy Commission under Grant No. At-(11-1)-1087.

REFERENCES

1. F. Basolo and R. G. Pearson, *Mechanisms of Inorganic Reactions*, John Wiley and Sons, New York, 2nd edition, 1967.
2. S. A. Johnson and F. Basolo, *Inorg. Chem.* **1**: 925 (1962).
3. (a) S. A. Johnson, F. Basolo and R. G. Pearson, *J. Am. Chem. Soc.* **85**: 1741 (1963); (b) A. J. Poe, K. Shaw, and M. J. Wendt, *Inorg. Chim. Acta* **1**: 371 (1967), and references therein.
4. A. Werner and A. P. Smirnoff, *Helv. Chim. Acta* **3**: 472 (1920).
5. S. Kida, *Bull. Chem. Soc. Japan* **39**: 2415 (1966).
6. M. Delepine, *Ann. Chim. (France)* **19**: 174 (1923); *Compt. Rend.* **236**: 559 (1953). M. Delepine and F. Lareze, *ibid*, **256**: 3912 (1963).
7. J. V. Rund, F. Basolo and R. G. Pearson, *Inorg. Chem.* **3**: 658 (1964); J. V. Rund, *ibid.* **7**: 24 (1968).
8. R. D. Gillard, J. A. Osborn, and G. Wilkinson, *J. Chem. Soc.* 4107 (1965); D. J. Baker and R. D. Gillard, *Chem. Comm.* 520 (1967).

9. R. A. Bauer and F. Basolo, *Chem. Comm.* 458 (1968).
10. R. A. Bauer and F. Basolo, *J. Am. Chem. Soc.* **90**: 2437 (1968).
11. A. W. Adamson, *J. Phys. Chem.* **71**: 798 (1967).
12. G. Gire, *Ann. Chim.* **9** (11): 278 (1938).
13. M. E. Baldwin, *J. Chem. Soc.* 4369 (1960).
14. W. A. Millen and D. W. Watts, *J. Am. Chem. Soc.* **89**: 6858 (1967).

COMPLEXES OF COBALT(II) AND NICKEL(II) WITH BIS(2-DIPHENYL-PHOSPHINOMETHYL)ETHER

A. T. Ram and H. B. Jonassen

Tulane University
New Orleans, Louisiana

INTRODUCTION

The most widely studied chelating diphosphines have been those where a saturated carbon chain links two or more phosphine groups. Linear saturated diphosphines[1-6] of the type $R_2P(CH_2)_nPR_2$ have been prepared where $R = CH_3$, C_2H_5 and C_6H_5 for $n = 0$, 1 and 2, and where $R = C_6H_5$ and C_6H_{11} for $n = 0$, 3, 4, and 5. In metal complexes of cobalt(II) and nickel(II) ions containing these ligands, as the number of CH_2 groups increases (when $R = C_6H_5$) from 0 to 5, a distinct change from monodentate to bidentate[1,2] coordination and from square planar to tetrahedral configuration has been reported.[3-6] When $n = 3$ and $R = C_6H_5$, planar $\leftrightarrow$ tetrahedral equilibrium in solution has been observed.[2]

Until recently, only diphosphines with different chain lengths of saturated carbon–carbon linkages have been studied. In 1965, Hansen and Aguiar[7] prepared a saturated diphosphine containing an ether link in the chain, namely *bis*(2-diphenylphosphinomethyl)ether, $Ph_2PCH_2OCH_2PPh_2$, (abbreviation: ppme). This paper reports the preparation, spectral, magnetic, and conductance measurements of some of the metal complexes prepared with this ligand. Two types of complexes have been studied:

(a) MLX_2: Where $M = Co(II)$ and $Ni(II)$; $X = Cl^-$, Br^-, NCS^- and NO_3^- and $L = $ ppme.
(b) ML_2X_2: Where $M = Co(II)$ and $Ni(II)$; $X = ClO_4^-$ and $L = $ ppme.

The recent work of Sacconi[5] with *bis*(2-diphenylphosphinoethyl)ether, (abbreviation: ppee), reports the characterization of metal complexes of the type $ML'X_2$, where $M = Co(II)$ and $Ni(II)$, $L' = $ ppee and $X = Cl^-$, Br^-, I^-, and NCS^-. These complexes possess four coordinate pseudotetrahedral symmetry with no evidence for the coordination of the central oxygen atom.

EXPERIMENTAL

Preparations

(a) *Bis*(2-diphenylphosphinomethyl)ether: This ligand was prepared according to the procedure given by Hansen and Aguiar.[7]

(b) *Bis*(2-diphenylphosphinomethyl)ether metal(II) chlorides, bromides, thiocyanates and nitrates: Hydrated metal [Co(II) and Ni(II)] salts (0.1 m conc) in absolute methanol (5 ml) was added to a hot 95 % ethanol solution of the ligand (0.2 m conc) slowly with stirring. A distinct color change was observed as soon as the metal solution was added. The solid complex precipitated out when concentrated and cooled to room temperature. Addition of dry ether hastens this separation. The solid product was filtered, washed several times with ether, and dried under vacuum over calcium chloride in a desiccator.

(c) *Bis*(2-diphenylphosphinomethyl)ether metal(II) perchlorates : These preparations are very similar to that of the 1 : 1 type complexes. $NiL_2(ClO_4)_2$ where L = ppme can be prepared either using ethanol or acetone solution of the ligand with methanol solution of the metallic salt. We obtained a slightly orange colored product in both cases with different magnetic moments, however. These complexes can be prepared from both ethanol and acetone solutions of the ligand but the complexes prepared from acetone yielded products which always had slightly higher paramagnetic values than those prepared from ethanol.

The physical properties and analytical data of these complexes are given in Table 1. The presence of solvent in some of these complexes were identified with the use of nuclear magnetic resonance spectra as well as with mass spectral measurements.

MAGNETIC SUSCEPTIBILITY MEASUREMENTS

These measurements were made by the Gouy method, at different temperatures, using a semi-microanalytical balance and an electro-magnet, at a field strength of approximately 6000 G. The Gouy tube was calibrated with $HgCo(NCS)_4$. Diamagnetic corrections were calculated from Pascal's constants[8] and from direct measurements on the ligands. The effective magnetic moment was calculated using the formula $\mu_{eff} = 2.84\sqrt{\chi_M^{corr}T}$ where χ_M^{corr} is the corrected molar magnetic susceptibility per metal atom.

Solution susceptibility in nitromethane, at two different concentrations, were made by a similar procedure to that for the solids, as described above. The Gouy tube was calibrated with the solvent, i.e., nitromethane, as well as with freshly prepared $(NH_4)_2Fe(SO_4)_2 \cdot 6H_2O$.[9] In very dilute concentrations, it was assumed that only one single species is present, since otherwise it would be necessary to have a knowledge of the correct concentrations of individual species present at such concentrations. Hence, only the susceptibility values at higher concentrations seem reasonable. Precaution was taken, as much as possible to avoid atmospheric oxygen contamination.

Table 1

Physical Properties and Analytical Data of Metal Complexes with
bis(diphenylphosphinomethyl)ether
$(L = (Ph_2PCH_2)_2O = ppme)$

Complex	% Carbon	% Hydrogen	% Phosphorus	% Halide Nitrogen	Color	M.P., °C
CoLCl$_2$						
Calc.	57.38	4.45	11.40	13.03	Grey	182°
Found	57.80	4.87	11.69	12.47		
CoLBr$_2$						
Calc.	49.29	3.79	9.77		Brown-	
Found	49.86	3.52	9.97		grey	220°
CoL(NCS)$_2$						
Calc.	57.05	4.08	10.53		Greenish	
Found	57.09	4.20	10.20		Yellow	165°
CoL(NO$_3$)$_2$**						
Calc.	50.84	4.84	9.38	4.23	Violet	180°
Found	50.48	4.13	8.72	3.92		
CoL$_2$(ClO$_4$)$_2$**						
Calc.	56.36	4.87	10.78	6.17	Yellow	128°
Found	56.94	5.42	10.81	6.12		
NiLCl$_2$*						
Calc.	56.29	4.86	10.77		Reddish	192– J
Found	56.14	4.47	11.19		brown	195°
NiLBr$_2$*						
Calc.	48.76	4.21	9.33	24.05	Dark	160– J
Found	47.64	4.02	9.45	24.74	Brown	162°
NiL(NCS)$_2$						
Calc.	57.06	4.08	10.53		Yellow	218– J
Found	56.78	4.22	10.32			220°
NiL(NO$_3$)$_2$						
Calc.	52.29	4.02	10.39		Pale	105°
Found	52.12	4.25	10.66		brown	
NiL$_2$(ClO$_4$)$_2$***						
Calc.	57.49	4.42	11.42		Yellow	235 – J
Found	57.72	4.97	11.10			237°

* [complex] · 1CH$_3$OH.
** [complex] · 2CH$_3$OH.
J decomposes.
*** NiL$_2$(ClO$_4$)$_2$ prepared in acetone and methanol, melts at 248°C.

Conductance Measurements

Conductance measurements were carried out using a Serfass Conductance Bridge Model RCM 1581 made by Industrial Instruments Co. (Cedar Grove, N.J.), with cells containing freshly coated platinum black. Conductance measurements and conductometric titrations were carried out with $NaB\phi_4$ in order to understand the mechanism of the disproportionation reaction.

Solution and Solid-State Spectra

Solution spectra in different solvents were measured using a Cary 14 recording spectrophotometer. The molar extinction coefficient, ε, was calculated at the absorption maxima and the spectra were measured from 350 to 2500 mμ. The diffuse reflectance spectra of complexes were measured using a Beckman DK reflectance spectrophotometer, with dry spectral grade lithium fluoride as the reference material.

The infrared spectra of the complexes were obtained on a Beckman Model IR-8 spectrophotometer, both in KBr and nujol mulls, over the 3 to 16 μ region. In certain compounds they were measured from 2.5 to 40 μ.

RESULTS AND DISCUSSION

Bis(diphenylphosphinomethyl)ether complexes were prepared by adding the respective metal salts to hot ethanol solutions of the ligand in a ratio of 1:2. Only for metal perchlorates did a 1:2 type complex result, all other salts gave 1:1 complexes.

1:1 Cobalt(II) complexes

The spectral data of these complexes in solution and solid phase are given in Table 2. The solution spectrum in chloroform and in nitromethane are nearly identical.

The splitting of the visible absorption band in solution (Fig. 1), the extinction coefficient and their positions are indicative of tetrahedral configuration in cobalt(II) complexes. A comparison of the spectra of these complexes with those reported for truly tetrahedral species such as $(CoCl_4)^=$ [10,11] and $[Co\{OP[N(CH_3)_2]_4\}]^{2+}$ [12] reveal that the bands observed in the visible region of the spectrum for these complexes are identical in intensity and position to the bands reported. Hence, the bands in the region of 13,500 to 18,000 cm^{-1} are assigned to the $^4A_2 \rightarrow {}^4T_1(P)$ transition. The fine structure is caused by spin–orbit coupling with the neighboring doublet levels (^2E and $^2T_1$). The second absorption band, around 7000 cm^{-1}, is assigned to the $^4A_2 \rightarrow {}^4T_1(F)$ state. Since the intensity of the absorption band,

Table 2

Spectroscopic Data for Co(II) Complexes with *bis*(diphenylphosphinomethyl)ether

Compound	State	Absorption max, cm^{-1} (ε molar for soln.)
$CoLCl_2$	Chloroform	27030 (410), 17330 (449), 16000 (331), 13790 (417), 7500 (broad) (90)
	Nitromethane	26880 (2310), 16900 (349), 15500 (390), 14710 (369), 7000 (broad) (100)
	LiF	26310 (v. high), 20000, 14710, 8474
$CoLBr_2$	Chloroform	23530 (364), 16400 (503), 14920 (425), 13480 (511), 8000 (broad) (80)
	Nitromethane	26880 (2000), 16130 (335), 14710 (456), 14290 (409)
	LiF	24390, 19230, 14920, 8190
$CoL(NCS)_2$	Chloroform	22990 (1250), 17390 (359), 16000 (452), 13790 (45), 900 (broad) (< 100)
	Nitromethane	26880 (v. high), 17480 (344), 15880 (539)
	LiF	24390, 19230 sh, 12990 broad, 5988 (broad max) (high intensity)
$CoL(NO_3)_2$*	Dimethylformamide	24690 (430), 18180 (45), 7273 (18)
	LiF	— 17540 (broad), 7300 (v. broad)
$CoL_2(ClO_4)_2$	Acetonitrile	23810 sh (301), 19280 sh (197)
	LiF	27000, 19230, 13330 sh, 7300

* Changes color to pink.

around 26,000 cm^{-1}, is highly dependent on the nature of the solvent, and is observed for all complexes (Table 2), in solution and in the solid phase, this seems to be a charge transfer transition band.

The solid-state spectrum in dry lithium fluoride (Fig. 1) shows only two absorption maxima in the visible region. The change in absorption maxima between the solution and solid phase clearly indicate the structural transformations. Two structural configurations seem possible from the nature of the spectrum in the solid state. These are (a) low spin trigonal bipyramidal structure with five coordination; and (b) square planar configuration. A transformation from a five-coordinate trigonal bipyramidal (TBP) configuration, in the solid state to a tetrahedral configuration in solution, would involve dissociation in solution and would most likely be incomplete. Since the solution spectrum does not retain the 20,000 cm^{-1} band, essentially only a single species (the tetrahedral form) is present in solution.

Considerably less is known about the spectral properties of square planar cobalt(II) complexes. All of the square planar complexes studied are of low spin configuration[13-17] with a ground state of $^2B_{2g}$ or $^2A_{1g}$. The spectral bands in the planar complex indicate two absorption maxima in the visible region, and one band around 8100 to 8600 cm^{-1}. The $^4A_2 \rightarrow {}^4T_1(F)$ transition in the tetrahedral form also appears around the region of 8000 cm^{-1}; thus, both low-spin square planar and high-spin tetrahedral forms exhibit an absorption band around this region.

The band envelopes are, however, quite different, the low-spin form possessing a characteristically narrow band, whereas the high-spin form

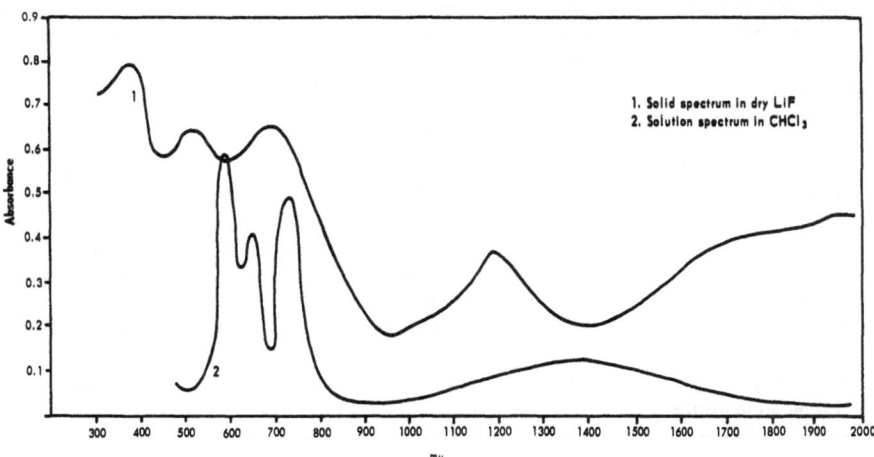

Fig. 1. Electronic absorption spectra of [Co(II)ppmeCl$_2$].

exhibits a broader feature around the same energy level.[14a] The solid-state spectrum of this complex more closely resembles the low-spin planar type complexes. The ligand field transition for this characteristic absorption band around 8500 cm^{-1} in low-spin planar cobalt(II) complexes has not yet been assigned due to the uncertainty in the ground state electronic configurations (i.e., $^2B_{2g}$ or $^2A_{1g}$).

The magnetic moment of these complexes (Table 3) is around 2.30 to 2.80 B.M. at room temperature in the solid state, and the magnetic susceptibility decreases with an increase in temperature. The magnetic moments of these complexes in nitromethane solution at two different concentrations are, however, much higher (Table 4).

The magnetic moment of the complex in the solid state is in agreement with the low-spin square planar structure indicated by the reflectance spectra, indicating the presence of one unpaired electron. Valid solution susceptibility information can be obtained if only one species is present. At the concentration measured, conductance measurements indicate the existence of one species. The observed magnetic moment also is in accordance with the tetrahedral configuration with three unpaired electrons. In more dilute solutions, the tetrahedral form is favored as indicated by the higher value of the magnetic moment.

The conductance measurements indicate that these complexes tend to behave like 1:1 electrolyte in 10^{-4} molar solution.

Table 3
*Magnetic Susceptibility Data of Co(II) and Ni(II) Complexes
with ppme in Solid Phase
(L = (Ph$_2$PCH$_2$)$_2$O = ppme)

Complex	Temp., °K	$\chi_m^{obs} \times 10^{-6}$	$\chi_m^{corr} \times 10^{-6}$	μ_{eff}, B.M.
CoLCl$_2$	291	2615	2962	2.60
CoLBr$_2$	300	2720	3087	2.74
CoL(NCS)$_2$	298	1937	2297	2.36
CoL(NO$_3$)$_2$	300	2814	3150	2.74
CoL$_2$(ClO$_4$)$_2$	303	2376	3100	2.76
NiLCl$_2$	291	315.9	660.5	1.26
NiLBr$_2$	308	810	1177	1.69
NiL(NCS)$_2$	293	−586.4	−226.6	Diamagnetic
NiL(NO$_3$)$_2$	293	2189	2525	2.48
NiL$_2$(ClO$_4$)$_2$ (Prepared in ethanol)	300	−1190	−466.4	Diamagnetic
NiL$_2$(ClO$_4$)$_2$ (Prepared in acetone)	300	1155	1968	2.19

*These measurements were made at different temperatures. It was noted that χ_m^{corr} decreases with increase of temperature.

Table 4
Magnetic Susceptibility Data of Co(II) and Ni(II) Complexes with ppme in Nitromethane Solution

$$(L = (Ph_2PCH_2)_2O = ppme)$$

Complex	Conc., M	Temp. 30°C		
		$\chi_m^{obs} \times 10^{-6}$	$\chi_m^{corr} \times 10^{-6}$	μ_{eff}, B.M.
CoLCl$_2$	2.94×10^{-3}	7347	7692	3.80
	1.89×10^{-3}	9004	9349	4.77
CoLBr$_2$	1.5×10^{-3}	8338	8705	4.60
	1.15×10^{-3}	9694	10051	4.94
CoL(NCS)$_2$	1.2×10^{-3}	7760	8120	4.45
	3.68×10^{-3}	4819	5179	3.55
CoL(NO$_3$)$_2$	INSOLUBLE			
CoL$_2$(ClO$_4$)$_2$	3.5×10^{-3}	4661	5385	3.62
	1.47×10^{-4}	11724	18148	6.65
NiLCl$_2$	3.86×10^{-3}	1414	1761	2.07
	1.05×10^{-3}	2821	3165	2.77
NiLBr$_2$	4.48×10^{-3}	2142	2509	2.14
	1.0×10^{-3}	3355	3722	3.01
NiL(NCS)$_2$	7.3×10^{-4}	1149	1509	1.19
NiL(NO$_3$)$_2$	4.25×10^{-3}	1166	1502	1.95
	1.16×10^{-3}	4452	4761	3.48
NiL$_2$(ClO$_4$)$_2$	1.62×10^{-3}	-3924	-3200	Diamagnetic
(prepared in EtOH)				
NiL$_2$(ClO$_4$)$_2$	1.42×10^{-3}	1090	1814	2.10
(prepared in acetone)	1.21×10^{-3}	1178	1902	2.15

Solution and solid-state spectra, and solution and solid-state magnetic susceptibility values of these complexes indicate the presence of the planar configuration in the solid state, which is transformed into a tetrahedral species in solution.

1:2 Cobalt(II) complexes

The only 1:2 complex synthesized was *bis*(diphenylphosphinomethyl)-ether cobalt(II) perchlorate. The spectral data of this complex in the solid phase (Table 2) show absorption bands around 27,000, 19,230, 13,330 sh and 7300 cm^{-1}, with weak intensities. However, the position of the 19,230 cm^{-1} and the shoulder at 13,330 cm^{-1} are in line with the corresponding 1:1 halide complexes, where a square planar configuration in the solid state is indicated. The 7300 cm^{-1} band is quite broad and has low intensity; hence, assignment is not possible.

In the solution phase, however, the bands appear as shoulders around 23,810 and 19,280 cm^{-1}. The intensity of these bands is less than expected for a true tetrahedral complex.[10,11] Therefore, it is more likely that solvent molecules coordinate with the metal ion, forming a pseudo-octahedral configuration. The intensity of such bands would be lowered, as they are Laporte forbidden d–d transitions.[18,19] Since the $^4T_{1g}(F) \rightarrow {}^4T_{1g}(P)$ transition in octahedral cobalt(II) complexes is of very low intensity, it was not observed in this case.

The magnetic moment (Table 3 and 4) of the complex in the solid phase, 2.80 B.M., is in line with a square planar configuration. It is difficult to interpret the solution susceptibility properly because of the dissociation revealed by conductance measurements. However, the much higher magnetic moments, 5.65 B.M. in dilute solution, indicate the probability of octahedral coordination, as the magnetic moment can vary from 1.9 to 5.2 B.M. in such configurations.

The conductance measurement indicates that the complex behaves as a 1:2 electrolyte. The infrared spectral data clearly indicate that the anion does not participate in the coordination, but is purely ionic. These data then indicate that this complex is planar in the solid state but, in solution, tends to attain pseudo-octahedral structure by the addition of two solvent molecules.

1:1 Nickel(II) complexes

The spectral data of these complexes in solution and in the solid phase are given in Table 5. The solution spectrum in chloroform and nitromethane

Table 5
Spectroscopic Data for Ni(II) Complexes with *bis*(diphenylphosphinomethyl)ether

Compound	State	Absorption max, cm^{-1} (ε molar for soln.)
NiLCl$_2$	Chloroform	— 21050 (737)
	Nintromethane	27320 (481), 23810 (163)
	LiF	27030, 21270, 13330 sh (Small), 7600 sh
NiLBr$_2$	Chloroform	— 20960 (644)
	Nitromethane	27390 (565), 23530 sh (235)
	LiF	27030, 20000, 13330 sh, 7600 sh
NiL(NCS)$_2$	Chloroform	— 22470 (1325)
	Nitromethane	27390 (775), 22990 (919)
	LiF	27200, 22,000, 13330 sh, 7600 sh
NiL(NO$_3$)$_2$	Nitromethane	25000 (92)
	LiF	27200, 22000, 13330 sh, 7563 sh
NiL$_2$(ClO$_4$)$_2$	Trifluroacetic Acid	21050 (590)
	Nitromethane	27250 (1386), 24390 (780)
	LiF	23810 (v. high)

are identical. The solution spectrum shows one band in the near ultraviolet region $(27,030 \, \text{cm}^{-1})$ and only one absorption band around $20,000 \, \text{cm}^{-1}$. The presence of a single absorption band in the visible region has been diagnostic for square planar nickel(II) complexes.[20,21] No bands are observed at energies less than $20,000 \, \text{cm}^{-1}$.

The solid-state spectrum in dry lithium fluoride is similar to the solution spectrum, indicating that no structural transformation has occurred in solution. However, weak shoulders or bands around $14,500$ and $7700 \, \text{cm}^{-1}$ (Fig. 2) are observed. The absorption maxima around $27,000 \, \text{cm}^{-1}$ is attributed to charge transfer transition, as it appears in all complexes of both cobalt(II) and nickel(II) with this ligand.

The magnetic moment (Table 3) has a subnormal value of 1.26 to 2.48 B.M. in the solid state. The complex $NiL(NCS)_2$ is diamagnetic. Square planar nickel(II) complexes should be either diamagnetic (low spin) or paramagnetic (high spin) with a magnetic moment around 2.8 to 3.2 B.M. Measurements of the magnetic susceptibility of the complexes with subnormal paramagnetism at different temperatures obey the Curie–Weiss Law. However, the curve does not uniformly decrease with increasing temperature. This probably indicates that traces of other structural isomers are present in the solid state. The solution susceptibility (Table 4) at concentrations around 10^{-3}M, indicates that no structural transformations between the solid and solution occur except for $NiL(NCS)_2$.

The conductance measurements show that these complexes tend to dissociate at dilute concentrations $(<1 \times 10^{-4} \, \text{M})$. The molecular weight

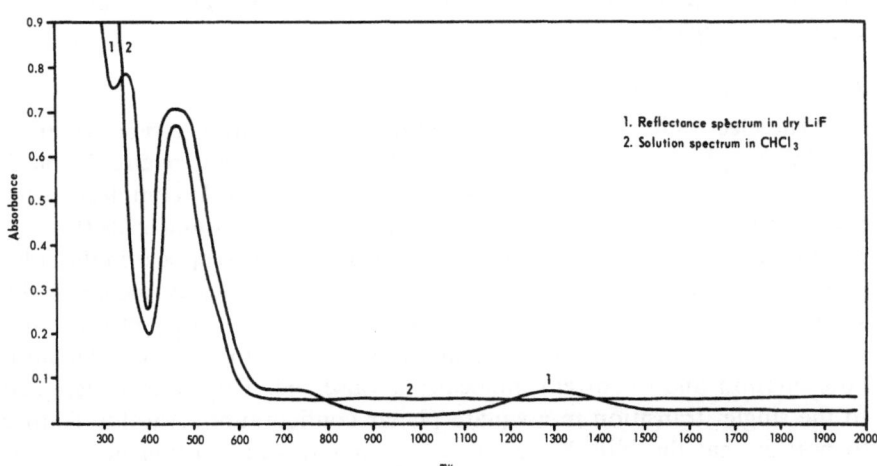

Fig. 2. Electronic absorption spectra of [Ni(II) ppme Cl$_2$].

of some complexes were determined in nitromethane. They appear to be monomeric and undissociated.

The absorption spectra, conductivity, and the molecular weight data, therefore, support a monomeric square planar configuration for all these complexes. The magnetic moment of the solid complex $NiL(NCS)_2$ indicates that in the solid state this complex exists as a low-spin square planar complex. The magnetic moments of the other 1:1 $NiLX_2$ complexes are intermediate between a true singlet and triplet state. Chatt and Booth[3] prepared a similar compound with subnormal moment $(NiCl_2(O_2PC_2H_4PO_2), \mu_{eff} = 1.54$ B.M.) to which they ascribed a cocrystallized paramagnetic and diamagnetic complex in a unit cell. Many other explanations for such subnormal nickel complexes are proposed in the literature such as spin state isomerization,[22-24] addition of solvent molecules to give six coordinated complexes,[25,26] and structural isomerization.[27-30]

The most likely explanation for the subnormal moment in these complexes is the type of cocrystallization postulated by Chatt.[3] The change in the magnetic value with temperature and the presence of shoulders in the spectra of the solids seem to be indicative of such a phenomenon.

Since these complexes are mainly planar, with C_{2v} symmetry,[2] the ligand field considerations predict four possible spin allowed $d-d$ transitions, of which $^1A_1 \rightarrow {}^1B_2$ occurs at the lowest energy. The observed band around 20,000 cm^{-1} is assigned to this transition. The three other predicted transitions, $^1A_1 \rightarrow {}^1A_2$, $^1A_1 \rightarrow {}^1B_2$, and $^1A_1 \rightarrow {}^1B_1$ are undoubtedly swamped by strong charge transfer bands which dominate at higher energies. The solutions and solid-phase spectrum and magnetic moment indicate the presence of square planar species admixed with some high-spin species for these 1:1 complexes.

1:2 Nickel(II) complexes

The electronic absorption spectral data of the only 1:2 type complex, bis(diphenylphosphinomethyl)ether nickel(II) perchlorate is given in Table 5. It is very similar to the corresponding 1:1 type nickel(II) complexes with this ligand in having one single absorption maxima around 22,500 cm^{-1}. It is important to note that no bands (not even shoulders) appear in the solid spectrum at energies lower than 22,500 cm^{-1}. This clearly indicates the square planar configuration of these complexes. The intensity of this band is not decreased in solution, again proving that no pseudo-octahedral coordination due to solvent molecules is observed. This band is assigned to the singlet transition in a square planar configuration. Another feature to note is that the NiP_2X_2 chromophore has this absorption maxima at slightly lower energies than the perchlorate complex (NiP_4 chromophore) which has the maxima around 23,000 cm^{-1}. Similar to the thiocyanate complex, the complex is diamagnetic both in the solid state and in solution—indicating the true singlet nature of the ground state.

However, if the complex is prepared in methanol and acetone, the product has a residual paramagnetism of about 1.80 B.M. Temperature dependent magnetic measurements of these two compounds indicated that the diamagnetic compound remains diamagnetic, while the other retains its paramagnetism. The solution susceptibility of the paramagnetic compound retains the paramagnetism (1.95 B.M.). This leads to the conclusion that the origin for the presence of structural isomerization is in the mode of preparation of these complexes.

Conductivity measurements at several concentrations show that the complex is a 1 : 2 electrolyte. The molecular weight determination corresponds to a 1 : 2 electrolyte in exhibiting one-third the actual molecular weight.

The complex has a square planar configuration in both solid and solution phase. The mode of preparation of the complex leads to the formation of both paramagnetic and diamagnetic species.

CONCLUSIONS

The *bis*(diphenylphosphinomethyl)ether complexes behave similarly to the *bis*(diphenylphosphinoethyl)ether complexes prepared by Sacconi[5] in forming four coordinate complexes. The data give no indication of fibe coordination, i.e., the central donor atom, oxygen, remains noncoordinated. Sacconi,[5] did not observe any structural transformation in the solid and solution phases with cobalt(II) *bis*(diphenylphosphinoethyl)ether complexes. The structural transformation in cobalt(II) complexes are more closely related to the nickel(II) complexes prepared by Van Hecke and Horrocks.[2]

ACKNOWLEDGMENTS

The authors are grateful and would like to thank Dr. Edward A. Boudreaux of Louisiana State University, New Orleans, for the use of Gouy Balance and the DK-2 reflectance spectrophotometer. We also appreciate the assistance provided by the National Science Foundation through chemistry research instrument grant GP 16090.

REFERENCES

1. K. Issleib and G. Schwager, *Z. Anorg. u. Allgem. Chem.* **310**: 43 (1961).
2. G. R. Van Hecke and W. D. Horrocks, *Inorg. Chem.* **5**: 1968 (1966).
3. G. Booth and J. Chatt, *J. Chem. Soc.* 3238 (1965).
4. K. Issleib and G. Hohlfeld, *Z. Anorg. u. Allgem. Chem.* **312**: 169 (1961).
5. L. Sacconi and J. Gelsomini, *Inorg. Chem.* **7**: 291 (1968).

6. C. E. Wymore and J. C. Bailar, Jr., *J. Inorg. Nucl. Chem.* **14**: 42 (1960).

7. K. Hansen, Ph.D. Dissertation, Tulane University, New Orleans, La., August 1967.

8. B. N. Figgis and J. Lewis, *Modern Coordination Chemistry*, Interscience Publishers, New York, 1960, p. 403.

9. D. M. Adams and J. B. Raynor, *Advanced Practical Inorganic Chemistry*, John Wiley and Sons, New York (1965), p. 150.

10. F. A. Cotton and M. Goodgame, *J. Am. Chem. Soc.* **83**: 1777 (1961).

11. Y. Tanabe and S. Sugano, *J. Phys. Soc. (Japan)* **9**: 753 (1954).

12. J. Chatt and B. L. Shaw, *J. Chem. Soc.* 705 (1950); 4021 (1959).

13. A. Davison, N. Edelstein, R. H. Holm and A. H. Maki, (a) *Inorg. Chem.* **2**: 1227 (1963); (b) *ibid.* **3**: 914 (1964); (c) *J. Am. Chem. Soc.* **85**: 2029 (1963); (d) *ibid.* **85**: 3049 (1963).

14. (a) G. W. Everett, Jr. and R. H. Holm, *ibid.* **88**: 2442 (1966); (b) *Inorg. Chem.* **7**: 776 (1968).

15. B. N. Figgis and R. S. Nyholm, *J. Chem. Soc.* **12**: (1954); 338 (1959).

16. A. B. P. Lever, J. Lewis, and R. S. Nyholm, *J. Chem. Soc.* 2552 (1963).

17. F. A. Cotton and D. L. Weaver, *J. Am. Chem. Soc.* **87**: 4189 (1965).

18. B. N. Figgis, *Introduction to Ligand Fields*, Interscience Publishers, New York, 1966.

19. P. W. Selwood, *Magneto-Chemistry*, Interscience Publishers, New York, 1956, p. 226.

20. G. Dyer, J. C. Hartley, and L. M. Venanzi, *J. Chem. Soc.* 1293 (1965).

21. G. Maki, *J. Chem. Phys.* **28**: 651 (1958).

22. G. Maki, *ibid.* **29**: 1129 (1958).

23. C. J. Ballhausen and A. D. Liehr, *J. Am. Chem. Soc.* **81**: 538 (1959).

24. H. N. Ramaswamy, H. B. Jonassen, and A. N. Aguiar, *Inorg. Chim. Acta* **8**: 141 (1967).

25. J. B. Willis and D. P. Mellor, *J. Am. Chem. Soc.* **69**: 1237 (1947).

26. H. C. Clark and A. L. Odell, *J. Chem. Soc.* 520 (1956).

27. R. H. Holm and K. Swaminattran, *Inorg. Chem.* **2**: 181 (1963).

28. L. Sacconi, *J. Chem. Soc.* 4608 (1963).

29. L. Sacconi and M. Ciampolini, *J. Am. Chem. Soc.* **85**: 1750 (1965).

30. C. R. C. Caussmaker, M. H. Hutchinson, J. R. Millard, L. E. Sutton, and L. M. Venanzi, *J. Chem. Soc.* 2705 (1961).

31. S. L. Holt, R. C. Bouchard, and R. L. Carlin, *J. Am. Chem. Soc.* **86**: 579 (1964).

CYANOAQUOCHROMIUM(III) COMPLEXES: ISOLATION, SPECTRA, AQUATION REACTIONS AND KINETICS*

W. B. Schaap, R. Krishnamurthy, D. K. Wakefield, and W. F. Coleman

Indiana University
Bloomington, Indiana

INTRODUCTION

The mechanisms of solvent substitution reactions of octahedral complex ions have been the subject of considerable study in recent years.[1-5] In these studies, chromium compounds have played an important role because the d^3 electronic configuration of the Cr^{+3} ion imparts properties to its compounds which make them especially valuable for studies involving kinetics, mechanisms, and the dependence of reaction rates on geometrical and electronic factors. In particular, chromium(III) compounds are sufficiently nonlabile so that intermediates can often be isolated and characterized, and there can be a stepwise replacement of one set of ligands by another set, with widely differing ligand field strength, without incurring a disruptive spin change.

Although the literature contains numerous descriptions of solvent-substitution reaction studies of various complexes of chromium(III), we know of no reported studies involving the hexacyanochromate(III) anion $Cr(CN)_6^{-3}$ prior to our own.[6]

Cyanide complexes are important to the theory of coordination compounds because of their type of bonding and their exceptional stability.

*This work was supported by the U.S. Atomic Energy Commission under Contract AT(11-1)-256 (Document No. COO-256-92). Address inquiries to W. B. Schaap.

Presented in part before the Division of Inorganic Chemistry at the 150th National Meeting of the American Chemical Society, Atlantic City, N.J., September 1965, at the 151st National Meeting, Pittsburgh, Pa., March 1966, and at the 153rd National Meeting, Miami Beach, Fla., April 1967. Also presented in part at the 9th International Conference on Coordination Chemistry, St. Moritz, Switzerland, September 1966.

Taken in part from theses submitted to the Graduate School of Indiana University in partial fulfillment of the requirements for the Ph.D. degree in Chemistry by: (a) R. Krishnamurthy, 1966; (b) Diane K. Wakefield, 1967; and (c) W. F. Coleman (1970).

Cyanide ion ranks highest of all the common ligands in the spectrochemical series; it stabilizes a variety of stereochemical configurations and exhibits one of the largest *trans* effects observed in substitution reactions of square-planar complexes of platinum(II) and nickel(II). For these reasons, we decided to exploit the observation that the hexacyanochromate(III) ion undergoes rapid, acid-catalyzed aquation and to carry out detailed kinetic studies of this sequence of reactions. This communication summarizes our studies on this system to date and includes procedures for isolating the intermediate cyanoaquochromium(III) complexes and the results of our spectral and kinetic investigations.

AQUATION REACTIONS*

Spectral Changes

If a dilute solution of potassium hexacyanochromate is made slightly acidic with perchloric acid, the original lemon-yellow color deepens and becomes deep orange within several minutes. The deep-orange color persists with little change for several hours but, if the acidity is increased to 1–2 M perchloric acid, the solution reddens, then becomes pink, and finally violet. Because the spectrum of the final violet solution is identical to that of the hydrated chromic ion, $Cr(H_2O)_6^{+3}$, the observations indicate that the hexacyanochromate(III) ion has undergone complete aquation via series of successive reactions involving relatively stable intermediates, i.e.,

$$Cr(CN)_6^{-3} + H_3O^+ \rightarrow Cr(CN)_{6-n}(H_2O)_n^{n-3} + n\,HCN$$

$(n = 1 \text{ to } 6 \text{ inclusive})$

Repetitive spectrophotometric scans recorded during the transition of the original yellow color of $Cr(CN)_6^{-3}$ to the semistable, deep-orange intermediate color show that the two absorption bands of $Cr(CN)_6^{-3}$, present initially and located at 307 and 377 mμ, drop down and disappear, and are replaced by another simple two-band spectrum having a more intense band at 467 mμ and a less intense band at 362 mμ. This final spectrum, characteristic of the deep-orange solution, persists virtually unchanged for several hours in a solution of pH $\sim$ 1, but disappears within a day.

It is noteworthy that no one isosbestic point persists throughout these spectral changes. The initial isosbestic points at 322, 350, and 393 mμ are soon lost. (Experiments done at lower acidities permitted more absorption curves to be recorded during the initial stages of the reaction and confirmed the existence of these isosbestic points.) After several minutes, a single new isosbestic point at 433 mμ starts to develop and is maintained until the transition to the deep-orange solution is complete. The behavior of the

*From the thesis submitted to the Graduate School of Indiana University in partial fulfillment of the requirements for the Ph.D. degree in Chemistry by R. Krishnamurthy, 1966.

isosbestic points[7] implies that the reaction is not of the simple A → B type, but that at least one intermediate must be involved between the parent ion, $Cr(CN)_6^{-3}$, and the relatively stable species giving rise to the deep-orange solution.

The frequencies of the absorption bands shift markedly toward lower energies as the aquation proceeds, while the charge transfer bands in the near ultraviolet shift toward higher energies. The directions of both shifts are consistent with those expected for the progressive substitution of water molecules for the coordinated cyanide groups. The intensities of the absorption bands reach their maximum with the stable, deep-orange intermediate solution, and then decrease again with broadening of the lowest-energy band as the aquation proceeds toward the hexaquochromium(III) ion. These further changes are consistent with the formation of additional intermediate mixed cyanoaquo complexes of lower symmetries.

Isolation of Intermediates

The intermediate species formed during the sequential aquation reactions were separated and identified by ion-exchange techniques. It was observed that the species contained in the deep-orange solution passed directly through both an anion and a cation-exchange resin column and that its spectrum was unchanged. Therefore, it was concluded that this species was probably the uncharged tricyanotriaquo molecule, $Cr(CN)_3(H_2O)_3$, and that it is essentially the only component of the persistent deep-orange solution. Analysis of the orange-colored effluent solution gave a CN^-/Cr ratio of 3.0 within experimental error.

If the aquation is quenched by raising the pH to 5.0 before conversion to the orange species is complete and if the resulting solution is passed through an anion-exchange column in the perchlorate cycle, a total of three species can be separated. First, the orange-colored, neutral species, which is not retained by the resin, appears in the effluent; then, after the concentration of sodium perchlorate in the eluting solution is increased, a deep-yellow band is eluted from the column which contains a ratio of four cyanide ions to one chromium(III) ion. A further increase in sodium perchlorate concentration removes the remaining, lemon-yellow band, the spectrum of which is identical to that of the parent ion, $Cr(CN)_6^{-3}$. No detectable amount of the pentacyano complex could be isolated and identified in this type of aquation experiment.

If the acidity of the solution is raised to 1–2 M in perchloric acid, aquation beyond the tricyano complex proceeds slowly. The color of the solution gradually reddens, then becomes pink, and finally the pale violet color of the hydrated ion. Again, no one isosbestic point persists throughout these spectral changes.

If this reaction is quenched before it is complete, the intermediate cationic species may be separated using a cation-exchange column in the sodium cycle. As before, the neutral tricyano complex is not retained by the column, but in this case a dark purple band is observed at the top of the column. Washing the column with dilute sodium perchlorate (pH = 3) causes the adsorbed band to split into three bands. The lowest band moves down the column quite rapidly and emerges as a peach-colored solution, the spectrum of which consists in a weak maximum at 378 mμ and a more intense band at 466 mμ having a distinct shoulder at about 540 mμ. This species has a CN^-/Cr ratio of 2.0.

The second band, which is pink and resembles permanganate ion in color, can be readily washed from the column with about 0.5 M sodium perchlorate (pH = 2–3). This species exhibits two broad, weak absorption bands of nearly equal intensities located at 393 and 525 mμ, which are identical to the bands attributed to $Cr(CN)(H_2O)_5^{+2}$ by Espenson, who prepared this complex by the reduction of a monocyanocobalt(III) complex with chromium(II) in an acidic aqueous solution.[8] This pink complex has a CN^-/Cr ratio of 1.0. Finally, the dark band at the top of the column can be eluted with a concentrated, acidified solution of sodium perchlorate. This species has a three-band spectrum identical to that of $Cr(H_2O)_6^{+3}$.

The ion-exchange separations of the species and their elution behaviors give no indication of the presence of multiple isomers of the intermediate cyanoaquo complexes. In addition, the neutral tricyano complex is much more stable than any of its precursors, and subsequent reactions in the second half of the overall aquation sequence are much slower. Under these circumstances, the aquation of $Cr(CN)_6^{-3}$ in acidic solutions always passes through an intermediate stage in which the solution contains essentially only $Cr(CN)_3(H_2O)_3$. If the reaction is quenched before this stage, only the neutral and anionic species are present; if the reaction proceeds beyond the stable intermediate, only the neutral and cationic species are in the solution. For these reasons the effort required to follow the reaction spectrophotometrically and to separate the species is greatly simplified, for no more than three or four species can coexist.

Stereochemistry

Because only one isomer is apparently formed at each step of the reaction sequence, the question of the stereochemical path of the reactions is intriguing. We believe that the relative inertness of the tricyano complex, $Cr(CN)_3(H_2O)_3$, and its successors, in comparison with its precursors, and the sharpness of the two-band absorption spectrum of $Cr(CN)_3(H_2O)_3$ argue convincingly for assigning to this species the *cis* or 1,2,3-configuration.

The complex $Cr(CN)_3(H_2O)_3$ can exist in two possible geometrical arrangements, *cis* (1,2,3-) or *trans* (1,2,6-). In the *cis* complex the mutual *cis* orientation of all three cyanide ligands would favor formation of strong π-bonds with the t_{2g} electrons of the metal atom. The d_{xy}, d_{yz}, and d_{xz} orbitals are of proper symmetry to form three equivalent, mutually orthogonal π-bonds with the cyanide groups. The remaining three positions *trans* to the three cyanide groups are occupied by water molecules which are not strongly involved in π-bond formation and do not compete favorably with cyanide groups opposite (*trans*) to them for the t_{2g} π-bonding electrons. As a result, each cyanide ligand *trans* to a water molecule is stabilized in comparison to a cyanide group *trans* to another cyanide ion. The latter are therefore more labile and, following protonation, are more readily replaced by water molecules. As the aquation of $Cr(CN)_6^{-3}$ proceeds, the incoming water molecules will be directed to positions which are *cis* to previously substituted water molecules. When the aquation has proceeded to the *cis*- or 1,2,3-tricyanotriaquo stage, there are no longer any cyanide ligands *trans* to other cyanide groups, and the rates of further substitution of water molecules decrease drastically at this point. This type of labilization arising from a *trans* effect has been observed by Wilkinson and coworkers[9,10] in the replacement of carbon monoxide in metal carbonyls.

The assignment of the *cis* (1,2,3-) configuration of the tricyano complex is also in line with the spectral features of this complex. The absorption spectrum of this species consists of two ligand-field bands which are symmetrical without apparent broadening or splitting. This is in agreement with the ligand-field predictions for the pseudo-cubic, trigonal (C_{3v}) field around the Cr^{+3} ion in the *cis* (1,2,3-) isomer.[11,12] This configuration is more symmetrical than the *trans* (1,2,6-) configuration because in the 1,2,3-complex the sums of the dipole moments of the ligands are equal along each of the three cartesian axes, so that the splitting pattern of the d orbitals is the same as it would be in an octahedral (O_h) field. The spectrum of the *trans* complex should exhibit some splitting and show broadened or definitely split absorption bands.

In view of these considerations, we propose the stereochemical path shown on page 182 for the overall aquation sequence.

The implication of this reaction sequence is that there is a very strong *trans*-labilizing effect operating in these cyanoaquo complexes, which is probably related to competition between opposite ligand pairs for the π-bonding electrons. A strong *trans*-effect is unusual for octahedral complexes of elements of the first transition series, but was recently reported by Moore, Basolo, and Pearson[13] on the basis of $H_2{}^{18}O$ tracer studies involving the iodopentaquochromium(III) ion. In the case of the cyanoaquo complexes, the *trans*-effect is evidently strong enough to cause the sequence of aquation reactions to be highly stereospecific.

(*cis*-diaquo)   (*cis*-(1,2,3-)isomer)   (*cis*-dicyano)
                                (stable Intermediate)

EXPERIMENTAL SECTION

Chemicals

Deionized water and reagent grade chemicals were used in all preparations and procedures. Stock solutions of sodium perchlorate used as eluants, with the ion exchange columns and to adjust the ionic strengths of solutions, were prepared either by neutralization of sodium carbonate with perchloric acid or from Fisher purified reagent, after tests showed it to be free of chloride ion. Potassium hexacyanochromate(III) was initially synthesized and purified by the method of Bigelow.[14] An improved, more rapid method of preparing this compound was developed during the course of these studies and is described below.

Ion Exchange Procedures

Analytical grade anion-exchange resin (Bio-Rad, AG 1-X4, 100–200 mesh or 200–400 mesh) was converted from its original chloride form to the perchlorate form by repeated equilibration with concentrated sodium perchlorate solution, first batch-wise in a beaker and finally in a column, until the effluent from the column gave a negative test for chloride. The column was then washed with distilled water to remove excess electrolyte.

Analytical grade cation-exchange resin (Bio-Rad, AG 50W-X4, 100–200 mesh), purchased in the hydrogen form, was converted to the sodium form by neutralizing with sodium hydroxide solution and then washing with acidified sodium perchlorate solution (pH $\sim$ 2) and finally with distilled water.

The ion-exchange resin beds were generally about 1.0 to 1.5 cm in diameter and about 20 to 25 cm in length. The columns were equipped with water jackets so that they could be operated either at 0°C, if ice water was circulated through the jackets, or at room temperature.

Portions of the appropriate hydrolysis mixtures, prepared as described below, were loaded onto the tops of the columns and the excess electrolyte, as well as all nonadsorbed complex species, removed by washing first with distilled water and then with 0.05 M sodium perchlorate. Adsorbed species were eluted individually from the column by gradually increasing the concentration of sodium perchlorate in the eluting solution. The effluent from the columns were collected in small fractions, usually from 15 to 25 ml, which were stored in the dark at 0° until used in subsequent experiments. Typical elution curves from the anion and the cation-exchange resin columns are shown in Figs. 1 and 2, respectively. Detailed procedures for the separation of individual species are given below.

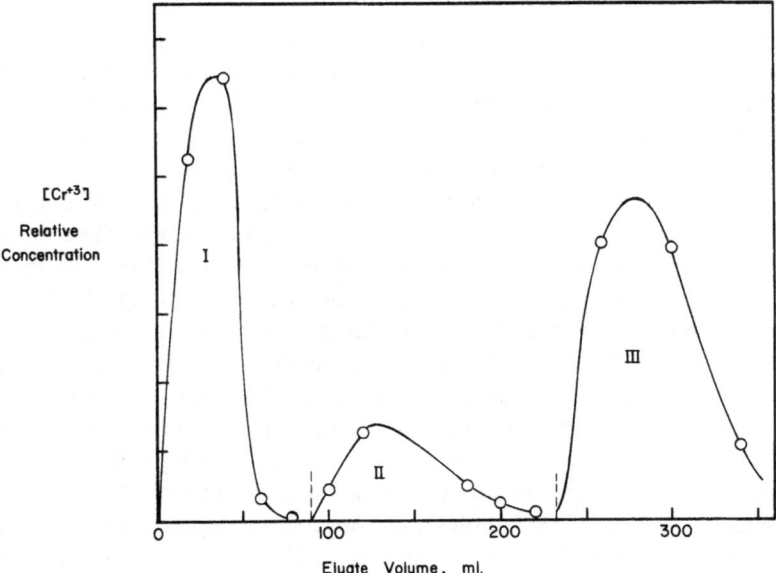

Fig. 1. Ion exchange elution curve showing separation of neutral and anionic cyano-aquochromium(III) complexes. I. $Cr(CN)_3(H_2O)_3$; II. $Cr(CN)_4(H_2O)_2^-$; III. $Cr(CN)_6^{3-}$.

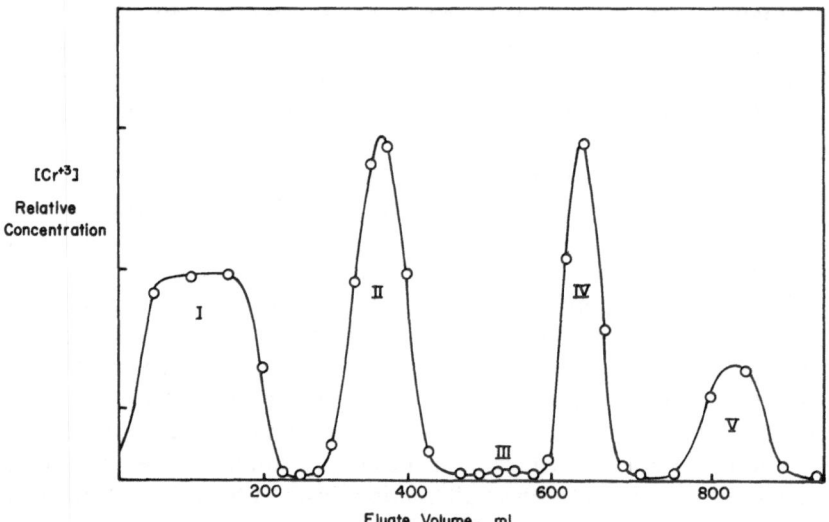

Fig. 2. Ion exchange elution curve showing separation of neutral and cationic cyano-aquochromium(III) complexes. I. $Cr(CN)_3(H_2O)_3$; II. $Cr(CN_2(H_2O)_4^+$; III. unidentified species; IV. $Cr(CN)(H_2O)_5^{2+}$ and V. $Cr(H_2O)_6^{3+}$.

Preparation of Hydrolysis Mixtures

Mixtures of the neutral and anionic cyanoaquo complexes can conveniently be prepared by mixing solutions of $K_3Cr(CN)_6$ and standard perchloric acid in a molar ratio of about 1 to $1\frac{1}{2}$. In a typical preparation, 25 ml of a solution containing 487 mg (1.5 mmoles) of the complex are added with stirring to 10 ml of 0.300 M perchloric acid. A slow stream of nitrogen gas is passed through the reaction mixture for about an hour to sweep out the molecular HCN liberated during the aquation steps. (The original yellow color of the solution rapidly changes to deep-yellow and then to orange, with a rise in pH to about 5.0, which does not change further with time.) The mixture is chilled to ice-temperature and the precipitated potassium perchlorate filtered off. The clear orange filtrate, containing a minimum of free electrolyte, is diluted and used to charge the anion exchange column.

A hydrolysis solution containing the neutral and cationic cyanoaquochromium(III) complexes can be prepared by combining potassium hexacyanochromate and perchloric acid in a molar ratio of 1:5. In a typical experiment, 0.65 g (2.00 mmoles) of potassium hexacyanochromate is dissolved in 250 ml of 0.04 M perchloric acid. The solution is kept in the dark at room temperature until the pH of the solution rises to 2.2, which takes approximately 40 to 45 hr. A stream of nitrogen is continuously bubbled through the solution to sweep out the liberated molecular hydrogen cyanide.

The deep red solution is cooled to 0° and the precipitated potassium perchlorate removed by filtration. The solution, which now contains only neutral and cationic cyanoaquo complexes, can be loaded onto the cation-exchange column.

Spectral and Kinetic Measurements

A Cary Model 14 Recording Spectrophotometer was used to record the spectra of isolated species and to follow the rates of the aquation reactions. The kinetic studies were carried out directly in a water-jacketed silica spectrophotometer cell, usually of 10-cm path length. As a usual procedure, an aliquot of the stock solution containing the cyanoaquo complex and sodium perchlorate, previously brought to the desired temperature, was mixed with another solution at the same temperature containing perchloric acid and sufficient sodium perchlorate to achieve the desired ionic strength. The mixed solution was quickly transferred to the water-jacketed cell and absorption measurements begun. In all experiments, the hydrogen ion concentration was at least 50 times greater than the concentration of the complex, which was kept low and in the range 5×10^{-5} to 3×10^{-3} M.

Reactions were followed at wavelengths appropriate to the various reactants or products. Molar absorbancies at the wavelengths used are listed in Table 1 or in the text. In some cases, the molar absorbancy index was observed to change slightly immediately upon addition of acid. In such cases, a corrected absorbancy index for the reactant was obtained by extrapolating the observed absorbancy measurements back to zero time.

All reactions were followed for three or four half-lives and gave good pseudo first-order plots of $\log(A_t - A_\infty)$ vs. time, where A_t is the absorbance at time t and A_∞ is the absorbance after more than eight half-lives. A computer program[15] which was adapted to the available CDC 3600 computer was used to calculate the least-squares-best slopes for the straight line plots of $\ln(A_t - A_\infty)$ vs. time and was used to obtain the pseudo first-order rate constants. The rate constants were calculated for the first, the first and second, and for the first, second and third half-lives in order to detect trends in the rate data.

Chemical Analyses

Quantitative analysis of various solutions or complexes for chromium was done spectrophotometrically following oxidation of chromium(III) to chromate by boiling with alkaline peroxide ($a_m = 4.83 \times 10^3$ l mole^{-1} cm^{-1} for CrO_4^{2-} at 372 mμ).[16]

Analysis for bound cyanide was carried out following decomposition of the complexes by titrating the liberated cyanide ion with standard silver nitrate solution. To do this, the solution to be analyzed (pH < 6) was first

Table 1

Absorption Spectral Data for the Cyanoaquochromium(III) Complexes[a]

Complex	λ_{max}, mμ	ν_{max}, cm^{-1}	ε_{max}, cm^{-1} liter mole^{-1}
$Cr(CN)_6^{3-}$	530[b]	18,900	0.4
	377	26,500	85.9
	307	32,600	59.5
$Cr(CN)_5(H_2O)^{2-}$	~420 (sh.)	23,800 (sh.)	80
	382	26,200	106
	311	32,200	50
$cis\text{-}Cr(CN)_4(H_2O)_2^-$	427	23,400	110
	335	29,900	50
$1,2,3\text{-}Cr(CN)_3(H_2O)_3$	687; 620[b]	14,600; 16,100	0.2; 0.3
	~590[b]	~17,000	~0.1
	467	21,400	115
	360	27,800	25.2
$cis\text{-}Cr(CN)_2(H_2O)_4^+$	~680[b]	14,700	~0.2
	~480[b]	20,800	~1
	~535 (sh.)	~18,700 (sh.)	30.2
	469	21,300	45.4
	378	26,500	25.8
$Cr(CN)(H_2O)_5^{2+}$	~670[b]	~14,900	~0.2
	525	19,000	26.0
	393	25,400	20.5
$Cr(H_2O)_6^{3+}$	575	17,400	13.4
	406	24,600	15.4
	256	39,100	8.8

[a]The data for the charge transfer absorption bands are not included in this table.
[b]Weak, spin-forbidden band.

well purged with nitrogen to remove dissolved HCN and then made alkaline by addition of excess sodium hydroxide solution and boiled. This treatment serves to decompose the cyanoaquo complexes (but not $Cr(CN)_6^{3-}$) and to cause precipitation of chromic hydroxide, which can be filtered off. Cyanide ion is determined in the filtrate by titration with silver nitrate solution in the presence of NH_4NO_3 and KI, according to the modified Liebig method.

The ionic strengths of solutions used in the kinetic studies were determined by converting the cations present in the solutions to hydrogen ion by means of a cation-exchange resin in the hydrogen form and then titrating the total acid. The concentration of free hydrogen ion originally present in the solutions was similarly determined by standard titration procedures.

PREPARATION OF CYANOAQUOCHROMIUM(III) COMPLEXES

Neutral and Anionic Complexes*

Preparation of a hydrolysis mixture by addition of less than a 3:1 ratio of perchloric acid to potassium hexacyanochromate(III) (see above) should yield a solution which contains only neutral and anionic cyanoaquochromium(III) complexes, providing that the mixture is stored at 0° or is not allowed to remain at room temperature for more than a few hours. After filtration of the hydrolysis mixture to remove precipitated potassium perchlorate, it is advantageous to pass the mixture through a short cation-exchange column (5 to 10 cm) in the sodium form to remove the remaining potassium ion (which otherwise might precipitate subsequently as potassium perchlorate) and to remove any cationic complexes possibly present in the solution. The effluent solution from the cation-exchange column is then passed through an anion column in the perchlorate cycle to adsorb the anionic complexes. A typical elution curve is shown in Fig. 1.

$Cr(CN)_3(H_2O)_3$

When the effluent solution from the cation-exchange column is loaded onto the anion-exchange column, an orange-colored solution passes directly through the column along with the solvent. This color is presumably due to the presence of an uncharged species which is not retained by either the cation or the anion-exchange resin. Rinsing the column with water and then with 0.05 M sodium perchlorate solution (pH ~ 5) serves to remove completely the uncharged species from the column. The anionic complexes are quantitatively adsorbed by the resin and form a bright yellow band at the top of the column.

In our experiments the orange-colored effluent solution from the anion-exchange column was collected in small fractions and stored in ice. The absorption spectra of the successive fractions were qualitatively identical in that the ratio of the absorbancies of the maxima observed at 467 and 360 $m\mu$ was found to be close to 4:1 in all fractions. Moreover, the molar extinction coefficients calculated for all fractions of the band, including the leading and trailing edges, were found to be constant within experimental error. The spectrum of the uncharged complex (shown in Fig. 3) is independent of pH in the range of 2.0 to 5.0 and has absorption maxima (in $m\mu$) and molar absorbancy indices (in l mole^{-1} cm^{-1}) as follows: 687 (0.2), 620 (0.3), 467 (115), 360 (25.2), 227 (2.1 × 10^3), and 236 (2.1 × 10^3).

*See footnote on p. 178.

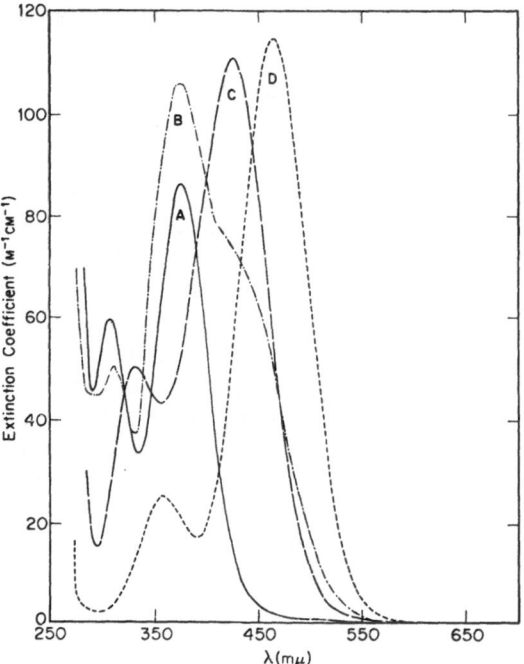

Fig. 3. Absorption spectra of anionic and neutral cyano-aquochromium(III) complexes. A. $Cr(CN)_6^{3-}$; B. $Cr(CN)_5(H_2O)^{2-}$; C. cis-$Cr(CN)_4(H_2O)_2^-$ and D. 1,2,3-$Cr(CN)_3(H_2O)_3$.

Analysis of the orange solutions to determine ligand cyanide and chromium(III) concentrations gave CN:Cr ratios of 3.00 within experimental error. This result, together with the ion exchange behavior, indicates that the uncharged, orange species is the tricyanotriaquo complex $Cr(CN)_3$-$(H_2O)_3$. The sharpness of the bands in the absorption spectrum and its constancy in the various fractions are proof that only one of the two possible geometrical isomers is present. For reasons developed below, this species has been identified as 1,2,3-$Cr(CN)_3(H_2O)_3$.

Relatively concentrated solutions of the tricyanotriaquo complex can be readily prepared from solid $K_3Cr(CN)_6$ by dissolving a weighed amount of the crystalline salt in water and then adding sufficient perchloric acid to give exactly a 3 to 1 ratio of acid to salt. The reaction reaches completion in six to ten hours, at which time the pH of the deep orange solution will have risen to 4 or 5. The solution is cooled in ice, purged for 1 hr with nitrogen gas and then filtered to remove the precipitated potassium perchlorate. The solution at this point contains essentially only the uncharged tricyano complex. If desired, it can be purified by passing it successively through an anion column in the perchlorate cycle, and a cation column in the sodium cycle.

Some attempts were made to investigate the aqueous solution chemistry of this neutral species. Addition of three equivalents of hydroxide ion per mole of $Cr(CN)_3(H_2O)_3$ results in an instantaneous change in color of the solution from orange to deep pink, which then slowly changes to green and finally decomposes forming a grey-green precipitate of chromic hydroxide. If, however, three equivalents of cold perchloric acid are added to the pink solution to neutralize the previously added base before decomposition has occurred, the original orange color is immediately and reversibly restored. The spectrum of this neutralized solution was found to be identical to that of the original complex. These color changes, together with a pH titration of the acidic protons of the water ligands with NaOH, indicate the following conjugate acid-base equilibria

$$[Cr(Cn)_3(H_2O)_3] \xrightleftharpoons{\ pK_1 \simeq 6.8\ } [Cr(CN)_3(H_2O)_2(OH)]^- + H^+$$

$$\xrightleftharpoons{\ pK_2 \simeq 9\ } [Cr(CN)_3(H_2O)(OH)_2]^{-2} + H^+ \xrightleftharpoons{\ pK_3 \simeq 10.5\ } [Cr(CN)_3(OH)_3]^{-3} + H^+$$
$$\text{(pink)}$$

The tricyano complex is extremely soluble in water, but forms insoluble orange polymers when solutions are concentrated by evaporation, by freeze drying techniques, or even by repeated freezing and thawing. The optimum pH range for storing the tricyano complex appears to be 3.5 to 4.5.

$Cr(CN)_4(H_2O)_2^-$

After removing the orange, uncharged species, washing the column with 0.2 to 0.3 M sodium perchlorate solution causes the adsorbed yellow band to move down the column and to divide into two yellow bands. The lower, more-rapidly moving band can be completely removed from the column as a deep-yellow solution before the breakthrough of the second band. (See Fig. 1)

The spectra of successively collected portions of the effluent were found to be qualitatively identical, with well-defined absorption maxima located at 427 and 335 mμ, having absorbancy indices of 110 and 50.0 l mole^{-1} cm^{-1}, respectively. The determination of cyanide-to-chromium ratios in the various fractions gave values between 4.0 and 4.2, consistently. These values are probably within experimental uncertainty of 4.00, considering that small amounts of free cyanide not removed by purging with nitrogen gas might have some tendency to adsorb on the anion-exchange column and would, therefore, be eluted with sodium perchlorate.

The results indicate that a single absorbing species is contained in the effluent sample and that it is the anionic complex, $Cr(CN)_4(H_2O)_2^-$. The spectrum of this complex ion is shown in Fig. 3. It has been identified as the cis-tetracyanodiaquo complex.

Although this complex can be isolated and characterized by the above procedure, it can be seen from the elution curve (Fig. 1) that the concentration of this species is small relative to that of either the neutral complex or $Cr(CN)_6^{-3}$. It can be prepared, however, at fairly high concentrations from available $K_3Cr(CN)_6$ by the following procedure.

A relatively concentrated solution, say 0.1 M, of $Cr(CN)_3(H_2O)_3$ is prepared from solid $K_3Cr(CN)_6$ as described previously. A 10-fold excess of solid potassium cyanide is now added to the solution causing an immediate color change from orange to pink, the color of the conjugate base of tricyanotriaquo complex. As the solution stands, the color changes gradually to deep yellow and the spectrum becomes similar to that of $Cr(CN)_4(OH)_2^{3-}$. After several hours, the solution is cooled in ice and cold dilute perchloric acid is added slowly until the pH is again about 5. The perchloric acid removes the excess potassium cyanide from the solution and greatly decreases the total electrolyte concentration. The molecular hydrogen cyanide formed is removed from the slightly acidic solution by bubbling with nitrogen gas for an hour and the precipitated potassium perchlorate is removed from the cooled solution by filtration. Following a final readjustment of the pH to 5.0 to 5.5, the resulting solution has a spectrum essentially identical to that of tetracyanodiaquo complex eluted from the anion-exchange column.

$Cr(CN)_6^{3-}$

After removal of the tetracyano complex from the column, the concentration of sodium perchlorate is increased to about 1.0 M in order to remove the remaining yellow band from the column. The lemon-yellow species next removed was found to exhibit a spectrum with absorption maxima located at 377 and 307 mμ and identical to all wavelengths to that of the parent ion, $Cr(CN)_6^{-3}$ (Fig. 3). No fractions containing a species having a CN^-/Cr ratio of 5.0 were isolated by this type of preparative procedure. Although the pentacyanomonoaquo complex must be formed in the aquation sequence, it is relatively unstable and is present in such small concentrations that attempts to isolate it by ion-exchange techniques described above are difficult. An alternate method of preparing this complex is given below.

Improved Preparation of $K_3Cr(CN)_6$*

In a typical preparation, a solution of 56 g (0.12 moles) of $[Cr(H_2O)_6]$-$(ClO_4)_3$ in 500 ml of water is added dropwise to a boiling solution of 130 g (2.0 moles) of KCN in 500 ml of water. The rate of addition should be such

*From the thesis to be submitted to the Graduate School of Indiana University in partial fulfillment of the requirements for the Ph.D. degree in Chemistry by W. F. Coleman (1970).

that incipient boiling continues and the solution remains clear and deep red to orange in color. If any precipitate forms, the addition should be halted and the solution allowed to boil until the precipitate disappears. Care should be taken that the solution does not bump, which results in the irreversible formation of a great deal of chromic hydroxide. In general, the addition should be accomplished in as short a time span as possible.

As soon as addition is complete, the reaction mixture is cooled to ice temperature and filtered by suction through a fine sintered-glass filter. The filtrate, which at this point contains mostly the tetracyanodihydroxy complex, is then deaerated with purified nitrogen for 1 hr. Under a nitrogen atmosphere, 2 ml of freshly prepared 0.1 M $Cr(H_2O)_6^{2+}$ solution, obtained directly from a Jones reductor in which 0.1 M $HClO_4$ is the activating acid, is added to the reaction mixture to catalyze the anation reactions and produce the hexacyano complex. Nitrogen is bubbled through the reaction vessel for another half hour. Oxygen is then bubbled through the solution for several minutes to oxidize the chromous ion to chromium(III), which causes the color of the solution to change from red-orange to golden yellow. The $K_3Cr(CN)_6$ salt is thrown down from this solution by the addition of 3 to 5 volumes of methanol. The solid is filtered by suction and dried with anhydrous ethyl ether. This solid is redissolved in a minimum of water and precipitated again by addition of a large excess of methanol. This dissolution and reprecipitation process may be repeated as often as desired with minimal loss of product. Three such treatments have been found to yield product of greater than 99% purity in the form of a fine pale yellow powder. The remaining impurity appears to be principally KCN.

Following this procedure, a yield of greater than 90% based on K_3Cr-$(CN)_6$ is obtained. If a more crystalline product is desired, the solid obtained from the first alcohol precipitation may be recrystallized from warm (45°) water. The yield following this recrystallization is about 50%. Chemical analysis and comparison of visible and u-v spectra have shown that the products obtained by these two methods of purification are of about equal purity (>99%).

Preparation of $Cr(CN)_5(H_2O)^{2-}$

Because of its very fast rate of aquation in acidic solutions, only a minute amount of the pentacyanochromium(III) complex (<1%) exists in the hydrolysis solutions and is difficult to isolate from such solutions. The ion can be isolated in small amounts, however, if the acid-assisted aquation reaction is quenched at an early stage in the reaction and if most of the unreacted $Cr(CN)_6^{3-}$ is removed by precipitation prior to the ion exchange separation. The procedure is as follows.*

*See footnote on p. 190.

A 1.3 g sample of $K_3Cr(CN)_6$ (0.004 moles) is dissolved at 50 ml of 0.1 N perchloric acid at 0°C. The solution is allowed to react for about 5 min, during which time nitrogen is bubbled through the solution to remove hydrogen cyanide. The reaction is quenched by rapidly bringing the pH of the solution up to 6.5 to 7.5 by dropwise addition of 4 M potassium hydroxide solution. The reaction mixture is then filtered to remove the precipitated potassium perchlorate. Three volumes of anhydrous methanol cooled to 0° are added to thrown down the unreacted $K_3Cr(CN)_6$ and the solution filtered. (The pH of the filtrate should be checked at this point to ensure that it remains near 7.)

The reaction mixture is then passed through a previously prepared anion exchange column (perchlorate cycle, 20 to 30 cm in length) maintained at 0°. The column is washed first with water and then with 0.05 M sodium perchlorate to remove the uncharged $Cr(CN)_3(H_2O)_3$ complex. The eluting solution is then changed to 0.1 M sodium perchlorate to remove $Cr(CN)_4$-$(H_2O)_2^-$. After this band is removed, the concentration of the sodium perchlorate is raised to 0.25 M to remove the pentacyano anion. Several pale yellow 10-ml fractions are obtained. In a typical run, the following species were recovered in the effluent solution from the anion exchange column: $Cr(CN)_3(H_2O)_3$, 80%; $Cr(CN)_4(H_2O)_2^-$, 14%; $Cr(CN)_5(H_2O)^{2-}$, ~0.5%; and $Cr(CN)_6^{3-}$, 5%. These relative percentages are based on chromium analyses and assume that most of the unreacted $K_3Cr(CN)_6$ was removed by precipitation prior to the ion exchange separation.

The pentacyano complex has also been prepared in dilute solution by photochemical techniques.* Its spectrum is included in Fig. 3.

Cationic Complexes†

A solution containing only the neutral and cationic cyanoaquo complexes is obtained when more than a 3:1 ratio of perchloric acid is added to a solution of potassium hexacyanochromate(III). The chilled hydrolysis solution, prepared as described previously, can be first passed through a short anion-exchange column in the perchlorate cycle to remove any possible anionic complexes and then loaded onto the cation exchange column in the sodium cycle. Because the separation of the cationic species takes more time than the separation of the anionic complexes, due to the greater number of species present on the column, it is advantageous to use a water-jacketed column and to carry out the separations near 0°C.

As in the case of the anionic complexes, the orange uncharged tricyanotriaquochromium(III) molecule is not adsorbed by either the anion

---

*See footnote on p. 190.

†From the thesis submitted to the Graduate School of Indiana University in partial fulfillment of the requirements for the Ph.D. degree in Chemistry by Diane K. Wakefield, 1967.

or the cation exchange column and is the first species emerging from the cation exchange column. The last traces of this complex can be removed by eluting with 0.05 M sodium perchlorate, after which distinctly colored bands of adsorbed species remain at the top of the column. The lowest of these bands is peach colored, above this is a faint yellow one, next a pink-violet and finally a purple band at the top. A typical elution curve is shown in Fig. 2.

$Cr(CN)_2(H_2O)_4^+$

The peach-colored band containing this species can be eluted from the cation-exchange column with 0.10 M sodium perchlorate solution at pH 2.5. (Analysis of such solutions gave chromium to cyanide ratios of 1 to 2.00 within experimental error.) The visible and ultraviolet absorption spectrum of the dicyano complex is shown in Fig. 4 and is independent of pH in the region from 2.0 to 4.0. The position of the maxima (in mμ) and corresponding

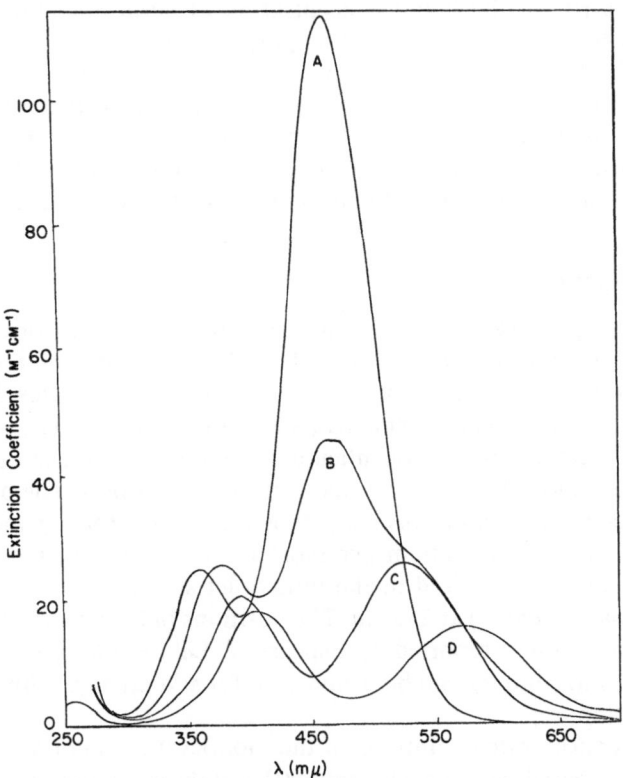

Fig. 4. Absorption spectra of cationic and neutral cyanoaquo-chromium(III) complexes. A. 1,2,3-$Cr(CN)_3(H_2O)_3$; B. *cis*-$Cr(CN)_2(H_2O)_4^+$; C. $Cr(CN)(H_2O)_5^{2+}$ and D. $Cr(H_2O)_6^{3+}$.

molar absorbancy indices (in l mole^{-1}cm^{-1}) in dilute sodium perchlorate are: 535 (30.2, shoulder), 474 (45.3); 464 (45.5), and 378 (25.8), with strong absorption in the ultraviolet. On freeze drying a solution of the dicyano complex a powder is obtained which is readily soluble in water, methanol and ethanol. The optimum storage pH for this complex appears to be about 3. For the hydrolysis procedure described above, about 30% of the product is in the form of the dicyanotetraaquochromium(III) ion.

Unknown Cationic Species

When the eluant concentration was increased in successive steps from 0.1 to 0.4 M sodium perchlorate a small concentration of an unknown species was eluted from the column which contained 0.6% of the total chromium present in hydrolysis mixture prepared as described previously. Separations carried out at various initial concentrations of potassium hexacyanochromate ranging from 10^{-3} to 10^{-1} showed that the amount of this species formed depends greatly on the initial concentration of hexacyanochromate used, which may be taken as evidence of its polymeric nature. The species is not produced in detectable amounts if very dilute solutions and long reaction times are used. The ultraviolet and visible spectral characteristics of the species are intermediate between those of the dicyano and tricyano complexes. Because it is eluted just before the cyanopentaaquo complex, we believe that the species probably has an overall charge of $+2$ and that it is most likely a dimer.

$Cr(CN)(H_2O)_5^{2+}$

This complex was found to constitute 20% of the hydrolysis mixture and can be eluted from the column with 0.6 M sodium perchlorate solution adjusted to pH 2.0. A chromium to cyanide ratio of 1:1.00 was determined by analysis for this pink, permanganate-colored species. The visible and ultraviolet spectrum of cyanopentaaquochromium(III) ion is independent of pH in the regions from 2 to 3.5 and is included in Fig. 4. The positions of the maxima (in mμ) and corresponding molar absorbancy indices (in l mole^{-1}cm^{-1}) in 0.6 M sodium perchlorate are: 525 (26.0), and 393 (20.5). The strong absorption band in the ultraviolet shows a point of inflection ($a = 268$ l mole^{-1}cm^{-1}) at 235 mμ. The positions of the absorption maxima are identical to those reported by Espenson.[8] Dilute solutions (pH $\simeq$ 2) of the monocyano complex can be stored at 0° for several days without serious decomposition.

More concentrated solutions of the monocyano complex can be conveniently prepared via the chromium(II)-catalyzed aquation of the tricyanotriaquochromium(III) complex, as suggested by Birk and Espenson.[17] According to this procedure, a relatively concentrated solution of the tricyanotriaquo complex is first prepared (see above). Dilute perchloric acid is added to the filtered solution so that the molar ratio of acid to complex is 2.0 to 1, or so that the solution is at pH 1.0, which ever is greater. The solution is placed in a flask equipped with a two-hole stopper and carefully purged with nitrogen (using a gas dispersion tube) for an hour to remove all oxygen. A small amount (1 to 2%) of chromium(II) ion, conveniently prepared with the aid of a Jones reductor, is admitted directly into the solution in the flask. With the nitrogen bubbling continuing, the catalyzed aquation is allowed to proceed at room temperature for about an hour, or until the solution attains the color of dilute permanganate. Oxygen or air is then bubbled through the solution for a few minutes to oxidize the chromous ion. At this point the solution contains principally $Cr(CN)(H_2O)_5^{2+}$ with $Cr(H_2O)_6^{3+}$ as an impurity. The monocyano complex can be purified by an ion-exchange separation if desired. Eluting this species with 1.0 M sodium perchlorate produces a more concentrated solution of the complex.

$Cr(H_2O)_6^{3+}$

The hydrated chromium(III) cation formed in the hydrolysis solution is strongly adsorbed on the resin, but can be removed from the column with a solution containing 0.5 M perchloric acid and 2.5 M sodium perchlorate. It is identified in the effluent by its typical three-band ultraviolet and visible absorption spectrum with maxima at 575, 406, and 256 mμ (Fig. 4), and comprises about 11% of the hydrolysis mixture.

Traces of hydrolytic polymers[18,19] are left on the column. The total amount of chromium(III) eluted from the cation-exchange column amounted to more than 98% of the amount loaded onto the column.

SPECTRAL PROPERTIES*

General Characteristics

The spectral absorption curves of the cyanoaquo complexes are shown in Figs. 3 and 4, along with the spectra of the cubic hexacyano and hexaaquo complexes. All of the complexes containing coordinated cyanide ion exhibit an intense charge transfer band, which begins to absorb between 250 and 300 mμ, as well as two other principal spin-allowed d–d bands, the lower energy one of which may show a distinct shoulder or unsymmetrical broadening. The third d–d band, which is clearly visible in the spectrum of $Cr(H_2O)_6^{3+}$ at 256 mμ, is undoubtedly hidden under the charge transfer band in the complexes containing cyanide. The wavelengths and frequencies of the absorption band maxima are collected in Table 1, together with their extinction coefficients.

*See footnote on p. 178.

The absorption edge of the charge transfer band shifts to shorter wavelengths as the number of coordinated cyanide ligands decreases. On the other hand, the absorption maxima of both principal d–d bands shift to longer wavelengths as cyanide ligands are replaced by water molecules, due to the much lower ligand-field strength of water.

The frequencies of the lowest energy spin-allowed bands agree quite closely throughout the series with the frequencies estimated by linear interpolation between the frequencies of the two unsubstituted complexes, $Cr(CN)_6^{3-}$ and $Cr(H_2O)_6^{3+}$. This agreement implies that the frequency of the first band is directly proportional to the average ligand-field strength calculated for a hypothetical average octahedral field.* For example, the first band observed for the tricyanotriaquo complex at 467 mμ (21,400 cm^{-1}) is very close in energy to the mean of the energies of the first bands of the hexacyano complex (26,500 cm^{-1}) and the hexaaquo complex (17,400 cm^{-1}). Furthermore, the frequencies of both spin-allowed bands of the tricyanotriaquo complex are very similar to those of the hexamminechromium(III) complex (21,500 and 28,500 cm^{-1}). This is significant because the ligand-field strength of ammonia ($Dq = 2150$ cm^{-1}) is also equal to the mean of the ligand-field strengths of cyanide ion ($Dq = 2650$ cm^{-1}) and water ($Dq = 1740$ cm^{-1}).

Spectral Assignments (Cubic Complexes)

The spectra of the partially substituted cyanoaquochromium(III) complexes may be most easily approached by reference to the unsubstituted cubic parent complexes, $Cr(CN)_6^{3-}$ and $Cr(H_2O)_6^{3+}$, and the intermediate 1,2,3-tricyanotriaquo complex, which can be considered "pseudo cubic." In the regular octahedral complexes of O_h symmetry, the spin-allowed absorption bands observable in the visible and near ultraviolet regions result from transitions from the ground state ($^4A_{2g}$) to higher energy quartet states, i.e.,

$$v_1 : {}^4A_{2g} \rightarrow {}^4T_{2g}$$

$$v_2 : {}^4A_{2g} \rightarrow {}^4T_{1g}$$

$$v_3 : {}^4A_{2g} \rightarrow {}^4T_{1g}$$

In addition to these quartet bands, at least two of which can usually be observed, one or two very weak transitions may be detected which arise from spin-forbidden transitions to doublet states. (See Fig. 6 at $Dt = 0$.)

*In the case of the dicyanotetraaquo complex, in which the first band exhibits a distinct shoulder, a 2:1 weighted mean is used for the comparison in order to account for the fact that the cubic parentage band is split into two components at 538 mμ (^4E) and one component at 467 mμ ($^4B_2$). The agreement is not very good for the pentacyano complex, which also has a distinct shoulder.

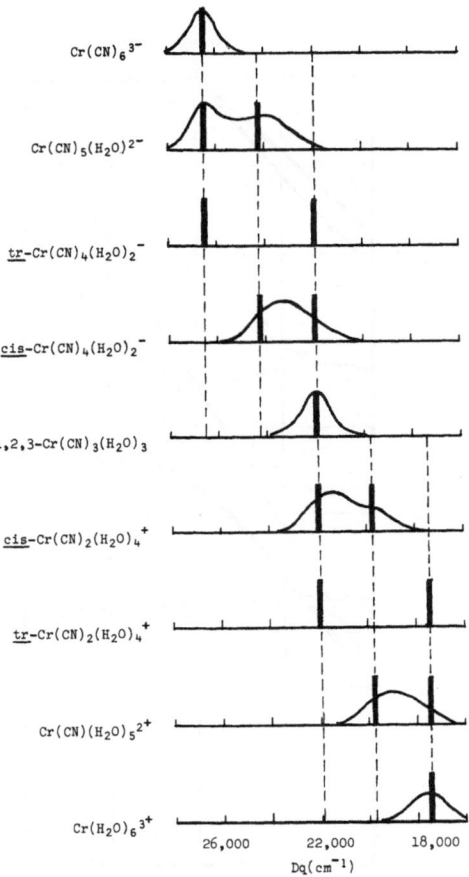

Fig. 5. Comparison of observed frequencies and splittings of the first d–d bands in cyanoaquochromium(III) complexes with values calculated neglecting configuration interaction. ($Dq_{CN} = 2650\,cm^{-1}$; $Dq_{H_2O} = 1740\,cm^{-1}$; splitting = 35/4 Dt, where $Dt = \mp 1/7$ ($Dq_{CN} - Dq_{H_2O}$) for single substitution.)

The assignments of the spectral bands of d^3 complexes with regular octahedral fields have been made in terms of the ligand field parameter Dq and the Racah parameters B and C. For this configuration, the frequency of the lowest energy, spin-allowed band is equal to 10 Dq, while the value of B can be obtained from the frequencies of the first (v_1) and second (v_2) spin-allowed absorption bands according to the following equation:[20]

$$B(cm^{-1}) = (2v_1^2 + v_2^2 - 3v_1 v_2)/(15v_2 - 27v_1)$$

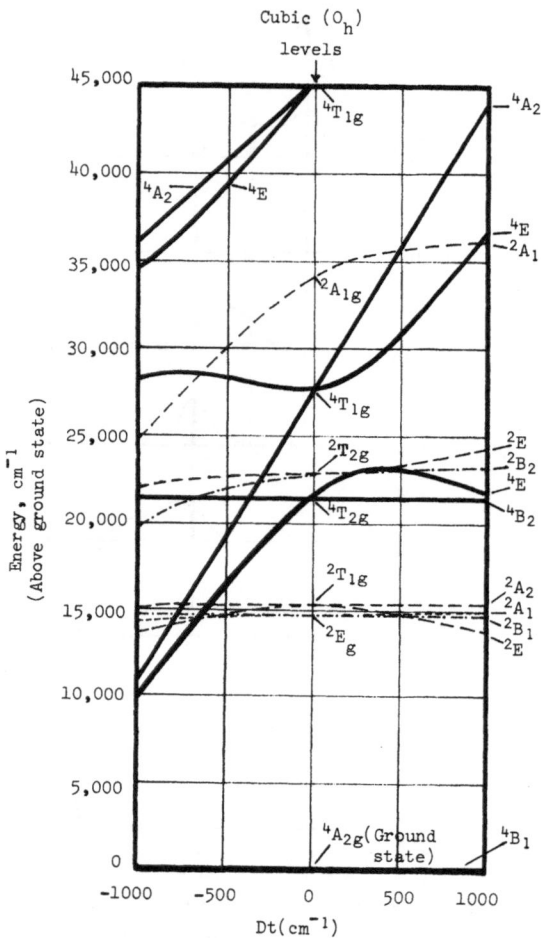

Fig. 6. Energy level diagram for the d^3 configuration showing the cubic levels (at $Dt = 0$) and the splitting of these levels in tetragonal fields as a function of Dt. (Heavy solid lines represent quartet states and the lighter broken lines are doublet states. $Dq = 2140 \text{ cm}^{-1}$; $B = 575 \text{ cm}^{-1}$, $C/B = 6$ and $Ds/Dt = 3$).

The value of C can be obtained from the energy of the weak, low-energy, spin-forbidden bands (^2E_g, $^2T_{1g}$), which are located approximately $(9B + 3C)\text{cm}^{-1}$ above the ground state, or from the $^2T_{2g}$ band which is $15B + 5C$ above the $^4A_{2g}$ ground level.

For the cubic complexes parentally related to the cyanoaquo complexes, the values of the Dq parameter are 1740 cm^{-1} for $Cr(H_2O)_6^{3+}$ and 2650 for

$Cr(CN)_6^{3-}$. The value of the B parameter varies inversely with Dq for these complexes, being $700 \, \text{cm}^{-1}$ for $Cr(H_2O)_6^{3+}$ and $550 \, \text{cm}^{-1}$ for $Cr(CN)_6^{3-}$. The ratio C/B obtained by fitting is 4.0 for the hexaaquo complex and 5.2 for the hexacyano species. Reasonably good values of these parameters for the intermediate cyanoaquo complexes may be obtained by interpolation between the values for the terminal parent complexes.[11,12] Energy level diagrams of the Tanabe–Sugano type have been constructed using parameters appropriate for octahedral chromium(III) complexes.[20]

The absorption spectrum of the 1,2,3-tricyanotriaquo complex is quite similar to that of the cubic hexacyano complex in general appearance. It consists of two relatively narrow bands, the lower energy one of which appears to be symmetrical and without splitting or broadening. The energy of this first band ($21,400 \, \text{cm}^{-1}$) is equal to the mean of the energies of the first bands of the hexacyano complex and the hexaaquo complex, as mentioned above. This implies that the 1,2,3-trisubstituted complex approximates a regular octahedral complex in which each of the six ligands has a ligand-field strength equal to the mean of the ligand-field strengths of cyanide and water, i.e., $(Dq)_{\text{ave}} = (2650 + 1740)/2 = 2195 \, \text{cm}^{-1}$. In justification of this view of the 1,2,3-tricyano complex as pseudo cubic, it may be noticed that the ligands CN^- and H_2O are located opposite each other along each of the Cartesian axes and that the sums of the dipole moments (or ligand fields) along each of the axes are equal. The d orbitals are therefore split in the same $3:2$ pattern that they are in cubic fields.

As a consequence of its pseudo-cubic symmetry, the spectrum of the 1,2,3-tricyano complex may be predicted from Fig. 6 at $Dt = 0$, or from Tanabe–Sugano diagrams for cubic d^3 complexes. In the trigonal (C_{3v}) field of the 1,2,3-tricyano complex, however, the term designations given for the energy levels in cubic symmetry transform as follows: $^4A_{2g} \rightarrow \, ^4A_2$; $^2E_g \rightarrow \, ^2E$; $^2T_{2g} \rightarrow \, ^2A_1, \, ^2E$; $^4T_{2g} \rightarrow \, ^4A_1, \, ^4E$; and $^4T_{1g} \rightarrow \, ^4A_2, \, ^4E$. It is evident from the spectrum of the tricyano complex that the trigonal field components of the cubic parentage bands remain very nearly degenerate.

Spectral Assignments—Substituted Complexes

The substituted 6-coordinate cyanoaquochromium(III) complexes can be treated as slightly distorted derivatives of the parent cubic or pseudo-cubic complexes considered above.[11,12,20] All of the possible geometrical isomers except one (the 1,2,6-tricyanotriaquo complex) can be treated formally by the same approach, i.e., by considering the substituting ligand(s), either one or two, to occupy positions along the Z-axis (four-fold axis of symmetry). The other four ligands are therefore undisturbed and may be considered to be the four ligands lying in the X–Y plane. Thus, the substituted complexes all maintain a four-fold or tetragonal axis of symmetry in this treatment.

In addition to the parameters Dq, B, and C, cited previously to describe a cubic field, the application of a perturbing axial field requires the addition of two more parameters, Ds and Dt. According to the simple crystal field theory, Dt is proportional to the difference in ligand field strengths of the original ligand X in MX_6 and the substituting ligand Y in MX_5Y, i.e.,

$$Dt = -2/7(Dq_X - Dq_Y) \qquad (Dq_Y > Dq_X)$$

For the *trans*-disubstituted complex, MX_4Y_2, the perturbation would be just twice as much as for single substitution, or

$$Dt = -4/7(Dq_X - Dq_Y)$$

The value of the Ds parameter is not so easily deduced and is usually obtained by fitting the observed spectral bands to the theory. For the purposes of the present discussion, however, and in line with a number of recent experimental results, it can be assumed that the value of the ratio Ds/Dt lies in the range 2 to 3.

In tetragonal fields the energies and symmetry designations of the one-electron d orbitals are as follows:

$$d_{z^2}(a_1): 6\,Dq + 2\,Ds + 6\,Dt$$

$$d_{x^2-y^2}(b_2): 6\,Dq - 2\,Ds + Dt$$

$$\left.\begin{array}{l} d_{xz} \\ d_{yz} \end{array}\right\}(e): -4\,Dq + Ds - 4\,Dt$$

$$d_{xy}(b_2): -4\,Dq - 2\,Ds + Dt$$

These expressions can be used to compute the splitting of the cubic e_g and t_{2g} orbitals upon substitution along the Z-axis if factors such as spin-orbit coupling and configuration interaction are neglected. Note that if the parameters Ds and Dt are equal to zero, the original cubic e_g orbitals are separated in energy by $10\,Dq$ units from the t_{2g} orbitals. In tetragonal fields, the $d_{x^2-y^2}$ orbital is also $10\,Dq$ units above the d_{xy} orbital, regardless of the values of Ds and Dt. Transitions between these two states can therefore be easily recognized because they occur at the same energy as the d–d transition of their unsubstituted cubic parent complexes.

In the case of the cyanoaquo complexes, therefore, the d–d spectral transitions of $Cr(CN)(H_2O)_5^{2+}$ and *trans*-$Cr(CN)_2(H_2O)_4^{+}$ can be assigned by comparing their spectra with that of their cubic parent $Cr(H_2O)_6^{3+}$. Similarly, the spectra of $Cr(CN)_5(H_2O)^{2-}$ and trans-$Cr(CN)_4(H_2O)_2^-$ are compared with the unsubstituted ion $Cr(CN)_6^{3-}$. Finally, cis-$Cr(CN)_4(H_2O)_2^-$ is considered to be a singly substituted tetragonal derivative of the pseudo-cubic intermediate complex 1,2,3-$Cr(CN)_3(H_2O)_3$ in which one CN^- has replaced a water molecule; and cis-$Cr(CN)_2(H_2O)_4^+$ is related to $Cr(CN)_3$-$(H_2O)_3$ by replacement of one cyanide ligand by H_2O. Because $Dq_{CN^-} = 2650$ cm^{-1} and $Dq_{H_2O} = 1740\ cm^{-1}$, the value of the parameter Dt is equal to $+260\ cm^{-1}$ for substitution of one CN^- for one H_2O and -260 for the reverse substitution. (In the sign convention used here, Dt is positive for increased axial field and is negative for a decreased axial field.) The value of Dt for the trans-disubstituted complexes will be $\pm 520\ cm^{-1}$.

For tetragonal complexes, one component of the first d–d band ($^4B_2$) will retain the energy (10 Dq) of the parent, while the other (^4E) will be separated from it by 35/4 Dt, if configuration interaction is neglected. For the cyanoaquo complexes listed above, the splitting is predicted to be 35/4 $(260\ cm^{-1}) = 2275\ cm^{-1}$ for single substitution, according to simple crystal field theory, and twice that amount for disubstituted trans complexes. These predictions for the frequencies of the components of the first d–d band are compared in Fig. 5 with the observed frequencies of the band components for the entire series of cyanoaquochromium(III) complexes.

It will be noted from Fig. 5 that the agreement between the predicted and observed frequencies is good only for one half of the complexes, i.e., those with negative values of Dt in which the original ligand of the cubic parent is replaced by another of lower ligand-field strength. This situation obtains because of the existence of configuration interaction. The ^4E tetragonal-field component of the lowest energy cubic level ($^4T_{2g}$) interacts with a ^4E component of the higher energy cubic $^4T_{1g}$ level. This interaction depresses the lower ^4E level and acts to *increase* the splitting if Dt is negative, and to greatly *decrease* the splitting if Dt is positive, as shown in Fig. 6.

Figure 6 is an energy level diagram which gives the energies of the lower energy doublet and quartet term states with respect to the ground state for the d^3 configuration as a function of the tetragonal field parameter Dt. The values of Dq, Ds, and C/B are chosen to be appropriate to the intermediate pseudo cubic complex $Cr(CN)_3(H_2O)_3$, but will apply fairly well over the entire series if changes in the Dq parameter are taken into account. At $Dt = 0$, the levels are those of a cubic field. Positive values of Dt correspond to replacement of water by cyanide ion, and negative values of Dt to the replacement of cyanide by a water molecule. Regardless of the value of Dt, the energy of the $^4B_2$ component of the $^4T_{2g}$ cubic parentage level remains 10 Dq units above the tetragonal $^4B_1$ ground state. The spectra of cyanoaquo complexes are discussed in detail in the reference cited.[11]

KINETIC STUDIES*

Kinetic studies of the aquation reactions of $Cr(CN)_6^{3-}$ and all of the cyanoaquo complexes formed in its complete aquation sequence have been carried out at 25° and in some cases at several temperatures. All of these aquation reactions exhibit acid catalysis and some possess, in addition, a detectable acid-independent pathway. (At higher pH values Birk and Espenson have detected a base-catalyzed aquation reaction for the mono-cyano complex.[17]) Although some of the reactions may be accurately de-scribed with simpler rate laws, the following general rate law applies in acidic solutions (pH < 3) and includes the possible coexistence of the acid in-dependent pathway and pathways involving single and double protonation of cyanide ligands:†

$$-\frac{d(C_t)}{dt} = \frac{(k_0 + k_1 K_1[\text{H}^+] + k_2 K_1 K_2[\text{H}^+]^2)(C_t)}{1 + K_1[\text{H}^+]}$$

In this rate law, C_t is the total concentration of the complex in protonated and non-protonated forms, k_0 is the rate constant for the acid independent path, K_1 and K_2 are the formation constants for the addition of the first and second protons, respectively, and k_1 and k_2 are the corresponding rate constants for the protonated species. The values for these rate parameters are collected in Table 2. Also listed in Table 2 are the values of the activation parameters for those reactions for which they have been evaluated.

In acidic solutions, the three pathways correspond to aquation of unprotonated, singly protonated and doubly protonated species. For example, in the case of the *cis*-dicyano complex these would be as follows:

(1) $Cr(CN)_2(H_2O)_4^+ + H_2O \xrightarrow{k_0} Cr(CN)(H_2O)_5^{2+} + HCN + OH^-$

(2) $Cr(CN)_2(H_2O)_4^+ + H^+ \xrightleftharpoons{K_1} Cr(CN)(CNH)(H_2O)_4^{2+}$

   $Cr(CN)(CNH)(H_2O)_4^{2+} + H_2O \xrightarrow{k_1} Cr(CN)(H_2O)_5^{2+} + HCN$

(3) $Cr(CN)(CNH)(H_2O)_4^{2+} + H^+ \xrightleftharpoons{K_2} Cr(CNH)_2(H_2O)_4^{3+}$

   $Cr(CNH)_2(H_2O)_4^{3+} + H_2O \xrightarrow{k_2} Cr(CNH)(H_2O)_5^{3+} + HCN$

The importance of the aquation of the unprotonated complex (the acid-independent pathway) to the overall aquation process appears to decrease as the number of cyanide ligands increases. This pathway is important only for

*See the footnote on p. 192 and Refs. 6, 17, and 21.

Table 2
Rate and Activation Parameters for the Aquation Reactions of the Cyanoaquochromium(III) Complexes[a]

A. Rate Parameters at 25°C						
Complex	Ionic Strength[b]	$10^5 k_0$, sec^{-1}	$10^3 k_1$, sec^{-1}	K_1, $l\,mole^{-1}$	$10^3 k_1 K_1$, $sec\,M^{-1}$	$10^3 k_2 K_2$, $sec\,M^{-1}$
$Cr(CN)_6^{3-}$	1.0	0.0	4.0	6.0	24	—
$Cr(CN)_5(H_2O)^{2-}$	1.0	—	(500)	(5)	2500	—
cis $Cr(CN)_4(H_2O)_2^-$	1.0	—	8.1	3.3	27	—
1,2,3-$Cr(CN)_3(H_2O)_3$	2.0	<0.3	0.58	1.1	0.66	0.57
cis $Cr(CN)_2(H_2O)_4^+$	2.0	<0.6	1.5	0.55	0.84	0.42
$Cr(CN)(H_2O)_5^{2+}$	2.0	1.1	3.2	0.19	0.59	—

B. Activation Parameters				
Complex	Ionic Strength	Parameter	$\Delta H^{\pm}$ $k\,cals\,mole^{-1}$	$\Delta S^{\pm}$, $cal\,mole^{-1}deg^{-1}$
1,2,3-$Cr(CN)_3(H_2O)_3$	2.0	k_1	16 ± 6	−19 ± 9
		$k_1 K_1$	19.0 ± 0.8	−6.4 ± 1.1
		$k_2 K_2$	22 ± 1	0.4 ± 1.3
cis $Cr(CN)_2(H_2O)_4^+$	2.0	k_1	16 ± 5	−18 ± 7
		$k_1 K_1$	20.0 ± 0.4	−5.5 ± 0.7
		$k_2 K_2$	22 ± 1	1.4 ± 3
$Cr(CN)(H_2O)_5^{2+}$	2.0	k_0	26.9 ± 0.3	8.9 ± 1.1
		k_1	16 ± 2	−16 ± 4
		$k_1 K_1$	20.2 ± 0.1	−5.5 ± 0.3

[a]General form of rate law assumed to be $-\dfrac{d(C_t)}{dt} = \dfrac{(k_0 + k_1 K_1[\mathrm{H}^+] + k_2 K_1 K_2[\mathrm{H}^+]^2)C_t}{1 + K_1[\mathrm{H}^+]}$.

[b]Ionic strength adjusted with sodium perchlorate.

the aquation of monocyanochromium(III) ion[21] and for the cis-dicyano-chromium(III) ion at 35.0°. Because of its comparatively positive entropy of activation, it was concluded that the aquation of monocyanochromium(III) ion via the acid-independent pathway involves an intramolecular proton transfer from a coordinated water molecule to the cyanide ligand yielding hydroxochromium(III) ion and hydrogen cyanide in the transition state. The rate of an aquation involving such a proton transfer mechanism should depend upon the acidity of the coordinated water molecules as well as upon the basicity of the ligand involved. Although the basicity of bound cyanide, as measured by the equilibrium constant for protonation, increases gradually as the number of coordinated cyanide groups increases (see Table 2), the acidity of coordinated water molecules is known to decrease sharply with decreasing positive charge in the case of chromium(III) complexes. Therefore, aquation via a proton transfer mechanism should become less important as the number of coordinated cyanide groups increases, which is in agreement with the observed results.

All the cyanoaquochromium(III) complexes studied exhibit an acid-catalyzed aquation pathway which involves first the protonation of a cyanide ligand, described by the equilibrium constant K_1, and then the aquation of the singly-protonated species, described by the rate constant k_1. In Fig. 7, the logarithm of the product $k_1 K_1$, which is proportional to the observed reaction rates and which is accurately known for each reaction, is plotted as a function of the number of ligand cyanide groups in the complex. Also, included in the figure are plots of $\log K_1$ and $\log k_1$ vs. the number of co-ordinated cyanides.

If one considers the aquation sequence beginning with the pentacyano-chromate(III) anion, the product $k_1 K_1$ is seen to decrease sharply until the neutral tricyano complex is reached, after which it levels off and remains

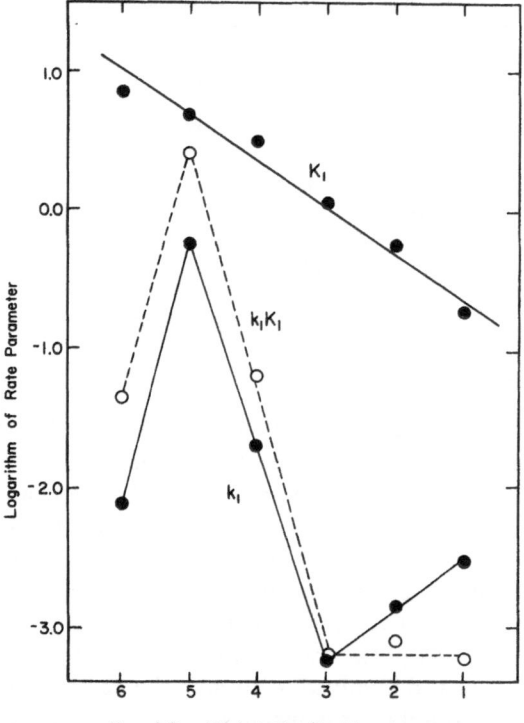

Fig. 7. The dependence of the logarithm of the rate parameters k_1, K_1 and $k_1 K_1$ for aquation of cyano-aquochromium(III) complexes on the number of co-ordinated cyanide ligands.

fairly constant for the remaining cationic complexes. The logarithm of the equilibrium constant $(-pK_1)$, on the other hand, varies nearly linearly throughout the series, decreasing as the number of coordinated cyanide groups decreases and the positive charge increases. Thus, the large decrease in the overall aquation rate, represented by the large decrease in the value of $k_1 K_1$ in going from the pentacyanochromate(III) ion to the neutral tricyanochromate(III) complex is primarily due to a large change in the first-order rate constant (k_1) for the aquation of the protonated species. These large differences in rate constants between the complexes with three or less cyanide ligands compared with those having four or more cyanides is a striking illustration of the *trans*-labilizing effect of cyanide ligands mentioned earlier.

In the aquation sequence, the precursors to the uncharged molecule 1,2,3-$Cr(CN)_3(H_2O)_3$ all have cyanide groups *trans* to another cyanide, whereas, in the tricyano complex and its daughters, the cyanide ligands are always opposite water molecules. The magnitude of the *trans* effect is sufficient to make the reaction essentially stereospecific, so that only *cis* isomers are formed in the aquation sequence.

The aquation of the hexacyanochromate(III) ion appears to be anomalous compared to the rest of the series. Not only is this complex stable indefinitely in neutral and basic aqueous solution (no acid-independent or base-catalyzed reactions), but also the rate of the acid-assisted reaction is much smaller than would be expected for a complex with all of its ligands in positions *trans* to other cyanide groups. We believe that the relative inertness of the hexacyano complex is a result of the absence of any water molecules in the innermost coordination sphere. Water molecules could be involved in the intimate aquation mechanism either by attracting and stabilizing the incoming water molecule through hydrogen bonding, or else by facilitating the transfer of a proton to the leaving cyanide ligand. The involvement of coordinated water molecules in the intimate aquation mechanism might also be invoked to explain the stereospecificity of the reaction, i.e., the incoming water molecules (after the first) always go to positions *cis* to previously coordinated water. This explanation, however, does not account for the drastic change in rate when the aquation product attains the composition 1,2,3-$Cr(CN)_3(H_2O)_3$.

The kinetics of the individual aquation reactions are discussed in more detail in the papers published elsewhere.[17,21]

REFERENCES

1. F. Basolo and R. G. Pearson, *Mechanisms of Inorganic Reactions*, John Wiley and Sons, Inc., New York, 2nd edition, 1967.

2. J. Lewis and R. G. Wilkins (eds.), *Modern Coordination Chemistry*, Wiley–Interscience, New York, 1960, Chapter 2 by D. R. Stranks.

3. R. K. Murman (ed.), *Mechanisms of Inorganic Reactions*, Advances in Chemistry Series, No. 49, American Chemical Society, Washington, D.C., 1965.

4. C. H. Langford and H. B. Gray, *Ligand Substitution Processes*, W. A. Benjamin, Inc., New York, 1965.

5. J. O. Edwards, *Inorganic Reaction Mechanisms*, W. A. Benjamin, Inc., New York, 1964.

6. R. Krishnamurthy and W. B. Schaap, Abstracts, 150th National Meeting of the American Chemical Society, Atlantic City, N.J., Sept. 1965. See also *Chem. and Eng. News*, Sept. 27, 1965, p. 44.

7. For a discussion of isosbestic points in relation to equilibria and kinetics, see H. L. Schläfer and O. Kling, *Angew. Chem.* **68**: 667 (1956); O. Kling and H. L. Schläfer, *Z. Electrochem.* **65**: 142 (1961).

8. J. H. Espenson, *J. Am. Chem. Soc.* **87**: 3280 (1965).

9. E. W. Abel, G. B. Hargreaves and G. Wilkinson, *J. Chem. Soc. (London)*, 3149 (1958).

10. E. W. Abel and G. Wilkinson, *ibid.* 1501 (1959).

11. R. Krishnamurthy, W. B. Schaap, and J. R. Perumareddi, *Inorg. Chem.* **6**: 1338 (1967).

12. J. R. Perumareddi, *J. Phys. Chem.* **71**: 3144, 3155 (1967).

13. P. Moore, F. Basolo and R. G. Pearson, *Inorg. Chem.* **5**: 223 (1966).

14. W. C. Fernelius (ed.), *Inorg. Syn.* **2**: 203 (1946).

15. K. B. Wiberg, *Physical Organic Chemistry*, John Wiley & Sons, Inc., New York, 1964, p. 554.

16. G. W. Haupt, *J. Res. Natl. Bur. Std.* **48**: 414 (1952).

17. J. P. Birk and J. H. Espenson, *J. Am. Chem. Soc.* **90**: 2266 (1968).

18. M. Ardon and R. A. Plane, *J. Am. Chem. Soc.* **81**: 3197 (1959).

19. J. A. Laswick and R. A. Plane, *ibid.* **81**: 3564 (1959).

20. J. R. Perumareddi, *Coordination Chemistry Reviews*, in press, 1969.

21. D. K. Wakefield and W. B. Schaap, *Inorg. Chem.* in press, 1969.

COORDINATION COMPOUNDS CONTAINING FLUOROPHOSPHINE LIGANDS

R. W. Parry

University of Michigan
Ann Arbor, Michigan

DISCUSSION

Interest in fluorophosphines as ligands was initiated when Chatt[1,2] established the formula $Pt(PF_3)_2X_2$ for a compound described years ago by Schutzenberger and Fontaine[3] and later by Moisson.[4] As expected, $Pt(PF_3)_2X_2$ has square planar *cis* geometry. The ability of the PF_3 unit to serve as a coordination ligand put severe strains on existing coordination theory, since electronegative fluorines attached to the phosphorus would be expected to pull back the phosphorus electron pair and seriously reduce its δ-bonding ability. Chatt rationalized the existence of $Pt(PF_3)_2X_2$ by suggesting that d electrons from the platinum were donated back to the d orbitals of the phosphorus to give $d\pi-d\pi$-bonding. The electronegative fluorines attached to the phosphorus would build up the effective positive charge on the central phosphorus atom, and lower the energy of the $3d$ orbitals enough to permit their use in bonding.

One of the predictions of Chatt's theory was that atoms without d electrons should not form complexes with PF_3. In particular, PF_3 complexes with boron and aluminum would not be expected. In preliminary studies Chatt found no complex formation when PF_3 was passed over hot aluminum chloride and no reaction when PF_3 was passed into a solution of aluminum bromide in cyclohexane. Similarly, PF_3 complexes with BF_3 could not be prepared. The foregoing observations are in agreement with Chatt's hypothesis, but subsequent developments demanded modifications of the model.

Bissot,[5] working in the University of Michigan laboratories, succeeded in preparing H_3BPF_3 as a volatile material existing in equilibrium with PF_3 and B_2H_6 at room temperature, even though boron in the borane group has no d electrons for π-bonding. It was suggested by Graham and Stone[6] that hyperconjugation of the electrons in the B—H bonds could provide π-bonds to the phosphorus, thus accounting for the existence of the compound. In support of this postulate, it was noted by Graham and Stone that neither Bissot nor anyone else had been able to prepare F_3PBF_3 where hyperconjugation would not be expected. In fact, the only known boron unit forming a complex with PF_3 was BH_3. Further, no aluminum complexes with PF_3 were known.

In 1959 Alton,[7] working in the University of Michigan Laboratories, examined the reactions between PF_3 and aluminum compounds in some detail. His most significant observations can be summarized as follows. In a sealed tube at 25°C, PF_3 under 8 atm of pressure reacts over a 4-hr period with rigorously dry Al_2Cl_6 to give quantitative yields of PF_3AlCl_3. Over a shorter period of time, the reaction is incomplete. Over longer periods of time, or in the presence of certain catalysts, a halide shift will occur:

$$nF_3PAlCl_3 \rightarrow nPCl_3 + [AlF_3]_n$$

The shift will also occur immediately, at temperatures above −20°C if a large excess of PF_3 is not present. Proof of complex formation was established by: (a) the stoichiometry of the reaction, (b) the displacement of PF_3 by $N(CH_3)_3$ at low temperatures to give $(CH_3)_3NAlCl_3$ and free PF_3, and (c) by the determination of the molecular weight of the complex using the vapor pressure depression technique in liquid PF_3 at −112°C. A molecular weight value of 202 ± 20 was obtained. Theory for F_3PAlCl_3 is 221. Some halogen interchange was observed even at −112°.

Alton[7] also examined the systems $Al_2(CH_3)_6$-PF_3, BF_3-PF_3, Al_2Cl_6-PCl_3, and $Al_2(CH_3)_6$-CO. The data showed unequivocally that there is a reaction between PF_3 and $Al_2(CH_3)_6$, but it is not simple. In view of the stepwise nature of the reaction between Al_2Cl_6 and PF_3, it is not unreasonable to believe that a complex $(CH_3)_3AlPF_3$ might form initially then decompose. No reaction was detectable between BF_3 and PF_3 or between $Al_2(CH_3)_6$ and CO even though extreme experimental conditions were used. Similarly, no reaction between Al_2Cl_6 and PCl_3 could be detected, although some slight solubility of Al_2Cl_6 in liquid PCl_3 was observed.

The existence of F_3PAlCl_3 does not indicate that Chatt's arguments are incorrect. In the formation of $(F_3P)_2PtCl_2$ $d\pi$–$d\pi$-bonding could still be a dominant factor. It does, however, indicate rather clearly that neither $d\pi$–$d\pi$-bonding nor hyperconjugation is necessary to interpret the formation of a complex between F_3P and metals such as aluminum. The ligand F_3P must have some sigma bonding propensity. If indeed this fact is true, how can one rationalize the fact that F_3P will form complexes with H_3B and $AlCl_3$ but not with BF_3 or AlF_3? The arguments that follow are from the thesis of Alton at the University of Michigan. His answer is based on peculiarities in the acid, not the d orbitals of the base PF_3. In order to invoke a minimum number of assumptions, the arguments will be presented, where possible in terms of experimentally determined parameters. Thus the validity of the arguments which follow does not depend upon the specific electronic model used for their rationalization. It does, however, eliminate some electronic arguments.

As a simplifying assumption let us first analyze complex formation of BF_3 and BH_3 by starting in both cases with planar molecules. The slight modification in the analysis required by the dimerization of BH_3 will be considered later. When planar BF_3 is coordinated, the molecule must be forced into a pyramidal shape and the B—F distance must be lengthened. In BF_3 the B—F distance is about 1.30 Å: in BF_4^- it is about 1.43 Å. It is clear that the work done in changing interatomic distances by 0.13 Å and in changing angles must be done at the expense of the energy released in the coordination process. Similar arguments apply to BH_3. High deformation energy in both cases makes coordination more difficult. The magnitude of such energy changes can be estimated in a number of ways. One procedure which gives some measure of the difference between the deformation energies of BH_3 and BF_3 may be gained from an examination of the potential-energy curves for the stretching of B—F and B—H bonds.

Approximate curves can be constructed from a Morse function:

$$U_{(r)} = D[1 - e^{-a(r-r_e)}]^2$$

where U_r = potential energy of system for an internuclear distance r, r_e = equilibrium distance, D = dissociation energy, and a = constant for a given molecule and may be expressed as

$$a = 1/2\sqrt{\frac{2k}{D}}$$

where k = the force constant for the stretching vibration.

The experimental data needed to make estimates of the deformation energies for BF_3 and BH_3, are: (a) the stretching force constants for the B—X bond in BF_3 and BH_3, (b) the equilibrium B—X distances in each of the species, and (c) the dissociation energies of the B—X bond in BF_3 and BH_3. Values for each of these quantities are available or can be calculated from available literature data. Values are shown in Table 1. Using the values shown, potential-energy curves can be constructed which give an order of magnitude estimate of the stretching deformation energies of BH_3 and BF_3. Values obtained by Alton using the data shown are about 30 kcal/mole for BF_3 and about 10 kcal/mole for BH_3. While these values are of limited absolute accuracy and do not include bending distortions, they do provide an experimentally based comparison between BF_3 and BH_3. It is apparent that less energy is required to distort BH_3 than to distort BF_3, hence BH_3

Table 1

Parameters Used in Estimating The Form of The Potential-Energy Curves for B—H and B—F Bonds. Stretching Deformation.

Species	Stretching force constant for B—X, dyne/cm × 10^{-5}	B—X bond energy, kcal/mole	Equil. bond distance, cm × 10^{-3}	Morse function a, cm^{-1} × 10^{-8}
$BH_{3(g)}$	3.21[a]	100	1.20	1.49
$BF_{3(g)}$	7.27[b]	164	1.30	1.88
$BH_{4(g)}^-$	2.77[c]	100	1.25	1.41
$BF_{4(g)}^-$	5.28[d]	164	1.40	1.52

[a] For H_3BCO [Sundaram and Cleveland, *J. Chem. Phys.* **32**: 166 (1960), 3.206 md/Å] and for H_3BPF_3 [R. C. Taylor, unpublished data, 3.224 md/Å].
[b] Wilson, Decius, and Cross, *Molecular Vibrations*, McGraw–Hill Book Co., New York (1955), p. 178.
[c] Emery and Taylor, *J. Chem. Phys.* **28**: 1029 (1958).
[d] Goubeau and Bues, *Z. anorg. allgem. Chem.* **268**: 221 (1952).

should be able to coordinate with weaker bases than BF_3. This fact alone could enable us to rationalize the existence of F_3PBH_3 and our inability to prepare F_3PBF_3 by the same techniques. The observation also would rationalize the fact that BH_3 dimerizes while BF_3 does not. The energy of interaction between BH_3 units would be larger than the deformation energy, but the energy of interaction between BF_3 units would be smaller than the deformation energy. When B_2H_6 reacts instead of BH_3, the energy required to displace a BH_3 unit would be of importance rather than the energy of deformation. For B_2H_6 this value is estimated to lie between 14 and 19 kcal/BH_3 unit. The best estimate is the lower value.

Other factors may be of major importance in considering complexes of PF_3. For example, the most stable coordination compounds of phosphorus are formed with metal ions having a high, positive field strength (high ionization potential). Stability of phosphorus complexes is particularly sensitive to field strength of the positive ion, hence to the internuclear distance between cation and coordinated phosphorus. Phosphorus ligands are more sensitive to this distance than are nitrogen ligands.

In addition, electronegative fluorines attached to the phosphorus atom would cause a build-up in effective positive charge on the phosphorus nucleus. This, in turn, would result in a contraction of the free electron pair of the phosphorus atom. Because of the contraction in the electron pair, a sigma bond could be formed only by Lewis acids which could approach closely. In BH_3 such close approach would be associated with a relatively small increase in deformation energy. On the other hand, in BF_3 close approach would require such a large increase in deformation energy that complex formation with BF_3 could not occur. Because of the strong dependence of phosphorus coordination on internuclear distance, small alterations in distance of approach would have a profound effect upon complex stability.

In 1967 Kuczkowski and Lide[8] reported the results of a microwave study of F_3PBH_3. Their results can be interpreted as supporting this model. The B—P bond distance in F_3PBH_3 is the *shortest yet found* (1.836 ± 0.012 Å); the H—B—H bond angle is 115° ± 1 almost comparable to that found in $(CH_3)_3NBH_3$ (112 ± 2), and the B—H distance is very slightly longer (1.207 ± 0.006 Å) than that found in BH_3CO (1.194 Å) where less BH_3 deformation would be required.

The formation of Cl_3AlPF_3 would be easily understood in terms of the above model. The larger size of aluminum and the ease with which it assumes a coordination number of four, as in Al_2Cl_6, imply a small deformation energy. If deformation energy is small, coordination with PF_3 might be expected. At least in the case of elements of group III, utilization of d orbitals of the P in PF_3 is not required.

In recent years, other fluorophosphines of the general form PF_2X have been prepared. One of the more interesting members of this family is PF_2H which turns out to be an unexpectedly strong Lewis base toward BH_3. Rudolph[9] found that this compound gives a much more stable BH_3 complex than does either PH_3 or PF_3. In terms of the foregoing model, the P—B bond in F_2HPBH_3 should be the shortest P—B bond known, presumably because of the H atom replacing the larger F in PF_3.

Another interesting member of the F_2PX family is F_2PNR_2. This ligand, first studied by Fleming[10] at the University of Michigan, presents some interesting data. The geometry of the free ligand as established by Morris and Nordman[11] is shown in Figure 1. The two CH_3-groups, the nitrogen, and the phosphorus all lie in a plane, suggesting that the free electron pair of the nitrogen is involved in a double bond to phosphorus. Such double bonding should enhance the sigma bonding propensity of the phosphorus atom somewhat. Fleming showed that when allowed to react with F_2PNR_2, BH_3 will coordinate to the phosphorus rather than the nitrogen to give a stable complex $H_3BF_2PNR_2$. On the other hand, when BF_3 reacts with F_2PNR_2, initial coordination is through the nitrogen, presumably because coordination to nitrogen is less distance-dependent than is coordination to phosphorus. In view of this fact, a smaller deformation of the BF_3 group will bring it into coordination range. Perhaps partly as a result of destruction of the N—P double bond in coordination of PF_3 to nitrogen, the coordinated parent ligand, $F_2PN(CH_3)_2$, can undergo facile decomposition to give F_3P and $F_2BN(CH_3)_2$. A minimum of bond rearrangement is required when BF_3 is bonded directly to nitrogen rather than phosphorus. Apparently the same factors which controlled the interaction of PF_3 with BH_3 and BF_3

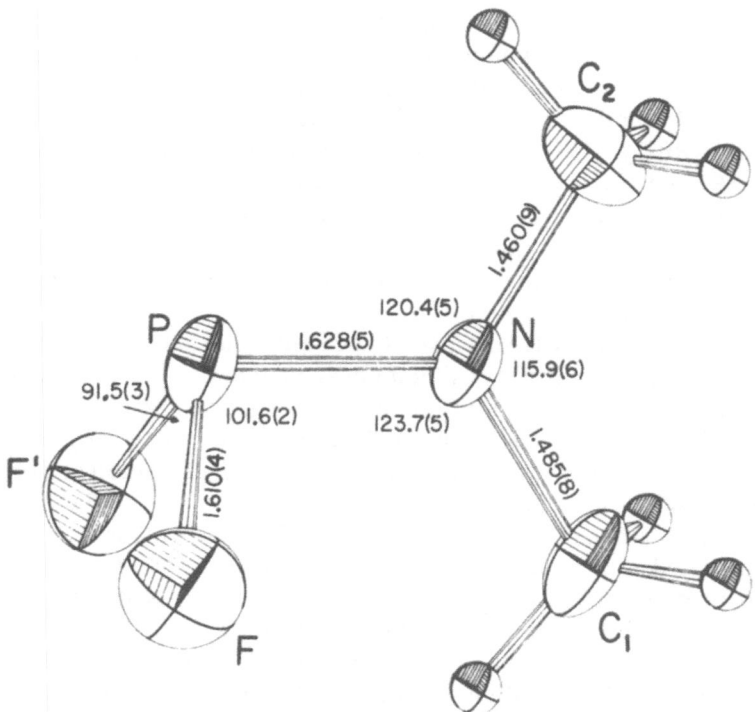

Fig. 1. ORTEP drawing of the $(CH_3)_2NPF_2$ molecule with 50% thermal motion
ellipsoids for the nonhydrogen atoms. Bond lengths, in Å, and angles, in degrees,
are given with standard deviations expressed in units of the last significant figure.
Reproduced from Morris and Nordman[11] with permission of authors and of
Inorganic Chemistry.

are involved in determining the interactions of F_2PNR_2 with these Lewis
acids.

Finally, what happens when the interaction of F_2PNR_2 with other
metal halides is studied? When Al_2Cl_6 is used as the reference acid, arguments
drawn from the case of BH_3 suggest that coordination should be through the
phosphorus. Bond,[12] working in the Michigan Laboratories, found that
$AlCl_3$ reacts with F_2PNR_2 to give a complex set of products including PF_3,
PCl_2NR_2, etc. Kopp[13] established that the initial coordination of F_2PNR_2
to $AlCl_3$ is through the phosphorous and that in the presence of excess
F_2PNR_2 rearrangement of the complex follows to give the products found
by Bond. The intermediate might be a five coordinate aluminum complex.
Appropriate steps in the process involve disproportionation of the complex
to give PF_3:

$$2F_2PN(CH_3)_2 + AlCl_3 \rightarrow AlCl_3 \cdot 2F_2PN(CH_3)_2$$
$$\rightarrow [(CH_3)_2N]_2FPAlCl_3 + PF_3$$

Halogen exchange gives $(CH_3)_2NPCl_2$. An appropriate equation is:

$$2nAlCl_3 \cdot 2F_2PN(CH_3)_2 \rightarrow AlClF_2 + Cl_2PN(CH_3)_2$$
$$\qquad\qquad\qquad \hookrightarrow AlCl_3 + AlF_3.$$

Finally, Cohn[14] examined systematically the reactions of $(CH_3)_2NPF_2$ with a series of halides taken from the first row of transition elements. The metal halides studied were $TiCl_4$, $FeCl_3$, $NiBr_2$, $CoBr_2$, $CuCl_2$, $CuCl$, and $GaCl_3$ as well as NaCl. No reaction was observed with NaCl. Four types of reactions were identified when $(CH_3)_2NPF_2$ was mixed with metal halide systems. These are:

1. The production of PF_3,
2. Halogen interchange indicated by $(CH_3)_2NPCl_2$ production,
3. Formation of a stable complex with $(CH_3)_2NPF_2$,
4. Stabilization of the lower oxidation states of the metal.

Reactions 1 and 2—disproportionation and halogen interchange—were observed with $TiCl_4$ and $AlCl_3$, metals which have small, highly charged cations and which contain no d electrons. In $GaCl_3$ the central metal has d electrons, although somewhat lower in energy than the d electrons of earlier elements in the series. Further the Ga^{+3} metal cation is also small and highly charged (0.62) and has a high field strength. In the presence of excess F_2PNR_2 disproprtionation occurs to give PF_3

$$GaCl_3 + 2Me_2NPF_2 \rightarrow GaCl_3 \cdot 2F_2PN(CH_3)_2 \rightarrow$$

$$GaCl_3[(CH_3)_2N]_2PF + PF_3$$

but the rate is slower than with $AlCl_3$. Halogen exchange to give $GaCl_2F$-$[(Me_2N)PCl]$ occurs only at higher temperatures.

Cobalt II bromide, $CoBr_2$, reacts with F_2PNR_2 to give a complex $CoBr_2 \cdot 2F_2PN(CH_3)_2$ which has a dissociation pressure of 15 mm at 22°C. The complex dissociates to give F_2PNR_2 and $CoBr_2$. No products of disproportionation nor halogen exchange were identified. Similarly, Gilje[15] prepared $NiBr_2 \cdot 2F_2PN(CH_3)_2$. The compound underwent relatively facile dissociation to $NiBr_2$ and free ligand. No evidence of ligand disproportionation or of halogen exchange was found.

With CuCl, which has a d^{10} metal cation, a complex $CuCl \cdot F_2PNR_2$ with a dissociation pressure of about 2 mm at 27° was formed. The complex is a tetramer,[16] $[CuCl \cdot F_2PNR_2]_4$. It is of interest that the Cu^+ and the Ga^{+3} ions are isoelectronic. The prime difference between the two seems to be one of ion charge and size $[Cu^+(0.96), Ga^{+3}(0.64)]$. Increasing charge on the ion promotes PF_3 production and halogen exchange. This is seen by examining $GaCl_3$ complexes with F_2PNR_2 at higher temperatures. Finally, Gilje[15] found that F_2PNR_2 reacts with $Ni(CO)_4$ at room temperature to give $Ni(CO)_2(F_2PNR_2)_2$. The compound is more stable than the $NiBr_2 \cdot 2F_2$-PNR_2 complex. The observation was confirmed by Schmutzler.[17] The corbonyl complex had no significant dissociation pressure since several weeks under high vacuum at 25°C gave no detectable loss of $F_2PN(CH_3)_2$.

OBSERVATIONS ON METAL HALIDES

Some results of the foregoing studies are summarized in Table 2. The following facts stand out. If the charge-size ratio of the central ion is high (high "ionic potential"), the predominant reaction is destruction of the ligand through disproportionation and halogen interchange. As the charge-size ratio falls, (and d electrons become available) the stability of the F_2PNR_2 complex increases. Maximum stability is reached with a carbonyl type of complex formed by metals with d orbitals.

SUMMARY

In a systematic survey of fluorophosphine coordination compounds the following points can be made.

1. Both H_3B and Cl_3Al will form complexes with PF_3, hence PF_3 must have some σ bonding propensity.

2. The difference in coordinating power between BH_3 and BF_3 can be understood in terms of differences in deformation energy. Semiquantitative estimates indicate that the deformation BF_3 requires at least 3 times as much energy as does comparable deformation of BH_3.

3. The unusually high base strength of PF_2H can be interpreted in terms of a reduction in non bonded repulsion and a closer approach between acid and base.

4. Using $F_2PN(CH_3)_2$ as a ligand and metal halides as acids trends in chemistry can be established.

5. Metal ions of small size and high charge which have no d electrons will coordinate to F_2PNR_2, but will promote its decomposition through disproportionation and halogen exchange.

6. So called "zero valent" metals such as one sees in carbonyls will form stable coordination compounds with $F_2PN(CH_3)_2$.

7. The electron-rich configuration of the atoms of the transition elements of group VIII seems to be required for formation of a complex which shows little tendency to rearrange. This may well be attributed to the use of d electrons in bonding. On the other hand, data do not support the conclusion that the fluorophosphine ligands are devoid of sigma bonding propensity.

Table 2

Coordination Compounds of $F_2PN(CH_3)_2$

Compound	Metal Ion	Configuration of Ion	Ion Radius	Charge/Radius	Complex	Observations
$TiCl_4$	Ti^{+4}	d^0	0.64 Å	6.3	Not isolated	Decomposes rapidly; PF_3 found.
$AlCl_3$	Al^{+3}	d^0	0·55 Å	5.5	$AlCl_3 \cdot 2F_2PN(CH_3)_2$	Decomposes rapidly. Disproportionation;·halogen exchange. PF_3 found.
$GaCl_3$	Ga^{+3}	d^{10}	0.62 Å	4.8	$GaCl_3 \cdot 2F_2PN(CH_3)_2$	Decomposes slowly. Disproportionation. PF_3 found.
$NiBr_2$	Ni^{2+}	d^8	0.78 Å	2.6	$NiBr_2 \cdot 2F_2PN(CH_3)_2$	Dissociates to give $NiBr_2$ and ligand. Dissoc. press significant.
$CoBr_2$	Co^{2+}	d^7	0.82 Å	2.4	$CoBr_2 \cdot 2F_2PN(CH_3)_2$	Dissociates to give $CoBr_2$ and ligand. Dissoc. press. = 15 mm.
$CuCl$	Cu^+	d^{10}	0.96 Å	1.04	$CuCl \cdot F_2PN(CH_3)_2$	Dissoc. to give $CuCl$ and F_2PNR_2 Diss. Press = 2 mm.
$Ni(CO)_4$	Ni^0	$d^8 s^2$	1.24 Å	0	$Ni(CO)_2(F_2PNR_2)_2$	No dissoc. at room temp.

8. The charge-size ratio for a number of ions seems to give a fairly good parameter for correlating currently available data on characteristics of the $F_2PN(CH_3)_2$ coordination compounds of the halides of the first transition group.

ACKNOWLEDGEMENT

My dependence upon my co-workers is clearly apparent in the manuscript. The assistance of my co-workers is gratefully acknowledged. Much of the work described herein was supported by the National Cancer Institute under Grant No. R10CA07989.

REFERENCES

1. J. Chatt, *Nature* **165**: 637 (1950).
2. J. Chatt and A. A. Williams, *J. Chem. Soc.* 3061 (1951).
3. P. Shutzenberger and C. Fontaine, *Bull. Soc. Chim., France* **17**: 482 (1891).
4. H. Moisson, *Bull. Soc. Chim., France* **5**: 454 (1891).
5. R. W. Parry and T. C. Bissot, *J. Amer. Chem. Soc.* **78**: 1524 (1956).
6. W. A. G. Graham and F. G. A. Stone, *J. Inorg. Nuclear Chem.* **3**; 164 (1956).
7. Earl R. Alton, *The Reactions of Phosphorus Trifluoride with Aluminum Chloride and Related Studies*, Ph.D. Dissertation, the University of Michigan, 1960.
8. R. L. Kuczkowski and D. R. Lide, Jr., *J. Chem. Phys.* **46**: 357 (1967).
9. R. W. Rudolph and R. W. Parry, *J. Amer. Chem. Soc.* **89**: 1621 (1967).
10. Sister Mary Albert Fleming, *The Coordination Chemistry of Dimethylamido-phosphorus difluoride and Bisdimethylamidophosphorus fluoride*, Ph.D. Dissertation, the University of Michigan, Ann. Arbor, Mich. 1963.
11. Morris, Jr., Earl D. and C. E. Nordman, *Inorganic Chemistry*, 1969 (in press).
12. A. C. Bond and R. W. Parry, unpublished results, 1961.
13. R. W. Kopp, "Complexes of $(CH_3)_2NPCL_2$ with $AlCl_3$", Ph.D. Dissertation, University of Michigan, Ann Arbor, Mich. (1966).
14. K. C. Cohn, "Reactions of Dimethylaminodifluorophosphine with Metal Halides, Ph.D. Dissertation, the University of Michigan, Ann Arbor, Mich., 1966.
15. John Gilje and R. W. Parry, unpublished results, the University of Michigan, 1962.
16. Kim Cohn and R. W. Parry, *Inorg. Chem.* **7**: 46 (1968).
17. R. Schmutzler, *Inorg. Chem.* **3**: 415 (1964).

THE BASE HYDROLYSIS OF *TRANS*-DIBROMOBIS(ETHYLENEDIAMINE)-CHROMIUM(III) CATION

M. S. Nozari and J. A. McLean, Jr.

University of Detroit
Detroit, Michigan

INTRODUCTION

A considerable amount of kinetic and mechanistic data has been reported for the base hydrolysis of *cis*- and *trans*-dihalobis(ethylenediamine)-cobalt(III) cations.[1] On the other hand, the only chromium(III) analogs which have been studied are *cis*- and *trans*-dichlorobis(ethylenediamine)-chromium(III) cations.[2] The present study was undertaken to gain additional base hydrolysis data for dihalobis(ethylenediamine)chromium(III) complexes, which will permit a more meaningful comparison with their cobalt(III) analogs.

EXPERIMENTAL

The synthesis of *trans*-dibromobis(ethylenediamine)chromium(III) bromide is described elsewhere.[3] Carbonate-free sodium hydroxide standard solutions were used in all runs. All other materials were reagent grade. When deliquescent salts were used in ionic strength studies, weighings were performed in a Dry Box under dry nitrogen. The components in the reaction mixture were separated by ion-exchange column chromatography according to the method of Garner and Quinn.[4]

Kinetic Measurements

Temperature was maintained constant within $\pm 0.04°$ by a thermostatically controlled bath. Light was routinely excluded by using actinic glassware.

Titrimetric data were obtained for bromide release and also for the disappearance of OH^-. A typical bromide release run consisted of placing the weighed complex and the standardized NaOH (along with weighed $NaClO_4$ or $LiClO_4$ in ionic-strength studies) in separate flasks in the temperature bath. When these reactants reached thermal equilibrium, they were mixed and vigorously stirred to achieve complete dissolution (about 30 to 900 sec, depending on the concentration and the temperature). The time of mixing was taken as $t = 0$. Samples were withdrawn at appropriate times with a fast flowing, thermostatted, jacketed-pipet, and discharged into a quenching solution consisting of dilute HNO_3, acetone and 5 drops of a non-ionic detergent. The time that the pipet was half empty was recorded as the reaction time t. The released bromide was titrated potentiometrically with standard $AgNO_3$. In some cases, the aliquot was quenched in ice-cold dilute HNO_3, passed through a jacketed cation exchange column, diluted with acetone and titrated as above. Data obtained from both methods were in excellent agreement. The disappearance of OH^- was measured by quenching the aliquot in a known volume of standard ice-cold HCl solution and titrating the quenched solution at $0°$ with standard NaOH using phenolphthalein as an indicator. The second-order rate constant, k_2, for bromide release was determined graphically from the expression

$$k_2 = \frac{2.303}{t(b - a)} \log \frac{a(b - x)}{b(a - x)}$$

where b = initial $[OH^-]$, a = initial [complex], and x = amount of Br^- released at time t. In runs where $a = b$, $1/(a - x)$ was plotted $vs.$ t to determine the second-order rate constant. With one exception, all hydroxide runs and many bromide runs were treated this way where $a - x = [OH^-]_t$ or $[Br^-]_t$. A few bromide release runs were carried out in the presence of H_2O_2 according to the method of Edgington and Pearson.[5] When the ratio of $[OH^-]$:[complex] was 2:1 in the disappearance of OH^- studies, the magnitude of the rate constant for the second step in base hydrolysis could be estimated by the method of Frost and Schwemer.[6]

In spectrophotometric runs, (a was equal to b) data for rate constants were obtained by quenching the aliquots in ice-cold HNO_3 and (a) scanning the spectra of these cold solutions from 640 to 340 mμ, or (b) separating the unreacted original complex by ion exchange and measuring absorbance of the eluted $trans$-dibromo complex. The rate constants in the first method were obtained graphically from the expression; $(A_0 - A_t)/(A_t - A_\infty) = k_2 at$, where A_0 and A_t are the absorbancies initially and at time t, and A_∞ is the absorbance calculated for the bromoaquo product. This method was limited by 600 mμ where cis- and $trans$-bromoaquo have a favorable isosbestic point. In the second method, the disappearance of the original complex was followed directly and the rate constant was obtained graphically from the expression: $1/[\text{complex}] = k_2 t$.

RESULTS AND DISCUSSION

Chan and Tobe[7] used fast-flowing pipets to titrimetrically measure the rates of base hydrolysis of cobalt(III) complexes whose half-lives are as small as 45 sec. Complex concentrations used in the present study are in the range of 9.0×10^{-4} to 5.0×10^{-3} M. The ratio of $[OH^-]:[complex]$ was varied from 1:1 to 20:1 at 0.0° but was held at 1:1 for all other temperatures. Under these conditions, the half-lives were far in excess of the lower limits imposed by Chan and Tobe. Nevertheless, the data obtained at 0° is considered to be the most accurate in that the sampling errors are minimized and contributions from competing acid hydrolysis are negligible at this temperature. All rate constants which are reported involve averages based on 4 to 12 runs. The second order rate constants were calculated from the extrapolated zero-time slopes of rate plots which were linear over two half-lives for titrimetric and spectrophotometric data.

Cobalt(III) complexes undergo base hydrolysis according to the following rate law[1]

$$-\frac{d[\text{complex}]}{dt} = k[\text{complex}][OH^-]$$

No deviations from this rate law have been reported for ethylenediaminecobalt(III) complexes. A recent report by Chan[8] that ion pairing causes halopentamminecobalt(III) complexes to deviate from a first-order dependence on $[OH^-]$ at higher $[OH^-]$ has been reinvestigated by Sargenson et al.[9] who found no deviation.

The Debye–Huckel limiting law (DHLL) predicts that the rate constant will decrease as the ionic strength increases in reactions between ions of different charge according to the following equation

$$\log k_\mu = \log k_0 + 2Az_Az_B(\mu)^{1/2}$$

where k_μ and k_0 are the rate constants at ionic strengths μ and 0, z_A and z_B are the charges, and the constant A has a value of 0.509 at 25°.[10] The effect of ionic strength on the observed bromide release rate constant in the present study is given in Table 1. The equation predicts that plots of k_μ vs. $(\mu)^{1/2}$ should be linear with slopes nearly equal to unity at 25°. When the data in Table 1 were plotted according to the DHLL, linear plots were obtained whose slopes were very close to the theoretical values. These data were also plotted according to the extended Debye–Huckel equation and values were obtained which agreed with the DHLL values within experimental error. The rate constants at $\mu = 0$ were obtained by extrapolation of the DHLL plots and are included in parentheses at the bottom of Table 1. These values were used to calculate Arrhenius parameters for this study. The plot of $\log k$ vs. $1/T$ was linear over the entire temperature range and the activation

Table 1
Ionic Strength Effect on the Second-Order Rate Constant

	k in $M^{-1} sec^{-1}$ at Various Temperatures			
μ	25°	20°	15°	0°
0.2000	—	—	—	0.05
0.1000	—	—	—	0.07
0.0500	8.03	—	1.34	0.08
0.0400	—	3.50	1.39	—
0.0300	—	3.70	1.56	—
0.0200	9.53	3.96	1.64	0.12
0.0100	10.23	4.42	1.81	0.13
0.0050	10.90	—	1.92	0.13
0.0020	—	—	—	0.15
0.0019	11.56	5.44	—	—
0.0018	—	—	—	0.15
0.0000	(12.60 ± 0.2)	(6.10 ± 0.2)	(2.30 ± 0.04)	(0.17 ± 0.01)

energy was found to be 27.5 ± 0.4 kcal/mole and ΔS^* is 36.6 ± 1.3 e.u. The bromide release constants were also checked spectrophotometrically and by the disappearance of OH^-. At 0° and ionic strengths of 0.01, these methods gave rate constants of 0.14 ± 0.01 which were in excellent agreement with the bromide release rate constant in Table 1. The method of Frost and Schwemer[6] was applied to hydroxide disappearance data to estimate the magnitude of the rate constant for release of the second bromide. The results of this method seemed inconclusive and are not in agreement with our preliminary measurements of the base hydrolysis of the *trans*-bromohydroxo cation.

Usually all experimental observations on rates of base hydrolysis can be explained equally well by either of two mechanisms.[1] Using the compound in the present study as a kinetic model, it may react by an S_N2 mechanism where OH^- is the nucleophile or by an S_N1CB mechanism where

$$[Cr(en)_2Br_2]^+ + OH^- \rightleftarrows [Cr(en)(en-H)Br_2] + H_2O \quad \text{(fast)}$$

$$[Cr(en)(en-H)Br_2] \rightarrow [Cr(en)(en-H)Br]^+ + Br^- \quad \text{(slow)}$$

$$[Cr(en)(en-H)Br]^+ + H_2O \rightarrow [Cr(en)_2(OH)Br]^+ \quad \text{(fast)}$$

Pearson and Edgington[5] have proposed a kinetic method to distinguish between these two mechanisms. This method predicts that the addition of excess H_2O_2 to a reaction mixture of substrate and OH^- will lead to an increased reaction rate by a factor which may be as large as 10^4 for an S_N2 mechanism. On the other hand, a reduction in rate by a factor as large as 150 is predicted for 1 M H_2O_2 in the S_N1CB mechanism. If R_2 is the rate with added H_2O_2 and R_1 is the rate without H_2O_2, the ratio of the rates is given by

$$R_1/R_2 = 1 + 150[H_2O_2]$$

Table 2 shows that the relative rates obtained at 15° using this mechanistic test strongly support an S_N1CB mechanism.

Table 2
The Effect of H_2O_2 on the Reaction Rate
at 15° and [Complex] = [OH⁻]

OH⁻, M × 10⁴	H_2O_2, M	R_1/R_2 Obs.	R_1/R_2 Calc.
9.81	0.018	3.4	3.7
9.39	0.0106	2.6	2.6

No reaction theory is complete unless it explains the stereochemistry of the reaction. Data collected by Garner[4] and Tobe et al.[11] have shown that cis-[M(en)₂X₂]⁺ complexes (where M = Co or Cr and X = Cl or Br) aquate with complete retention of configuration. The trans isomers of chromium(III) also give predominantly trans products. No steric course studies have been carried out for the base hydrolysis of chromium(III) complexes, but for cobalt(III) complexes there is extensive rearrangement for both cis and trans isomers. Basolo and Pearson[12] have used an S_N1CB mechanism to explain these rearrangements. These authors point out that the statistical factor for base hydrolysis of trans complexes predicts a mixture of 66.6% cis and 33.4% trans product. However, steric and electrostatic factors will be expected to alter these idealized proportions. In the present study we attempted to follow the steric course of the reaction in two ways. The first method consisted of scanning the spectra of acid-quenched aliquots which had been removed from the reaction mixture at various times. It was assumed that the bromoaquo intermediates were more stable than the hydroxobromo intermediates produced in the reaction mixture. This rationale has been previously utilized in similar base hydrolysis studies of cobalt(III) complexes.[7] The spectra obtained by this method were quite complex and spectral similarities of the possible geometrical isomers precluded the resolution of the components by simultaneous equations. The second method consisted of separating the various components in the quenched reaction mixture by ion exchange chromatography. Unfortunately, the method does not cleanly separate cis- and trans-bromoaquo complexes. Nevertheless, conditions could be chosen which would cleanly separate trans-dibromo, the bromoaquo mixture, and the bisaquo mixture. It was then possible to spectroscopically resolve the bromoaquo mixture in terms of its components at time t. The extent of reaction was determined by titrating aliquots from a second identical sample, under the same conditions and at the same time t. Samples analyzed between 20–40 % reaction indicated that the bromoaquo mixture consisted of greater than 60 % *trans*-bromoaquo product. This information was useful in showing that the *trans*-dibromo complex does undergo steric change in base hydrolysis. The mixture of bromoquo isomers cannot be explained in terms of their interconversion because this process has been shown to be too slow to be significant in this case.[4] Presently, we are investigating the base hydrolysis rates for the *cis*- and *trans*-bromoaquo intermediates. These rate data will be used to calculate the *cis*:*trans* ratio of bromoaquo intermediates formed at $t = 0$ in the base hydrolysis of the *trans*-dibromo complex.

REFERENCES

1. For collected references see F. Basolo and R. G. Pearson, *Mechanisms of Inorganic Reactions*, John Wiley and Sons, Inc., New York, 1967.
2. R. G. Pearson, R. A. Munson and F. Basolo, *J. Am. Chem. Soc.* **80**: 504 (1958).
3. A. M. Weiner and J. A. McLean, Jr., *Inorg. Chem.* **3**: 1469 (1964).
4. L. P. Quinn and C. S. Garner, *Inorg. Chem.* **3**: 1348 (1964).
5. R. G. Pearson and D. N. Edgington, *J. Am. Chem. Soc.* **84**: 4607 (1962).
6. A. A. Frost and W. C. Schwemer, *J. Am. Chem. Soc.* **74**: 1268 (1952).
7. S. C. Chan and M. L. Tobe, *J. Chem. Soc.* 4531 (1962).
8. S. C. Chan, *ibid.* 1124 (1966).
9. D. A. Buckingham, I. I. Olsen, and A. M. Sargeson, *Inorg. Chem.* **7**: 174 (1968).
10. D. R. Stranks, in: J. Lewis and R. G. Wilkens, ed., *Modern Coordination Chemistry*, Interscience Publishers, Inc., New York, 1960, p. 102.
11. S. C. Chan and M. L. Tobe, *J. Chem. Soc.* 5700 (1963).
12. R. G. Pearson and F. Basolo, *Inorg. Chem.* **4**: 1522 (1965).

CALCIUM COMPLEXING BY PHOSPHORUS COMPOUNDS

C. F. Callis, A. F. Kerst, and J. W. Lyons

Monsanto Company,
Inorganic Research Department
St. Louis, Missouri

INTRODUCTION

The ability of the polyphosphates to form water-soluble complexes with calcium has been known for a long time. The phenomenon of binding alkaline earth ions in soluble complexes, and thus preventing the formation of undesirable precipitates, is generally called "sequestration." This was the subject of earlier patent literature.[1] The role of these calcium-polyphosphate complexes in softening of water was described in Chapter 23 of the book edited by Bailar.[2] The complexing of a number of metal ions with the various phosphates was reviewed in detail in 1958.[3]

Upon completion of this compilation of the available literature, it was evident that significant deficiencies existed in the available information. Studies of metal complexes have been continued in the Monsanto laboratories to obtain the expanded knowledge needed to achieve a goal of more efficient complexing agents.

Perhaps the most significant finding reported in a series of articles,[4-13] describing these research results, was that a large positive entropy change was found to accompany the association of calcium with polyphosphates. This positive entropy change was attributed to the release of waters of hydration upon association.[4] The enthalpies of association for calcium complexes of the polyphosphates were found to be small.

The objective of the work described in the present paper was to develop a general picture for the calcium complexes of phosphorus compounds, in which both the molecular weight and structural parameters were varied. This necessitated the examination of compounds recently synthesized in laboratories along with re-examination of previously reported compounds under constant experimental conditions. The pleasing result of this effort has been the development of more efficient complexing agents, as described in the information that follows.

223

EXPERIMENTAL

Chemicals

The syntheses of nitrilotris(methylenephosphonic acid) (NTMP),[14] 1-hydroxyethylidene-1,1-diphosphonic acid (HEDP),[15] bis(phosphonyl-methyl) phosphinic acid (PMPA)[16] and tris(phosphonylmethyl) phosphine oxide (PMPO)[17] have been reported.

Crystalline 1-aminoethylidene-1,1-diphosphonic acid (AEDP) was prepared by reacting PCl_3 with acrylonitrile. Analysis: Calculated for $C_2H_9NP_2O_6$: C, 11.71; H, 4.38; N, 6.84; Found: C, 11.95; H, 4.23; N, 6.45.

Sodium polyphosphonate was prepared by dehydrating disodium methylenediphosphonate at 300° for $2\frac{1}{4}$ hr. Viscosity and titration data indicate this material is a low molecular weight, linear polymer possessing a backbone structure containing $(-\overset{\text{O}}{\underset{\text{O}}{\text{O}}}PCH_2\overset{\text{O}}{\underset{\text{O}}{\text{P}}}-)_n^{-2}$ units with $n > 5$. The ^{31}P NMR exhibited a broad signal -9.0 ppm downfield from H_3PO_4. Strong acid hydrolysis of this polymer yielded methylene diphosphonate.

The polyphosphinic acid was prepared by polymerization of $ClCH_2P-(OC_2H_5)_2$, followed by oxidation and acid hydrolysis. This material exhibits a ^{31}P NMR signal at -29.1 ppm relative to H_3PO_4 and a much smaller signal at -16.5 ppm. These signals correspond to the phosphinate and phosphonate phosphorus atoms, respectively.

Tetramethylammonium bromide (TMA^+Br^-) was reprecipitated twice from methanol-ether. Reagent grade chemicals were used unless otherwise indicated.

All compounds used in this study exhibited appropriate ^{31}P and ^1H NMR spectra. The sample of bis(phosphonylmethyl) phosphinic acid (PMPA) showed a trace of ethyl absorption in the proton NMR. This corresponded to less than 5% incomplete hydrolysis of the pentaethyl ester to the free acid.

Method

A nephelometric method has been used in this work for determination of dissociation constants of polyphosphonates at very low free-calcium ion concentrations $(10^{-6}$ M). This method provides direct determination of the complex-calcium to ligand ratio. This direct determination of the calcium complex model is very important since anticipated models or extrapolation of the calcium complex model determined from other methods can lead to serious error. Model refers to the type of metal complex being detected as indicated by the mole ratio of metal to ligand. For example, the calcium ion activity electrode (Orion Research Inc., Cambridge, Mass.)

detects only a 1:1 calcium complex of nitrilotriacetic acid (NTA). However, nephelometric data in the presence of an oxalate precipitating ion indicate nearly equivalent amounts of a 1:1 and 1:2 calcium-NTA complex are present at the turbidimetric end point.[18] This finding casts some doubt upon the reported[19] extrapolation of data obtained, using an electrode specific for calcium ion, to determine the calcium to ligand ratio under the conditions of a nephelometric method.

The nephelometric method used for determining the complex stabilities has been described previously.[9] These measurements were carried out under nitrogen at pH 11.0. Consideration of the pK_a values for the compounds being investigated indicates only calcium complexes of the unprotonated polyphosphonates need be considered at pH 11. The pH was monitored using a Beckman Zeromatic pH meter. The pH meter was calibrated at pH 10.00 with Sargent standard buffer solutions. Temperature compensation was made as indicated by the manufacturer. All pH adjustments were made with tetramethylammonium hydroxide. Ionic strength was maintained at 0.10 M using tetramethylammonium bromide.

The nephelometric titrations were carried out in 250 ml of a 10^{-3} to 10^{-2} M solution of complexing agent. A specific amount of oxalate ion was added to the solution and the ionic strength adjusted to 0.10 M. Solutions were studied at three different oxalate concentrations over a 1.11×10^{-2} M to 5.55×10^{-2} M concentration range. The solution was allowed to equilibrate at the desired temperature, and the pH adjusted to 11.0. The titrant was 0.10 M $Ca(NO_3)_2$ solution. The complexing agent was measured quantitatively by removing aliquots for titration from a standard stock solution.

Nuclear Magnetic Resonance

The ^{31}P NMR measurements were obtained at 24.288 Mc with a Varian high-resolution spectrometer. The chemical shifts are reported relative to 85% phosphoric acid. A Varian A-60 spectrometer was used to obtain ^1H NMR data. The instrument was calibrated with tetramethylsilane and chloroform. The measurements were made on samples prepared with constant ligand concentrations.

CALCULATIONS

Calcium complex dissociation constants were calculated following a modification of a previously described method.[4] The dissociation constant β of the calcium complex as defined by equation (1) may be calculated using equation (2). Equation (2) was derived assuming the concentration of free calcium ions is negligible at the turbidimetric end point.

$$\beta = \frac{[Ca^{+2}][L^{-p}]}{[CaL^{-p+2}]} \tag{1}$$

$$\beta = \left(\frac{[L^{-p}]_t A}{\theta YZ} - 1\right)\frac{K_{sp}}{[C_2O_4^{-2}]} \tag{2}$$

$$K_{sp} = [Ca^{+2}][C_2O_4^{-2}] \tag{3}$$

In equation (2), $[L^{-p}]_t$ is the total ligand concentration in moles per liter (or in case of polymers the moles of phosphorus per liter), A is the milliliters of solution at the nephelometric end point, θ is the moles of ligand (or in the case of polymers, the moles of phosphorus) per bound calcium ion, Y is the number of ml of Z molar calcium nitrate solution added to the end point, $[C_2O_4^{-2}]$ is the molar concentration of oxalate anion, and K_{sp} is the solubility product of calcium oxalate as given by equation (3). The literature values[4] for K_{sp} at 0.10 M ionic strength are 1.32×10^{-8} at 25° and 4.32×10^{-8} at 50°. A value of 2.08×10^{-8} was used for K_{sp} at 35°. This value was interpolated from a plot of the log of reported values of K_{sp} at 25°, 37° and 50° vs. $1/T$. The number of negative charges of the unbound ligand or monomer unit is given by $-p$.

The value of θ, which is the ratio of ligand to metal in the calcium complex, was determined using equation (4).

$$\theta = \frac{\left(\frac{[L]_{t1}A_1}{Y_1Z_1}\right)[C_2O_4^{-2}]_2 - \left(\frac{[L]_{t2}A_2}{Y_2Z_2}\right)[C_2O_4^{-2}]_1}{[C_2O_4^{-2}]_2 - [C_2O_4^{-2}]_1} \tag{4}$$

This expression is readily derived from equation (2) assuming β is constant and determining Y_1 and Y_2 at two different oxalate concentrations, $[C_2O_4^{-2}]_1$ and $[C_2O_4^{-2}]_2$.

Equation (2) is valid for 1:1 calcium-ligand complexes and can be used in the case of polyanions where the binding sites are assumed to be mutually independent. For nonpolymeric compounds where θ does not equal 1, higher or lower calcium-ligand complexes are indicated. A θ value less than one indicates the presence of complexes with more than one calcium ion complexed per molecule of complexing agent, L^{-p}. In such instances an additional equilibrium expression must be considered as given by equation (6).*

$$CaL^{-p+2} \overset{K_1}{\rightleftharpoons} Ca^{+2} + L^{-p} \tag{5}$$

$$Ca_2L^{-p+4} \overset{K_2}{\rightleftharpoons} CaL^{-p+2} + Ca^{+2} \tag{6}$$

*Waters of hydration have not been indicated in these equations.

The equilibrium expressions are given by equations (7) and (8). The total ligand concentration, $[L^{-p}]_t$, and total calcium ion concentration, $[Ca^{+2}]_t$

$$\beta_{CaL} = K_1 = \frac{[Ca^{+2}][L^{-p}]}{[CaL^{-p+2}]} \tag{7}$$

$$\beta_{Ca_2L} = K_2 = \frac{[Ca^{+2}][CaL^{-p+2}]}{[Ca_2L^{-p+4}]} \tag{8}$$

at the turbidimetric end point are given by equations (9) and (10), respectively.

$$[L^{-p}]_t = [L^{-p}] + [CaL^{-p+2}] + [Ca_2L^{-p+4}] \tag{9}$$

$$[Ca^{+2}]_t = [Ca^{+2}] + [CaL^{-p+2}] + 2[Ca_2L^{-p+4}] \tag{10}$$

At the end point $[Ca^{+2}] \ll [CaL^{-p+2}] + 2[Ca_2L^{-p+4}]$, therefore, equation (10) may be written as

$$[Ca^{+2}]_t = [CaL^{-p+2}] + 2[Ca_2L^{-p+4}] \tag{11}$$

Since θ is defined as the moles of ligand complexed per mole of calcium ion, equation (12) is valid.

$$\theta = \frac{[CaL^{-p+2}] + [Ca_2L^{-p+4}]}{[CaL^{-p+2}] + 2[Ca_2L^{-p+4}]} \tag{12}$$

By defining $X = [Ca_2L^{-p+4}]/[CaL^{-p+2}]$, the total ligand concentration, $[L^{-p}]_t$, and total calcium ion concentration, $[Ca^{+2}]_t$, given by equation (9) and (11) may be expressed as a function of X according to equations (13) and (14).

$$[L^{-p}]_t = [L^{-p}] + [CaL^{-p+2}](1 + X) \tag{13}$$

$$[Ca^{+2}]_t = [CaL^{-p+2}](1 + 2X) \tag{14}$$

These relationships can be rearranged to give quantities for $[L^{-p}]$ and $[CaL^{-p+2}]$ which may be substituted into equation (7) to give equation (15).

$$\beta_{CaL} = \frac{K_{sp}}{[C_2O_4^{-2}]} \left\{ \frac{[L^{-p}]_t(1 + 2X)A}{YZ} - 1 - X \right\} \tag{15}$$

In equation (15), the quantity $[Ca^{+2}]_t$ has been substituted by the experimental value YZ/A and $[Ca^{+2}]$ has been given in terms of K_{sp} according to equation (3). The value for β_{Ca_2L} is obtained by expressing equation (6) in terms of equation (3) and X which is a function of θ.

$$\beta_{Ca_2L} = \frac{K_{sp}}{X[C_2O_4^{-2}]} \tag{16}$$

Table 1
Data Used in Calculating the Dissociation Constants for Several Phosphorus Containing Calcium Complexes[a]

Compound	°C	[Phosphonate] $\times 10^3$	ml 0.10 M Ca^{+2} solution to end point for given concentration of C$_2$O$_4^{-2}$		
			1.11×10^{-2} M C$_2$O$_4^{-2}$	2.22×10^{-2} M C$_2$O$_4^{-2}$	4.44×10^{-2} M C$_2$O$_4^{-2}$
N(CH$_2$PO$_3$H$_2$)$_3$ [NTMP]	25	3.40	8.18	7.26	6.16
	25	1.70	3.98	—	—
	35	3.40	8.28	—	7.19
	50	3.40	8.38	—	—
OP(CH$_2$PO$_3$H$_2$)$_3$ [PMPO]	25	3.05	8.07	7.66	6.32
	25	1.52	4.05	—	—
	35	3.05	8.45	—	8.32
	50	3.05	9.30	—	—
HO$_2$P(CH$_2$PO$_3$H$_2$)$_2$·2H$_2$O [PMPA]	25	3.04	5.97	5.66	4.79
	25	1.52	2.93	—	—
	35	3.04	6.82	—	—
	50	3.04	8.21	—	6.51
NH$_2$ | CH$_3$C(PO$_3$H$_2$)$_2$ [AEDP]	25	3.27	6.57	5.81	4.57
	25	1.64	3.23	—	—
	35	3.27	7.18	—	—
	50	3.27	8.54	—	6.00

Concentration of polymers is given as moles P/liter

		[Phosphorus] $\times 10^3$		(10.33 in presence 5.55×10^{-2})	
				(13.07 in presence 3.33×10^{-2})	
HO $CH_3\overset{\mid}{C}(PO_3H_2)_2$ [HEDP]	25	5.28	12.97		
	50	5.28	18.09		11.72
$ClCH_2P[CH_2P]_nCH_2PO_3H_2$ Polyphosphinic $\quad\ \overset{\mid}{OH}\ \ \overset{\mid}{OH}$ $n = 9$ Acid	25	5.61	3.23	3.08	2.94
	25	11.22	6.54	—	—
	35	5.61	3.28	—	—
	50	5.61	3.80	—	3.46
$(-\overset{\mid}{\underset{\mid}{P}}-CH_2\overset{\mid}{\underset{\mid}{P}}-O-)_n$ Sodium $O^-Na^+O^-Na^+$ $n > 5$ Polyphosphonate	25	6.4	3.92	3.70	3.36
	25	3.2	1.92	—	—
	35	6.4	4.01	—	—
	50	6.4	4.67	—	4.41

*Runs were made with 250 ml of solution at 0.10 M ionic strength adjusted with TMA^+Br^- at pH 11.0.

For calcium-ligand complexes with greater than $1:1$ ratios, θ will be greater than 1 and the following equilibrium expressions must be considered.

$$CaL^{-p+2} \overset{K_1}{\rightleftharpoons} Ca^{+2} + L^{-p}$$

$$CaL_2^{-2p+2} \overset{K_2}{\rightleftharpoons} CaL^{-p+2} + L^{-p+2}$$

The dissociation constants are given by equations (17) and (18).

$$\beta_{CaL} = \frac{[Ca^{+2}][L^{-p}]}{[CaL^{-p+2}]} \tag{17}$$

$$\beta_{CaL_2} = \frac{[CaL^{-p+2}][L^{-p}]}{[CaL_2^{-2p+2}]} \tag{18}$$

By defining $W = [CaL_2^{-2p+2}]/[CaL^{-p+2}]$ and deriving the appropriate material balance expressions assuming $[Ca^{+2}]$ at the end point is much less than the concentration of complexed calcium ion, β_{CaL} and β_{CaL_2} may be calculated using equations (19) and (20). The development of these equations is directly analogous to that used in obtaining equations (15) and (16).

$$\beta_{CaL} = \frac{K_{sp}}{[C_2O_4^{-2}]}\left\{ \frac{[L^{-p}]_t A(1+W)}{YZ} - 1 - 2W \right\} \tag{19}$$

$$\beta_{CaL_2} = \frac{1}{W}\left\{ [L^{-p}]_t - \frac{YZ(1+2W)}{A(1+W)} \right\} \tag{20}$$

RESULTS

Experimental data used in calculating dissociation constants for several calcium-polyphosphonate complexes are given in Table 1. The values of θ determined according to equation (12) and used in the dissociation constant calculations are summarized in Table 2. Values of θ are generally accurate to within ± 0.05 units. The large value for bis(phosphonylmethyl) phosphinic acid (PMPA) (1.15 ± 0.05) probably reflects the effect of the small amount of partially hydrolyzed ester in this sample and not a higher calcium chelate complex. This suggestion is supported by the θ value of 0.85 at 50°, indicating the presence of a Ca_2L complex. A θ value of 1.15, if real, would indicate a CaL_2 complex. That both of these situations should occur over the small temperature change of the experimental determinations is very unlikely, and the data are interpreted in terms of CaL and Ca_2L complexes only.

To illustrate the complexing behavior of phosphorus based complexing agents, the compounds investigated in this work have been divided into structural types. The values of recently determined dissociation constants are presented with additional published data, where such data are available.

Table 2

Moles of Complexing Agent Per Bound Calcium Ion θ at 0.10 M Ionic Strength Adjusted With TMA^+Br^-

Compound	θ	
	25°	50°
$N(CH_2PO_3H_2)_3$	0.94 ± 0.03[a]	0.96
$OP(CH_2PO_3H_2)_3$	0.84 ± 0.02	0.78
$HO_2P(CH_2PO_3H_2)_2 \cdot 2H_2O$	1.15 ± 0.05	0.85
$\underset{\displaystyle\overset{\displaystyle NH_2}{\vert}}{CH_3C(PO_3H_2)_2}$	1.04 ± 0.06	0.81 ± 0.04
$\underset{\displaystyle\overset{\displaystyle OH}{\vert}}{CH_3C(PO_3H_2)_2}$	0.95	0.59
$ClCH_2\overset{\displaystyle O}{\overset{\displaystyle \Vert}{P}}[CH_2\overset{\displaystyle O}{\overset{\displaystyle \Vert}{P}}]_nCH_2PO_3H_2{}^b$ $\;\;\underset{\displaystyle OH}{\vert}\;\;\underset{\displaystyle OH}{\vert}$ $n = 9$	4.20 ± 0.10	3.52
$(-\overset{\displaystyle O}{\overset{\displaystyle \Vert}{\underset{\displaystyle \Vert}{\underset{\displaystyle O}{P}}}}CH_2\overset{\displaystyle O}{\overset{\displaystyle \Vert}{\underset{\displaystyle \Vert}{\underset{\displaystyle O}{P}}}}O^-)_n^{-2}Na_2^{+2b}$ $n > 5$	3.92 ± 0.04	3.38

[a]The values given are an average of duplicate or triplicate determinations.
[b]For polymers, θ is the moles of phosphorus/binding site.

Complexing Agents Containing Two Phosphorus Atoms

Values of dissociation constants for calcium complexes with pyrophosphate,[4] $P_2O_7^{-4}$, imidodiphosphate,[6] $O_3PNHPO_3^{-4}$, and methylene diphosphonate[10] have been reported. These data are presented in Table 3.

Careful examination of the published data on the above complexing agents, and others to be discussed later, leads to some discrepancies. This difference arises when θ values calculated using the published data are compared to the θ values used in obtaining the published dissociation constants. The θ values have been calculated from published data for those compounds, where such data were available, and are presented in Table 4. The value of θ calculated for $O_3PNHPO_3^{-4}$ is significantly greater than the value used in the published work. The calculated value of θ suggests that a mixture of 1:1 and 1:2 complexes of $O_3PNHPO_3^{-4}$ exists at the nephelometric end point. In fact, at 50° the 1:2, CaL_2, complex is predominant. The dissociation constants, β_{CaL} and β_{CaL_2}, given in Table 3 have been recalculated for

Table 3

Dissociation Constants for Complexing Agents Bearing Two Phosphorus Atoms
(Values were obtained at 0.10 M ionic strength adjusted with TMA^+Br^-

Compound	°C	$-\log \beta_{CaL}$	Ref.
$P_2O_7^{-4}$	25.0	5.39	4
	37.0	5.44	4
	50.0	5.39	4
$O_3PNHPO_3^{-4}$	25.0	5.36[a]	6
$O_3PCH_2PO_3^{-4}$	25.0	6.0	10
	50.0	5.9	10

[a]This value has been recalculated from published data correcting for the presence of a 1:2, CaL_2, complex ($-\log \beta_{CaL_2} = 3.98$).

Table 4

Moles of Complexing Agent per Bound Calcium Ion, θ, for Several Phosphorus Compounds at 0.10 M Ionic Strength Adjusted With TMA^+Br^-

Compound	θ[a]		Used in published calculations	Ref.
	25°	50°		
$(O_3PCH_2PO_3)^{-4}$	1.10	—	1.0	10
$(O_3POPO_3)^{-4}$	0.85	1.15	1.0	4
$(O_3PNHPO_3)^{-4}$	1.55	2.60	1.0	6
$\overset{O}{(O_3POPOPO_3)^{-5}}$	0.93	1.00	1.0	4
$\overset{O \quad\quad O}{(O_3PNHPNHPO_3)^{-5}}$	1.10	1.15	1.0	6

[a]The values of θ were calculated from the data reported in the references indicated.

$O_3PNHPO_3^{-4}$ at 25° using a θ value of 1.55 which corresponds to a ratio of 1.22 moles CaL_2 per mole CaL.

Another group of compounds which has been investigated is the substituted methylenediphosphonates. Values of the dissociation constants for the calcium complexes reported in the literature, and some recently determined values, are presented in Table 5.

The substituent effect on the values of β_{ML} has been correlated with Taft σ^* constants for the alkali metal ions[15] (Li^+, Na^+, K^+, and Cs^+) and for the alkaline earth metals[20] (Ca^{+2} and Mg^{+2}). These authors found that 1-hydroxy-1,1-ethylidenediphosphonic acid, HEDP, did not follow this correlation and ascribed this finding to participation of the hydroxyl group

Table 5
Values of Dissociation Constants for Calcium Complexes of Substituted
Methylene Diphosphonates

$$O_3^{-2}P-\underset{\underset{Y}{|}}{\overset{\overset{X}{|}}{C}}-PO_3^{-2}$$

Substituents	°C	Ionic strength	Added electrolyte	$-\log \beta_{CaL}$	$-\log \beta_{CaL_2}$	Method[a]	Ref.
X = Y = −H	25	0.10	TMA⁺Br⁻	6.0	—	N	10
	25	0.50	TMA⁺Cl⁻	4.70 ± 0.15	—	P	20
X = −CH₃,	25	0.50	TMA⁺Cl⁻	5.21 ± 0.07	—	P	20
Y = −H							
X = Y = −CH₃	25	0.50	TMA⁺Cl⁻	6.33 ± 0.12	—	P	20
X = −CH₃,	25	0.50	TMA⁺Cl⁻	5.74 ± 0.10	—	P	20
Y = −OH	25	0.10	TMA⁺Br⁻	7.09	—	N	—
	50	0.10	TMA⁺Br⁻	5.53	6.20	N	—
X = −CH₃,	25	0.10	TMA⁺Br⁻	6.71	—	N	—
Y = −NH₂	50	0.10	TMA⁺Br⁻	6.08	4.26	N	—

[a]N = nephelometry; P = potentiometry.

of HEDP in complex formation. The exceptional calcium complex behavior of HEDP is further indicated when the dissociation constants for the 1-amino analog, AEDP, are considered. Consideration of inductive effects would lead one to predict the amino compound AEDP would form a more stable calcium complex than the hydroxy substituted HEDP. The fact that this is not observed provides additional evidence that some interaction of the −OH group in the HEDP-metal complex must be considered.

Complexing Agents Containing Three Phosphorus Atoms

This class of phosphorus-containing complexing agents may be divided into two groups: (a) those compounds with a linear arrangement of the anionic binding sites, and (b) those with a branched arrangement of these sites.

Values of the dissociation constants for several linear phosphorus containing complexing agents are presented in Table 6. These values decrease in the order expected upon considering the distance between the phosphorus atoms. The distances between the atoms would increase in the order −CH₂− < −NH− < −O−, whereas the electronegativity of these groups increases in the same manner. Present understanding of these complexes does not permit indicating which of these phenomena are most important in determining the calcium complexing ability.

Table 6
Values of Dissociation Constants for Calcium Complexes of Linear Complexing Agents Bearing Three Phosphorus Atoms

$(0.10$ M ionic strength adjusted with $TMA^+Br^-)$

Compound	°C	$-\log \beta_{CaL}$[a]	Ref.
O $(O_3POPOPO_3)^{-5}$ O	25	6.41	4
	37	6.33	4
	50	6.27	4
O $(O_3PNHPNHPO_3)^{-5}$ O	25	6.74	6
	37	6.21	6
	50	6.01	6
O $(O_3PCH_2PCH_2PO_3)^{-5}$ O	25	6.87	—
	50	6.37[b]	—

[a] All constants were determined using nephelometric data.
[b] In addition to the CaL complex, a CaL_2 ($-\log \beta_{CaL_2}$ = 5.03) complex was present at the nephelometric end point at 50°.

Values of dissociation constants for phosphorus containing complexing agents with a branched arrangement of the $-PO_3^{-2}$ moieties are given in Table 7. This series of compounds illustrates an important feature of these phosphonates which differentiates them from their carboxyl analogs. Calcium complex formation by these materials is less sensitive to the bridging substituent. Values of the dissociation constants of the carboxylic acid analogs have been included in Table 7 for comparative purposes. The fact that the phosphonates are less sensitive than the carboxylates to the bridging substituent is not too surprising. The more basic doubly charged $-PO_3^{-2}$ group is so much more important in binding of the metal ions that it overshadows the effect of neighboring substituents. In the case of the singly charged $-CO_2^-$ group neighboring substituents play a more important role in the binding of metal ions.

In comparing the calcium complexing behavior of these polyphosphonates, some interesting points are raised. The most striking feature is the ability of tris(phosphonylmethyl) phosphine oxide (PMPO) to form Ca_2L complexes under conditions of the nephelometric studies where nitrilotris-(methylenephosphonic acid) (NTMP) does not. Perhaps interaction between the nitrogen atom and the $-PO_3^{-2}$ groups of NTMP decreases the number of sites available for complexing more than one calcium ion.

Table 7
Values of Dissociation Constants for Calcium Complexes of $X(CH_2PO_3H_2)_3$ Compounds and Their Carboxylate Analogs

Phosphonate compounds	°C	Ionic strength	Added electrolyte	$-\log \beta_{CaL}$	$-\log \beta_{Ca_2L}$	Method[a]	Ref.
$N(CH_2PO_3H_2)_3$	25	1.00	KNO_3	6.68	—	P	13
(NTMP)	25	0.10	TMA^+Br^-	6.87	—	N	—
	35	0.10	TMA^+Br^-	6.76	—	N	—
	50	0.10	TMA^+Br^-	6.56	—	N	—
$ON(CH_2PO_3H_2)_3$	25	1.00	KNO_3	5.69	—	P	13
(ONTMP)							
$OP(CH_2PO_3H_2)_3$	25	0.10	TMA^+Br^-	6.83	5.70	N	—
(PMPO)	50	0.10	TMA^+Br^-	6.80	5.30	N	—
Carboxylate Compounds							
$N(CH_2CO_2H)_3$	20	0.10	KCl	6.41	—	P	(b)
(NTA)	25	0.10	TMA^+Br^-	6.70	3.40	N	—
$ON(CH_2CO_2H)_3$	25	0.10	KNO_3	2.46	—	P	11
ONTA							

[a]N = nephelometry; P = potentiometry.
[b]This value was reported by G. Schwarzenbach et al., Helv. Chim. Acta. **38**: 1147 (1955).

One of the most surprising results is the contrast in complexing properties and pH titration curves of PMPO and the published[13] results for nitrilotris(methylenephosphonic acid) N-oxide (ONTMP). Whereas the titration curve for PMPO indicates all 6 protons are removed at pH 10.5, published data[21] for NTMP and ONTMP indicate these materials are monoprotonated at pH 10.5. The values of the pK for monoprotonated NTMP and ONTMP are 12.34 $\pm$ 0.14 and 12.05 $\pm$ 0.16, respectively, at 25° in the presence of 1.0 M KNO_3. Figure 1 shows the variation in ^{31}P NMR chemical shifts of PMPO upon neutralization with NaOD. The published data[13] for the effect of neutralization upon the chemical shifts of the ^{31}P NMR signals for ONTMP and NTMP are also given in Fig. 1.

The contrast in the curves for PMPO and ONTMP is interesting. The greatest change in the ^{31}P chemical shift for PMPO occurs upon adding from one to three equivalents of NaOD. In the case of NTMP and ONTMP the greatest change in ^{31}P chemical shift occurs upon addition of the sixth equivalent of NaOD. These results support the hypothesis of some special interaction between the phosphonate groups and the bridging groups in ONTMP and NTMP. The contrast in the properties of PMPO and the published results for ONTMP are unexpected and warrant further study.

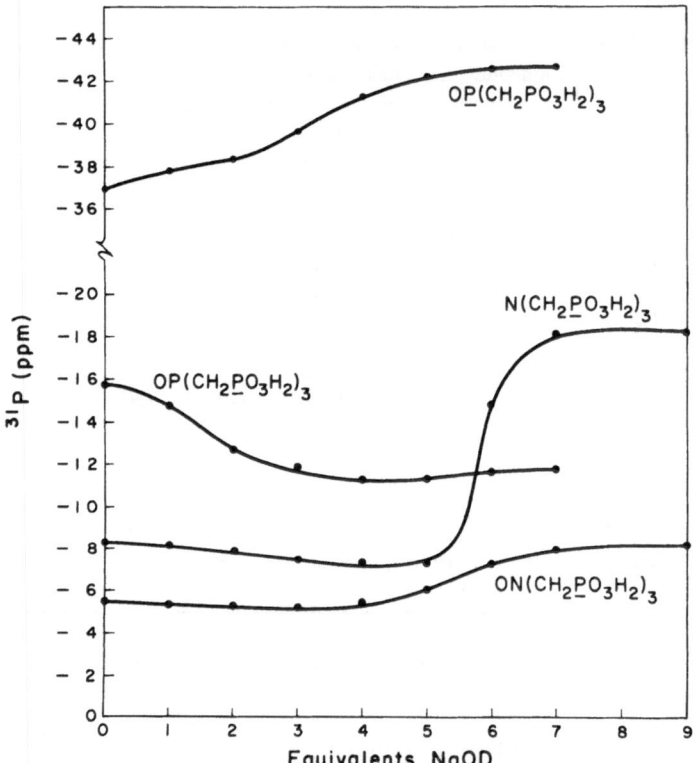

Fig. 1. Variation of ^{31}P chemical shifts with equivalents of NaOD for OP(CH$_2$PO$_3$H$_2$)$_3$ (0.70 M) at 25°. Values for N(CH$_2$PO$_3$H$_2$)$_3$ (0.33 M) and ON(CH$_2$PO$_3$H$_2$)$_3$ (0.10 M) were taken from Ref. 11.

Polymeric Phosphorus–Containing Complexing Agents

The accumulation of a large number of charges in a confined space leads to the interesting metal binding properties exhibited by polyelectrolytes.

In order to better understand the nature of the calcium polyphosphate complexes, polymers containing phosphorus atoms joined by —CH$_2$— groups and those joined by alternating —CH$_2$— and —O— substituents were investigated. The results of this investigation are presented in Table 8 along with the published data for the polyphosphates. In this series of compounds, the calcium complex stability increases as the connecting —O— atoms are replaced by —CH$_2$— groups. This same behavior was exhibited by the low molecular weight compounds previously discussed (Table 6).

Table 8

Calcium Complexes of Polymeric Phosphorus Containing Complexing Agents

(0.10 M ionic strength adjusted with TMA^+Br^-)

Compound	°C	$-\log \beta_{CaL(\theta)}$ [a]	θ [b]
$(-PO PO-)_n^{-2} Na_2^+$ $n = 3$ [c]	25	6.80	—
	37	6.71	4.3
	50	6.76	3.7
$(-PCH_2PO-)_n^{-2} Na_2^+$ $n > 5$	25	7.21	3.9
	50	7.13	3.4
$(-PCH_2PCH_2-)_n^{-2} TMA_2^+$ $n \sim 5$	25	7.38	4.2
	50	6.89	3.5

[a]All results are from nephelometric studies.
[b]In the case of polymers, θ is the moles of phosphorus/bound calcium ion.
[c]This data was taken from Ref. 4.

Thermodynamic Parameters

Although thermodynamic parameters are very helpful in understanding the forces in complex formation, we have been unable to determine these in all cases in the present work. This is because the number of sites per metal ion changes in many of the cases studied in moving from room temperature to 50°.

In the case of NTMP, triphosphate and diimidotriphosphate anions, meaningful thermodynamic data are available. These data are presented in Table 9. The diverse nature of the interaction of calcium ion with ligand is indicated by the variation in the values of ΔH and ΔS. A plot of ΔH vs. ΔS for these values gives a straight line, indicating a constant relationship between these quantities. Such a correlation has been observed in the case of many association reactions.[22] This observation has been discussed in terms of the effect upon ΔH and ΔS of the release and reorientation of water molecules during complex formation. Negative ΔH and ΔS values such as those observed for calcium complexing with the diimidotriphosphate anion suggest association of the calcium ion and ligand occurs with release of very little water from the hydration shells. In contrast, one would conclude more water molecules are released upon forming the calcium complex of the polyphosphates. It is also possible that some contribution to a decrease in ΔS for complexing of diimidotriphosphate arises from orientation of the $-NH-$ bridging groups.

Table 9

Values of ΔF, ΔH, and ΔS for Calcium Complex Formation by NTMP, Triphosphate and Diimidotriphosphate Anions at 0.1 M Ionic Strength

Compound	$\Delta F(25°)$, kcal/mole	ΔH, kcal/mole	ΔS, e.u.
$N(CH_2PO_3H_2)_3$	-9.4	-5.2 ± 0.2	14.1 ± 0.1
$(O_3POPOPO_3)^{-5}$ (with O above and below central P)	-8.8	-2.5 ± 0.2	21.1 ± 0.1
$(O_3PNHPNHPO_3)^{-5}$ (with O above and below central P)	-9.2	-12.8 ± 0.4	-12.3 ± 0.3

SUMMARY

There are consistencies in the data worth re-emphasizing. First, it is clear that chain length in the range studied is the most important variable (see Fig. 2). The stability of the calcium complexes increases in the order

$$\text{dimer} \ll \text{trimer} < \text{higher oligomers}$$

The decrease in the value of the dissociation constant in going from dimer to trimer is about one order of magnitude; from trimer to higher oligomers, about one-half an order of magnitude. The change in ionized sites from dimer to trimer involves the addition of only one charge per molecule. The higher oligomers may add several additional charged sites per molecule but with diminishing effect on the stability constant. In these higher oligomers, only 3.5 to 4 phosphorus atoms are utilized per cation even when more are available.

The effect of different atoms in the connecting link between phosphorus atoms is of somewhat lesser importance. The order for increasing stability of the calcium complex is

$$-O- \ \leq \ -N- \ < \ -C-$$

with about two to four-tenths of an order of magnitude in the value of the dissociation constant separating each pair. Whether the difference arises from specific chemical differences (inductive effects, etc.) or from geometrical effects (more compact molecules, and hence higher charge densities) is a question still unresolved. However, available thermodynamic data suggest that the effect of these connecting groups upon the solvation properties may be very important in determining the calcium complexing behavior of these

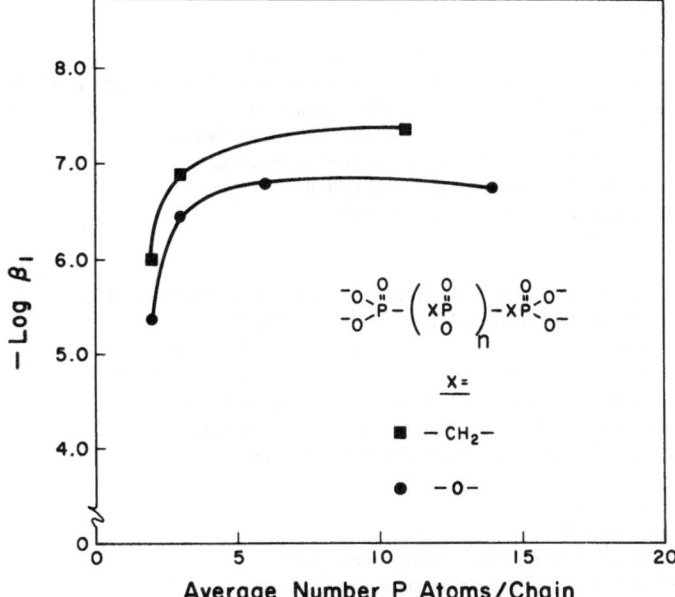

Fig. 2. Plot of −log of the thermodynamic dissociation constant *vs.*
the average number of phosphorus atoms per polymer chain. Values
were obtained at 25° at 0.10 M ionic strength.

compounds. The plot of ΔH *vs.* ΔS for these complexes fall on a straight line.
However, the data for the —NH— linked compounds do not fall between
the —O— and —CH$_2$— linked compounds as one might predict considering
that the relative strengths of the complexes decrease in the order in which the
connecting atoms appear in the periodic table.

Unexpected observations in this work were (a) the differences in the
combining ratios of metal to ligand for the different ligand structures, and (b)
the changes in the calcium complex model with relatively small variations
in experimental conditions. These findings warrant further investigation
with independent methods such as conductometric, polarographic, or
potentiometric techniques that are sensitive to low concentrations of metal
ion.

ACKNOWLEDGMENTS

The authors wish to thank Dr. M. M. Crutchfield for many stimulating
discussions and for providing us with the polyphosphonate sample. We are
also grateful to Dr. L. Maier for supplying us with samples of PMPA,
PMPO, and the polyphosphinic acid.

REFERENCES

1. R. E. Hall, U.S. Patent, 1,956,515 (1934); Reissue 19,719 (1935).
2. J. C. Bailar, Jr., *Chemistry of the Coordination Compounds*, Chapter 23, by R. D. Johnson and C. F. Callis, Reinhold Publishing Corp. New York, 1956.
3. J. R. Van Wazer and C. F. Callis, *Chem. Rev.* **58**: 1011–1046 (1958).
4. R. R. Irani and C. F. Callis, *J. Phys. Chem.* **64**: 1398 (1960).
5. R. R. Irani and C. F. Callis, *ibid.* **64**: 1741 (1960).
6. R. R. Irani and C. F. Callis, *ibid.* **65**: 296 (1961).
7. R. R. Irani and C. F. Callis, *ibid.* **65**: 934 (1961).
8. R. R. Irani, *ibid.* **65**: 1463 (1961).
9. R. R. Irani and C. F. Callis, *J. Am. Chem. Soc.* **39**: No. 3, 156–159 (1962).
10. R. R. Irani and K. Moedritzer, *J. Phys. Chem.* **66**: 1349 (1962).
11. M. M. Crutchfield and R. R. Irani, *J. Am. Chem. Soc.* **87**: 2815 (1965).
12. R. P. Carter, R. L. Carroll, and R. R. Irani, *Inorg. Chem.* **6**: 939 (1967).
13. R. P. Carter, M. M. Crutchfield, and R. R. Irani, *ibid.* **6**: 943 (1967).
14. K. Moedritzer and R. R. Irani, *J. Org. Chem.* **31**: 1603 (1966).
15. R. L. Carroll and R. R. Irani, *Inorg. Chem.* **6**: 1994 (1967).
16. L. Maier, *Angew. Chem.* Int. Ed. **7**: 384 (1968).
17. L. Maier, *ibid.* **7**: 385 (1968).
18. A. F. Kerst, unpublished data.
19. R. W. Cummins, *Detergent Age*, March 1968, p. 22.
20. R. L. Carroll and R. R. Irani (in press).
21. R. R. Irani and R. P. Carter, Jr., International Congress of Industrial Chemists, Brussels, Belgium, S. **26**: 606 (1967).
22. W. J. Hamer, *The Structure of Electrolytic Solutions*, Chapter 25 by J. F. Duncan and D. L. Kepert. John Wiley and Sons, Inc., New York, 1959.

THE POLY(METAL PHOSPHINATES)

B. P. Block

Pennsalt Chemicals Corporation
King of Prussia, Pennsylvania

For about the past ten years the research efforts of my group at Pennsalt have been devoted to that class of coordination polymers containing co-ordinate covalent bonds in the backbone. Bailar had undertaken a program on such materials with emphasis on bis(chelating) groups a few years after I left Illinois. At the time we started thinking about coordination polymers at Pennsalt, then, the bis(chelate) approach was obviously in good hands, so we concentrated on simpler bridging groups. Our study was confined to an examination of one-, two-, and three-atom bridging groups, i.e., bridging groups in which the shortest path from one coordination center to the next contains one, two, and three atoms, respectively. This limitation was made in order to avoid, inasmuch as possible, systems in which competition between bridging and chelation might be a problem; for a group with two donor sites can in principle function in either way.

After the usual false starts, three-atom bridges were eventually selected as the most promising. The phosphinate group $R_2PO_2^-$ appeared to be particularly desirable as a three-atom bridging group for the following reasons. It contains only two donor atoms, both oxygen; it has a charge of -1; and the central atom in the bridge is shielded from attack. These factors are desirable for the design of oxidatively, hydrolytically, thermally stable polymers. The most significant result of the early work with the phosphinate group was the recognition that it did indeed bridge but that cyclization posed a problem.[1] This led to the exploration of systems with two or more bridges per central atom, a study that still continues.

Several interesting polymers have emerged from the program, including the following types:

1. with tetrahedral coordination centers

$$M(OPRR'O)_2 \qquad (2, 3, 4, 5)$$

241

2. with five-coordinate centers

$$TiO(OPRR'O)_2 \qquad\qquad (6)$$

3. with octahedral coordination centers

$$M(a)(b)(OPRR'O)_2 \qquad\qquad (7, 8, 9)$$

$$M(AB)(OPRR'O)_2 \qquad\qquad (10)$$

$$M(OPRR'O)_3 \qquad\qquad (11)$$

$$M(OPRR'O)_4 \qquad\qquad (6)$$

Coordination centers studied include beryllium, lanthanum, cerium, titanium, zirconium, thorium, uranium, chromium, manganese, iron, cobalt, nickel, copper, zinc, cadmium, aluminum, tin, and lead. Many nonmetallic elements have been incorporated into the bridging groups and side groups, so that, in all, some 30 elements have been included in polymers of the phosphinate type. Of most interest are some of the polymers with tetrahedral and octahedral centers.

It is possible to make phosphinate polymers with tetrahedral centers by a number of different reactions. Much of the early work was done with fusion reactions in order to avoid any problems that might arise because of competition between solvent and bridging group for coordination sites. However, this did not prove to be a problem, and the most convenient laboratory-scale reaction is that of a metal acetate with a phosphinic acid in ethanol,[4] i.e.,

$$M(OOCCH_3)_2 + 2RR'P(O)OH$$
$$\xrightarrow{\;C_2H_5OH\;} M(OPRR'O)_2 \downarrow + 2CH_3COOH$$

The resulting polymers are powders which vary from soluble, readily fusible materials to insoluble, infusible materials. The side groups on the phosphorus determine the properties, apparently because they affect interchain attraction and the rigidity of the polymer backbone. Molecular weights as high as 50,000 have been reported for soluble polymers of this type.[12] It is possible to cast films of, and to mold, many of them; however, the fabricated specimens only have tensile strengths of 100 to 200 psi. According to thermogravimetric analysis decomposition temperatures vary from somewhat less than 300° to over 500°. Most stable are the diphenyl- and dimethylphosphinates. The initial volatile product in the decomposition of zinc diphenylphosphinate is benzene.[4] This suggests that the phosphorus-phenyl bond is the first to rupture. Others have made similar observations for beryllium diphenylphosphinate, although other products isolated suggest backbone degradation for beryllium dialkylphosphinates.[12]

The structures of these polymers are not known with any assurance. The reduced specific viscosities are extremely concentration dependent which makes it difficult to estimate intrinsic viscosities. This suggests that there is association at higher concentrations and in the solid which breaks down to some extent in extremely dilute solutions. Such behavior also has been observed for metal soaps.[13] Molecular weights of the phosphinates with tetrahedral centers determined by the usual extrapolation techniques, however, frequently exceed 10,000. Consequently, the phosphinates are not completely analogous to soaps. It appears that they must be linear polymers to explain their solubility and molecular weights. Two structures have been written, the double-bridged structure that one would intuitively expect[3,14] and a structure with alternating single and triple bridges.[15] The former is

presented as speculative, whereas the latter is claimed to be established by

x-ray structure work. The data published to date, however, do not indicate a very high degree of reliability for the structure determination, and the manner in which the single crystals were obtained raises questions about how representative they are. Further evidence is required to establish the structure in the bulk polymer.

Polymers formed by phosphinate bridges between chromium centers, i.e., octahedral centers, have proved most interesting. Much has been done with polymers containing two phosphinate groups per chromium. There are two classes of such polymers, one consisting of those polymers with chelating ligands, the other those with unidentate ligands. Historically, a representative of the former was found first.[10] It was made by the fusion of chromium(III) acetylacetonate with diphenylphosphinic acid:

$$Cr(AcCHAc)_3 + 2(C_6H_5)_2P(O)OH$$

$$\frac{1}{x}\{Cr(AcCHAc)[OP(C_6H_5)_2O]_2\}_x + 2AcCH_2Ac$$

Extraction of the fusion product with a sequence of solvents yielded a spectrum of products among which was an infusible, soluble polymer with a molecular weight greater than 10,000. It can be cast from solution with plasticizers. At about 300° it starts to disproportionate and lose weight. This is the temperature at which chromium(III) acetylacetonate starts to lose weight, so it is likely that the weak spot in the polymer is in the chromium-acetylacetonate linkage.

A double-bridged structure is suggested for the chelated structure, but again there is no proof. The structure of the related dimer $(AcCHAc)_2Cr[OP(C_6H_5)_2O]_2Cr(AcCHAc)_2$ has been determined, and the details of its double bridge are known.[16] Extension to the chelate polymer leads to the following repeat unit:

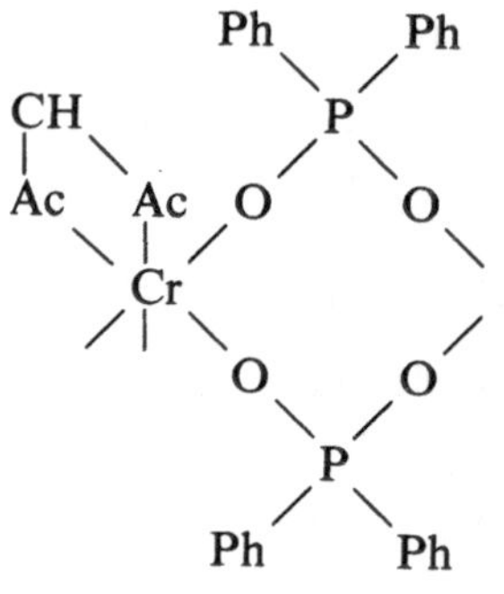

The intrinsic viscosity of this polymer can be determined, for the reduced specific viscosity is not markedly concentration dependent. Its value of over

$0.1 dl/g$ in conjunction with the solubility of the polymer suggests that the polymer is essentially linear. One interesting consequence of the structure indicated is that the backbone of the polymer must be zigzag or spiral. The presence of the chelating ligand, which must occupy cis positions, prevents the two double bridges to any center from lying in the same plane. The same conclusion holds if the backbone contains alternating single and triple bridges rather than double bridges.

The second class of polymers, containing two phosphinates per chromium, is represented by a family of polymers. $[Cr(H_2O)(OH)(OPRR'O)_2]_x$. A typical route to their preparation is[7]:

$$Cr(OOCCH_3)_2 \cdot H_2O + KOP(C_6H_5)_2O \xrightarrow{H_2O} Cr[OP(C_6H_5)_2O]_2 \cdot H_2O$$

$$Cr[OP(C_6H_5)_2O]_2 \cdot H_2O + O_2 \xrightarrow{H_2O} \{Cr(H_2O)(OH)[OP(C_6H_5)_2O]_2\}_x$$

In the case of the diphenyl polymer, the initial oxidation product has a relatively low molecular weight, but polymerization continues if the polymer is incubated in chloroform solution. The details of the oxidation process are very important, insofar as the growth is concerned. If the oxidation is not performed in a relatively homogeneous manner, a coordinatively unsaturated component, which leads to gel formation during the growth step, is formed.[9] In the discussion that follows, references to the diphenyl polymer are to samples made so as to avoid the unsaturated component. The polymers containing phosphinate groups other than diphenylphosphinate do not have as great a tendency to contain significant amounts of coordinatively unsaturated species.

Members of this family are soluble in some solvents and can be cast as films, some with and some without plasticizers. Tensile strengths and per cent elongations vary from 2000 to 500 psi and from less than 5 to 30%, respectively. Like the chelated polymers, however, they are infusible and cannot be molded. The aromatic phosphinates yield polymers decomposing a little under 400°, a surprisingly good stability for materials containing water. The alkyl-phosphinate polymers are less stable as expected.

In general intrinsic viscosities of almost $1 \, dl/g$ have been obtained, and a molecular weight of 180,000 corresponding to a degree of polymerization of about 345 was measured for one fraction of the diphenyl polymer. With these polymers the slope of the reduced specific viscosity-concentration curve is sufficiently small that extrapolation to intrinsic viscosity is straightforward. These solution properties suggest linearity. In contrast to the chelated polymers, geometric isomerism is possible for the polymers with unidentate ligands. If the unidentate ligands are *cis*, the polymer chain will have features similar to those of the chelated polymer, whereas a *trans* relationship would lead to a ribbon-like chain:

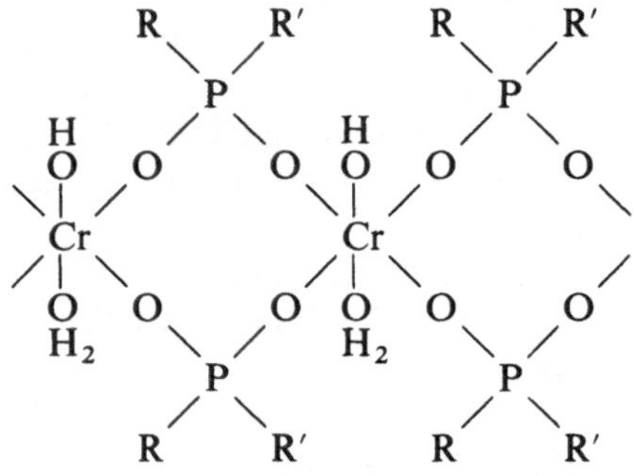

Single-triple bridging is not possible for the *trans* form. It is probable, however, that the configuration is primarily *cis*. The end groups appear to have an extra water and phosphinate coordinated to them, so the polymer most probably grows by elimination of water between end groups, i.e.,

$$-Cr(H_2O)(OH)(H_2O)(OPRR'O) + (H_2O)(OPRR'O)Cr(H_2O)(OH)-$$

$$\longrightarrow \quad -Cr(H_2O)(OH)(OPRR'O)_2Cr(H_2O)(OH)- \; + 2H_2O$$

The statistical chance for the formation of a *cis* as opposed to a *trans* configuration if there is random distribution of the four groups at each end is $5:1$. Interestingly enough, a *cis* double-bridged configuration has been established for $[Mn(CH_3COOC_2H_5)_2(OPCl_2O)_2]_x$ by x-ray work.[17] Since this system has no steric constraints on it, the *cis* configuration found for it supports in a sense a *cis* configuration for the chromium(III) bis(phosphinates) with unidentate ligands.

Recent work with systems containing three phosphinates per chromium has given surprising results.[11] Tris(phosphinates) ought to be crosslinked, intractable materials, and indeed tris(phosphinates) of chromium(III) prepared from chromium(III) salts and alkali metal phosphinates by metathesis appear to have such properties. However, films of tris(phosphinates) can be made by the treatment of solutions of $[Cr(H_2O)(OH)(OPRR'O)_2]_x$ with a phosphinic acid in stoichiometric amount. Once the film is made very little can be done with it, so it is apparently formed by reaction casting. These polymers are all insoluble and infusible. Tensile strengths and per cent elongations of the films vary from 8000 to 500 psi and from less than 1 to 120%, respectively.

The structure of these tris(phosphinates) is puzzling. Success in forming films appears to depend on starting with linear polymers. The physical properties do not clearly point to any simple structure formed by linking chains in a regular manner.

Work is now in progress on practical exploitation of various of these poly(metal phosphinates).

ACKNOWLEDGMENT

The work at Pennsalt was supported in part by the Office of Naval Research.

REFERENCES

1. B. P. Block, E. S. Roth, C. W. Schaumann, and L. R. Ocone, *Inorg. Chem.* **1**: 860 (1962).
2. B. P. Block, S. H. Rose, C. W. Schaumann, E. S. Roth, and J. Simkin, *J. Am. Chem. Soc.* **84**: 3200 (1962).
3. S. H. Rose and B. P. Block, *ibid.* **87**: 2076 (1965).
4. S. H. Rose and B. P. Block, *J. Polym. Sci.*, **Part A-1 4**: 573 (1966).
5. S. H. Rose and B. P. Block, *ibid.* 583.
6. G. H. Dahl and B. P. Block, *Inorg. Chem.* **6**: 1439 (1967).
7. A. J. Saraceno and B. P. Block, *J. Am. Chem. Soc.* **85**: 2018 (1963).
8. A. J. Saraceno and B. P. Block, *Inorg. Chem.* **3**: 1699 (1964).
9. K. D. Maguire and B. P. Block, *J. Polym. Sci.*, **Part A-1 6**: 1397 (1968).
10. B. P. Block, J. Simkin, and L. R. Ocone, *J. Am. Chem. Soc.* **84**: 1749 (1962).
11. A. J. Saraceno, J. P. King, and B. P. Block, *J. Polym. Sci.*, **Part B 6**: 15 (1968).
12. P. J. Slota, Jr., L. P. Freeman, and N. R. Fetter, *ibid.*, **Part A-1 6**: 1975 (1968).
13. A. E. Leger, R. L. Haines, C. E. Hubley, J. C. Hyde, and H. Sheffer, *Can. J. Chem.* **35**: 799 (1957).
14. G. E. Coates and D. S. Golightly, *J. Chem. Soc.* 2523 (1962).
15. F. Giordano, L. Randaccio, and A. Ripamonti, *Chem. Commun.* 19 (1967).
16. C. E. Wilkes and R. A. Jacobson, *Inorg. Chem.* **4**: 99 (1965).
17. J. Danielsen and S. E. Rasmussen, *Acta Chem. Scand.* **17**: 1971 (1963).

RECENT STUDIES ON COMPLEXES OF OXOVANADIUM(IV)

J. Selbin

Louisiana State University
Baton Rouge, Louisiana

INTRODUCTION

Our original interest in complexes of transition metal oxocations was aroused by the idea that a central *molecule-ion* would provide special means for studying coordination compounds, beyond those generally available for examining complexes of central *atom-ions*. This presumes that the oxocation metal–oxygen bond is terminal (i.e., not bridging) and of such strength (bond order higher than 1.5) that it will persist from one complex to another. There are over 50 such transition metal oxocations,[1,2] but only a few have received extensive investigation; these are UO_2^{2+}, VO^{2+}, MoO^{3+}, and ReO^{3+}. Somewhat less thoroughly studied, but nevertheless common species, are OsO_2^{2+}, ReO_2^+, NbO^{3+}, MoO_2^{2+}, and WO^{3+}. Our own oxocation research has dealt very extensively with VO^{2+} complexes and to a much lesser extent with UO_2^{2+}, UO_2^+, MoO^{3+}, and CrO^{3+}. Our efforts to isolate compounds of such often-reported species as UO^{2+}, ZrO^{2+}, and TiO^{2+} have met with only very limited success and we have concluded from both our own and the work of others that these oxocations will only exist under very unusual circumstances. Therefore, they do not meet our interest criterion of persistence from complex to complex. The purpose of this paper will be to report and summarize some of our most recent findings concerning complexes of VO^{2+}, a unique and probably the most stable and persistent diatomic molecule-ion.

The discussion of our recent work on complexes of VO^{2+} will be subdivided into the following sections: (a) New complexes with phosphorus and sulfur donor ligands; (b) polynuclear complexes and attempts to obtain VO^{2+}–Cu^{2+} bridged polymers; (c) equatorial π bonding and hydrogen bonding to the vanadyl oxygen; and (d) electronic spectral studies.

NEW COMPLEXES WITH PHOSPHORUS AND SULFUR DONOR LIGANDS

It is well known that VO^{2+} behaves as a *class a* acceptor or a *hard acid* ion. The many hundreds of complexes of this species with oxygen and nitrogen donor atom ligands, the much greater stability of the fluoro complexes relative to the chloro and bromo ones, and the almost total absence of any complexes with other donor atoms from the nitrogen and oxygen families attest to this behavior.[3,4] Although we made no effort to obtain any of the recently prepared strong multidentate ligands of P, As, and S, we, nevertheless, succeeded in the preparation and characterization of several complexes with phosphorus and sulfur donor ligands,[5–7] and we obtained numerous impure products with arsenic ligands and certain other phosphorus ligands.[5] These products, like so many other unusual or anomalous compounds once believed beyond isolation, are the result of simply picking the right combination of reactants or right reaction medium or right reaction conditions. Of course, it must be admitted that all of the complexes exhibit a larger degree of thermodynamic instability than the oxygen and nitrogen donor complexes.

The only phosphine complexes[6] to be obtained in pure form are the green products $VOCl_2 \cdot L \cdot H_2O$, $VOBr_2 \cdot L \cdot H_2O$, and $VOBr_2 \cdot M \cdot H_2O$ [where L = ethylene *bis*-(diphenylphosphine), $(C_6H_5)_2PCH_2CH_2P(C_6H_5)_2$ and M = methylene*bis*-(diphenylphosphine), $(C_6H_5)_2PCH_2P(C_6H_5)_2$]. The infrared frequencies assigned to $v(V=O)$ in these compounds are 1003, 992, and 1000 cm^{-1}, respectively; and particular attention was given to establishing the absence of $P=O$ frequencies in these compounds. This was aided by the intentional preparation of VO^{2+} complexes of the phosphine oxides of the foregoing ligands. Nothing unusual was noted about the electronic spectral bands of the phosphine complexes in the *d–d* or ligand field region, there being two normally placed maxima at ~ 13 kK and ~ 14.5 kK. This emphasizes the primary importance of the oxo-oxygen in setting up the order and splitting of the *d* orbitals. A lower energy third band than usually found (actually a shoulder at ~ 22 kK) is almost surely charge transfer (ligand-to-metal) in origin, as this band is normally found at ~ 25 kK or higher when there are more electronegative ligand atoms. Green complexes of the analogous bidentate arsenic ligands, as well as of tri-*n*-butylphosphine, triphenylphosphine, tri-*n*-octylphosphine, and triphenylarsine were obtained as impure products with both $VOCl_2 \cdot xH_2O$ and $VOBr_2 \cdot xH_2O$, suggesting that other starting materials, such as $VO(DMSO)_5Br_2$ or $VO(DMSO)_5$-$(ClO_4)_2$, may eventually yield pure products with these and other phosphorus and arsenic ligands not investigated by us.

We have also prepared several complexes with sulfur donor ligands,[5–7] including the following (of which 1–4 have not been previously reported):

1. $VO(mptt)_2 \cdot H_2O$

mptt = 5-mercapto-3-phenyl-1,3,4-thiadiazole-2-thione

2. $K_2VO(mb)_2 \cdot H_2O$ mb = o-mercaptobenzoate

3. $VO(Hmb)_2$
4. $VO(dmt)_2 \cdot H_2O$ dmt = 2,3-dimercaptotoluene

5. $VO(Pipdtc)_2 \cdot H_2O$	Pipdtc = piperidinedithiocarbamate $C_5H_{10}NCS_2^-$	
6. $VO(Et_2dtc)_2$	Et_2dtc = diethyldithiocarbamate $(C_2H_5)_2NCS_2^-$	
7. $VO(EtPhdtc)_2$	EtPhdtc = ethylphenyldithiocarbamate $(C_2H_5)(C_6H_5)NCS_2^-$	
8. $VO(Dicyhxdtc)_2 \cdot H_2O$	Dicyhxdtc = dicyclohexyldithio-carbamate $(C_6H_{10})_2NCS_2^-$	
9. $VO(Mecyhxdtc)_2 \cdot H_2O$	Mecyhxdtc = methyl(cyclohexyl)-dithiocarbamate $(CH_3)(C_6H_{10})NCS_2^-$	

Compounds 5–9 have been throughly investigated by IR, visible UV, and EPR spectroscopy,[7] in order to learn more about the electronic structure of the dithiocarbamate molecules while at the same time shedding more light

on the elusive complexes of VO^{2+} with sulfur-donor ligands. Similar work has recently been published by McCormick[8] on a series of four *dtc* complexes, the dimethyl, diethyl, diisopropyl, and pyrrolidine analogues, of which only the diethyl compound overlaps our series of five compounds. Our work is in essential agreement with his with regard to (a) the V=O stretching frequencies and their interpretation, (b) the $C \doteq N$ stretching frequencies, which imply that the most important canonical structure is

$$M \overset{\displaystyle S}{\underset{\displaystyle S}{\lessgtr}} C \doteq N \overset{\displaystyle R}{\underset{\displaystyle R'}{\lessgtr}}$$

and (c) the general stability of the complexes and the stability order of the ligands in regard to the electron-donating ability of the R groups. However, we are not in agreement with McCormick's interpretation of the electronic spectral data or his interpretation of what occurs when his *dtc* complexes are dissolved in the coordinating solvents pyridine and dimethylsulfoxide (DMSO). With regard to the former disagreement, we do not believe that the two visible bands, observed at ~ 17 and $\sim 18.5\,kK$, can be unambiguously assigned as the first two d–d transitions. One of these may be a charge transfer band, an assignment which we can justify by use of the recent Vanquickenborne and McGlynn (VM)[9] MO energy scheme for $VO(H_2O)_5^{2+}$, modified appropriately for the weaker σ-bonding sulfur ligands. Furthermore, the three bands which we find in the 20–30 kK region for our *dtc* complexes are most probably ligand-to-metal charge transfer bands (from filled four-centered π orbitals on the carbon, nitrogen and two sulfur atoms of the *dtc* molecule to the empty π antibonding orbital on the VO^{2+}). The band reported by McCormick in the 20–25 kK region and assigned by him to a d–d transition ($b_2 \rightarrow a_1^*$) is almost surely the first of these three charge transfer bands. Intraligand bands occur beyond 35 kK.

McCormick has interpreted his very different optical spectra obtained in pyridine and DMSO solvents as being due simply to the addition of a solvent molecule to the available sixth position in his *bis-dtc* complexes. However, we have obtained EPR spectra of the *dtc* complexes in DMSO which clearly point to at least dissociation, and possibly even redox decomposition, in these strong solvents. In less potent solvents, such as toluene, the complexes apparently remain intact.

POLYNUCLEAR VANADIUM COMPLEXES AND ATTEMPTS TO OBTAIN VO^{2+}–Cu^{2+} BRIDGED POLYMERS

We recently have become interested in obtaining heterodinuclear complexes containing the d^1 VO^{2+} ion bridged to the d^9 Cu^{2+} ion. All of the available evidence indicates that redox will not occur, but the electron exchange interactions should prove very interesting. Initial experiments involving $Cu(aca)_2H_2O$ (aca = acetylacetonate ion) led us to a study of simply the exchange interaction found between copper ions, when bromide ions were added to a CH_2Cl_2 solution of $Cu(aca)_2$.[10] Bromide-bridged dimers were clearly being formed and these seemed to be preferred to any possible sought after species such as $[VO(aca)_2\text{-Br-}Cu(aca)_2]^-$.

Our next efforts to bridge the two *aca* complexes using chloride ion in CH_2Cl_2 solution again failed to produce the desired heteronuclear species, but led instead to a new mixed valence compound of vanadium.[11] The deep-violet crystalline compound has the stoichiometry $[(C_2H_5)_4N]_2[V_5O_{14}H_2]$, with the mean formal positive charge on the vanadium being 4.8. Independent titrations by $KMnO_4$ and by $FeSO_4$ showed that the ratio of V^v to V^{iv} is indeed 4:1, certifying the mean oxidation state of 4.8. The compound was further characterized by its (a) infrared spectrum ($V{=}O$ stretches at 975, 925, and 900 cm^{-1}); (b) conductance in CH_3CN (2:1 electrolyte indicated); (c) differential thermal analysis (one very broad endothermic peak beginning at 300°); (d) thermogravimetric analysis (weight loss of 37% between 300 and 400°, which corresponds to a loss of $[(C_2H_5)_4N]_2O$); (e) electronic spectra in CH_3CN solution and in Nujol mull at 77° (bands at 12.1, 20.0, 30.8, 41.1, and 46.5 kK); (f) magnetic susceptibility (μ_{297}^{corr} mysteriously high at 1.9 B.M.); and (g) electron spin resonance spectrum (showing 15 lines, and thereby suggesting two interacting V^{iv} ions in equivalent sites). The latter led us to write the formula as $[(C_2H_5)_4N]_4[V_{10}O_{28}H_4]$. The compound may be prepared at room temperature from $VO(aca)_2$ (or other β-diketone complexes) in CH_2Cl_2 (or other similar inert solvents), containing $(C_2H_5)_4NCl$ (or certain other tetraalkylammonium salts), and $Cu(aca)_2$ (or $Zn(aca)_2$ or $Co(aca)_2$ or $Ni(aca)_2$ or other β-diketone complexes of Cu^{2+}). It is significant, but not at all understood, that each of the three components, vanadyl complex, tetralkylammonium salt, and some other complex, must be present in the inert solvent in order for the compound to form.

Finally, our most recent efforts to produce vanadyl–copper dimers have also failed in their main purpose but have yielded instead some new polynuclear chloro complexes of Cu^{2+}.[12] These species are $Cu_3Cl_{11}^{5-}$ [with $(CH_3)_4N^+$ and $(C_2H_5)_4N^+$] and $CuCl_3^-$ [with $(C_2H_5)_4N^+$ and $(n\text{-}C_4H_9)_4N^+$]. The new compounds have been characterized by IR, EPR, and electronic spectra as well as by their magnetic susceptibility and conductance data. The peculiar thing about these products is that the presence of certain vanadyl complexes appears to be necessary for their formation, although the role played by the vanadyl compounds is not known at present. The reactions leading to the new compounds utilized $CuCl_2 \cdot 2H_2O$ in methanol solution containing either $[(C_2H_5)_4N]_3VOCl_3(OSO_3)(H_2O)$ or $M_3VO(NCS)_5$

[$M = (CH_3)_4N^+$, $(C_2H_5)_4N^+$, or $(n\text{-}C_4H_9)_4N^+$]. The particular product obtained depended not only upon the vanadyl complex anion used, but on the organic cation present. Work is continuing along several new lines in efforts to produce vanadyl–copper dimers.

EQUATORIAL π BONDING AND HYDROGEN BONDING TO THE VANADYL OXYGEN

We have deduced evidence for the existence of equatorial π bonding in $VOSO_4 \cdot 4H_2O$ and $VOSO_4 \cdot 2CH_3OH$ from proton NMR spectra.[13] The evidence derives from NMR *contact shift* data on the water proton and the methanolic -OH proton, respectively. For example, the *downfield* shift observed for the water proton has been interpreted by us as being due to the transfer of spin density from the b_2 orbital, formally a half-filled nonbonding d_{xy} orbital in the Ballhausen–Gray[14] energy scheme, to the b_2 (π symmetry) orbital on the water molecule. The electron spin-nuclear spin coupling constant evaluated from the contact shift data is not as large as has been found[15] for other paramagnetic ions with unpaired t_{2g} electrons (Co^{2+}, Fe^{2+}, Mn^{2+}, and Fe^{3+}) in aqueous solutions; but this is to be expected since these ions all have more t_{2g} electrons, and hence higher spin transfer tendencies, that is, greater π-bonding tendencies. However, the fact remains that some π bonding is evidenced in the VO^{2+} case and so the Vanquicken-borne–McGlynn MO energy scheme,[9] which allows for this, is to be considered more satisfactory than the Ballhausen–Gray[14] MO scheme which ignores equatorial π bonding.

There are very few studies which deal specifically with hydrogen bonding by solvents to transition metal complexes. Most of the literature deals only incidentally with this subject as, for example, H bonding is often invoked to explain or account for certain physical data. Hydrogen bonding to certain oxocation complexes has been considered in only a few recent papers[16-20] and to oxovanadium(IV) complexes in two others[21,22]; but the evidence for it is relatively meager and generally ambiguous. We have, therefore, carried out a rather thorough investigation of this phenomenon as it occurs in β-ketoenolate complexes of VO^{2+}.[23] Having noted for our complexes that solvents of different H-bonding potential often have an effect on the position of electronic band maxima and always upon band intensities, we proceeded to examine H bonding by the methods normally employed for such studies; namely, solubilities, infrared spectra, and NMR spectra.

Clear evidence for H-bonding to $VO(aca)_2$ and specifically for stronger H bonding to the vanadyl oxygen than to the equatorial oxygens, comes from solubility studies. We picked the solvents CCl_4, CH_2Cl_2, and $CHCl_3$ because of their increasing H-bonding ability; CH_3CN, because of its high dielectric constant, making possible the estimation of dielectric effects; and benzene, both as another reference solvent and because it is used in so many solvent studies. The pertinent data is collected in Table 1. The solubility of VO(aca)$_2$ is seen to increase markedly from CCl$_4$ to CH$_2$Cl$_2$ and to increase further with CHCl$_3$. The respective dielectric constants are 2.2, 4.8, and 9.1, illustrating that there is no linear increase in solubility with dielectric constant. Furthermore, the dielectric constant for CH$_3$CN is a comparably high 38.8, whereas the solubility is actually lower in this solvent than in either CH$_2$Cl$_2$ or CHCl$_3$. Coordination to the sixth position in the VO(aca)$_2$ molecule is highly unlikely for CHCl$_3$ (but quite likely for CH$_3$CN) and so this must also be ruled out as an important solubilizing factor. Of course, solubility depends upon many factors, some quite subtle, and we would not place great importance on these results alone, but they do represent additional evidence to support that from IR and NMR studies. We should note that the increased solubility of the other *aca* complexes in H-bonding media is in line with the prediction made in 1965[24] that H bonding to β-ketoenolates is probably much larger and more significant than previously considered.

Table 1

Solubility of VO(aca)$_2$ and Other Inorganic Complexes in Five Selected Organic Solvents, in g/ml

Complex	CCl$_4$	CH$_2$Cl$_2$	CHCl$_3$	CH$_3$CN	C$_6$H$_6$	CHCl$_3$/CCl$_4$	CHCl$_3$/C$_6$H$_6$
VO(aca)$_2$	>0.0005	0.0644	0.0969	0.0140	0.0058	194	17
MoO$_2$(aca)$_2$	0.0056	0.0103	0.0175	—	0.0076	3.1	2.3
Th(aca)$_4$	0.0789	0.0921	0.1674	0.1081	0.0708	2.1	2.4
Al(aca)$_3$	0.0758	0.1011	0.2059	0.0706	0.1655	2.7	1.2
Cu(aca)$_2$·H$_2$O	0.0009	0.0235	0.0591	0.0003	0.0001	657	590

Chloroform was chosen as the solvent for infrared studies of H-bonding to VO(aca)$_2$. An increase in the absorption of the C—H stretching frequency may be taken as the rough measure of H-bond interaction in this situation.[24,25] Since the C—H stretches of the β-diketone groups may mask the C—H stretch of the CHCl$_3$, chloroform-d was employed. The C—D stretching frequency was observed at 2285 cm^{-1}, well removed from any interfering bands. The absorbance of pure CDCl$_3$ was subtracted from the absorbance found when the complex was present and the absorbance obtained in this manner (A_H) was assumed to be due primarily to H bonding. Then by the relationship below, ε_H can be obtained.

$$A_H = \varepsilon_H b C_H$$

(where ε_H = effective molar absorptivity of the complex due to H bonding, b = cell length in cm, and C_H = the molar concentration of the H-bonding complex). This kind of measurement did not give the accuracy or precision necessary to rank the various complexes according to their H-bonding ability, but ε_H values ranged from 23/M-cm to 270/M-cm, with values for the VO(aca)$_2$ in the neighborhood of 100/M-cm. Further evidence then was established for H-bonding, but again, standing alone it would not be at all convincing that it is the vanadyl oxygen which plays the major role in this effect rather than the *aca* oxygens.

Hydrogen bonding by an alcohol can be identified by the positions of the NMR signal (δ) for the hydroxyl proton.[25] For example, the position of the signal for that proton increases as the concentration of an alcohol in benzene increases. Presumably, at the higher concentrations the alcohol molecules are closer together and thus more H bonding can occur. The top line of data in Table 2 shows the dependence of δ on the concentration of CH$_3$OH in benzene. The δ-value at any given CH$_3$OH concentration can be shifted by addition of another solute which forms a stronger H bond. Thus, the remainder of Table 2 shows how δ varies with the addition of fixed concentrations (0.041 M) of various *aca* complexes including VO(aca)$_2$. The three mole ratios of complex to CH$_3$OH represented by the 1, 2, and 3 % CH$_3$OH solutions are 1:5, 1:10, and 1:15, respectively.

Table 2

Position of the NMR Signal (δ) of the Hydroxyl Proton of Methyl Alcohol in Various Concentrations (volume %) in Benzene and With Inorganic Complexes (0.041 M)

Complex	1%	2%	3%
No complex	0.42	1.08	1.65
VO(aca)$_2$	1.65	1.69	1.65
MoO$_2$(aca)$_2$	1.65	1.92	2.19
Th(aca)$_4$	1.01	1.40	1.80
Al(aca)$_3$	0.62	1.18	1.68
Be(aca)$_2$	0.50	--	--

It is seen from the data that the δ value for pure 1% CH$_3$OH is 0.42 and a very large increase, to 1.65, occurs upon the addition of VO(aca)$_2$ (in mole ratio 1:5). The fact that there is roughly no change in the value when the mole ratio is increased to 1:10 and 1:15 suggests, but does not prove, that the H-bonding capacity of VO(aca)$_2$ is nearly saturated in the 1% CH$_3$OH solution. Of course, it is pertinent to inquire whether the increase in δ is due to H bonding or to coordination to the sixth position in VO(aca)$_2$, since CH$_3$OH is certainly capable of both kinds of interaction. The analogous data for $MoO_2(aca)_2$, a molecule with no open sixth position, would appear to rule out the explanation in terms of coordination. The data suggest further that $MoO_2(aca)_2$ possesses even greater capacity for H bonding than does $VO(aca)_2$. This is not unreasonable particularly if one pictures the H bond as being a trifurcated bond (hydrogen atom roughly centered above a triangular face consisting of three oxygen atoms). The tetragonal pyramidal structure of $VO(aca)_2$[4] presents four such triangular surfaces, each composed of two *aca* oxygens and one vanadyl oxygen, to which the H bond can attach. On the other hand, in the roughly octahedral $MoO_2(aca)_2$, where the two oxo ligands are believed to lie *cis* to each other[26,27] there would be twice as many triangular surfaces: two with two molybdenyl oxygens and one *aca* oxygen; four with one molbdenyl oxygen and two *aca* oxygens; and two with three *aca* oxygens. Furthermore, the molybdenyl oxygens would be expected to be more highly negative and less sterically hindered than the *aca* oxygens. Thus, it appears reasonable to expect increased H bonding with $MoO_2(aca)_2$ over $VO(aca)_2$, as is deduced from the data in Table 2. The greater broadness of the signal for the hydroxyl proton when it is interacting with $MoO_2(aca)_2$ seems also to suggest that the H bonding is occurring to more than one type of surface. It should be noted further that the corresponding δ values for $MoO_2(aca)_2$ in CH_2Cl_2 with 1, 2, and 3 % CH_3OH added were 2.10, 2.11, and 2.12, respectively, whereas the values for CH_3OH alone in CH_2Cl_2 were 1.26, 1.48, and 1.73, respectively. The smaller effect on δ of the hydroxyl proton caused by adding more CH_3OH or $MoO_2(aca)_2$ in CH_2Cl_2 as solvent shows that both the alcohol and the complex interact with the H-bonding solvent.

Finally, the δ values for the three other *aca* complexes listed in Table 2, chosen for their three different structures and numbers of triangular *aca* oxygen faces, lend support to our suggestion that the H-bonding effect (at least for an alcohol) is greater with the multiple bonded oxygens than with the *aca* oxygens.

ELECTRONIC SPECTRAL STUDIES

Recent spectral studies in our laboratories have concerned *bis-β*-ketoenolate complexes of VO^{2+}, six such complexes having been examined in the 8000 to 30,000 cm^{-1} region[27] and seven such in the 20,000 to 50,000 cm^{-1} region.[28] Tentative assignments for the visible bands are made according to an empirical energy level scheme[27] which is different from the calculated MO schemes.[9,14] However, the latter are for the $VO(H_2O)_5^{2+}$ ion and it is very probable that no one energy level scheme will prove satisfactory for the great variety of complexes formed by this unique ion. Somewhat firmer assignments have been made[28] for the ultraviolet bands exhibited by the β-ketoenolate complexes. They include: *spin-forbidden*, π-π^*, transitions in the 20 to 25 kK region; *ligand-to-metal*, $\pi_3 \rightarrow d_{x^2-y^2}$ (a_1), transitions in the 25 to 28 kK region; *intraligand spin-allowed*, π_3-π_4^*, transitions in the 28 to 39 kK region; and higher energy *intraligand*, π-π^*, allowed transitions in the 40 to 50 kK region.

Most recently, we have examined[29] the low-temperature (77 and 4°K) spectra of some high-symmetry complexes of VO^{2+}. These include VOF_5^{3-}, $VOF_4(H_2O)^{2-}$, $VOCl_4^{2-}$, $VOCl_3(OSO_3)(H_2O)^{3-}$, $VO(NCS)_5^{3-}$, and $VO(NCS)_4(H_2O)^{2-}$. Band assignments have been made on a much firmer empirical basis than previously possible and the $(xy) \rightarrow (x^2 - y^2)$ transition is used to establish a spectrochemical series for a VOX_4-type complexes as follows:

$$Cl^- < DMSO \sim N_3^- < F^- < H_2O < -NCS^- < CN^-$$

The assignment of the first charge transfer transition (ligand-to-metal) leads us to the following ligand series

$$F^- < N_3^- < Cl^- < -NCS^- < CN^- \ll H_2O \text{ and DMSO}$$

The order in this series of F^-, Cl^-, and H_2O may be rationalized in terms of available valence state ionization energies for these species, which further justifies our assignment of the transition which produces the series.

REFERENCES

1. J. Selbin, *J. Chem. Ed.* **41**: 86 (1964); *Oesterr. Chem. Ztg.* **65**: 275 (1964).
2. J. Selbin, *Angew. Chem. Internat. Ed. Engl.* **5**: 712 (1966).
3. J. Selbin, *Chem. Rev.* **65**: 153 (1965).
4. J. Selbin, *Coord. Chem. Rev.* **1**: 293 (1966).
5. G. Vigee, Ph.D. Dissertation, Louisiana State University, Baton Rouge, La., 1968.
6. J. Selbin and G. Vigee, *J. Inorg. Nucl. Chem.* **30**: 1644 (1968).
7. G. Vigee and J. Selbin, *J. Inorg. Nucl. Chem.* (in press).
8. B. J. McCormick, *Inorg. Chem.* **7**: 1965 (1968); *Inorg. Nucl. Chem. Letters* **3**: 293 (1967).
9. L. G. Vanquickenborne and S. P. McGlynn, *Theoret. Chim. Acta* **9**: 390 (1968).
10. C. Heitner-Wirgiun and J. Selbin, *J. Inorg. Nucl. Chem.* **30**: 2403 (1968).
11. C. Heitner-Wirguin and J. Selbin, *J. Inorg. Nucl. Chem.* **30**: 3181 (1968).
12. O. Piovesana and J. Selbin, *J. Inorg. Nucl. Chem.* (in press).
13. G. Vigee and J. Selbin, *J. Inorg. Nucl. Chem.* **30**: 2273 (1968).
14. C. J. Ballhausen and H. B. Gray, *Inorg. Chem.* **1**: 111 (1962).
15. B. B. Wayland and W. L. Rice, *Inorg. Chem.* **5**: 54 (1966).
16. J. C. Ryan, *Inorg. Chem.* **2**: 348 (1963).
17. A. E. Baker and H. M. Haendler, *Inorg. Chem.* **1**: 127 (1962).
18. J. H. Beard, J. Casey, and P. K. Newman, *Inorg. Chem.* **4**: 797 (1965).
19. R. K. Murman and D. R. Forrester, *J. Chem Phys.* **67**: 1383 (1963).
20. S. J. Lippard and B. J. Russ, *Inorg. Chem.* **6**: 1943 (1967).
21. A. I. Rivkind, *Zh. Strukt. Khim.* **4**: 64 (1963).

22. F. A. Walker, R. L. Carlin, and P. H. Rieger, *J. Chem. Phys.* **45**: 4181 (1966).
23. R. T. Arndts, Ph.D. Dissertation, Louisiana State University, Baton Rouge, La., 1969.
24. J. P. Fackler, Jr., T. S. Davis, and I. D. Chawla, *Inorg. Chem.* **4**: 130 (1965).
25. G. C. Pimentel and A. L. McClellan, *The Hydrogen Bond*, W. H. Freeman and Co., San Francisco, Calif., 1960.
26. Y. A. Buslaer, Y. Y. Kharitanov, and R. L. Davidovich, *Izr. Akad. Nauk. SSSR* **3**: 589 (1967).
27. F. W. Moore and R. E. Rice, presented at the 154th National ACS Meeting, Chicago, Ill., Sept. 1967.
27. J. Selbin, G. Maus, and D. L. Johnson, *J. Inorg. Nucl. Chem.* **29**: 1735 (1967).
28. D. Ogden and J. Selbin, *J. Inorg. Nucl. Chem.* **30**: 1227 (1968).
29. O. Piovesana and J. Selbin, *J. Inorg. Nucl. Chem.* **31**: 433 (1969).

SOME REACTIONS OF THE DIBROMOIODIDE ION WITH PLATINUM(II) COORDINATION COMPOUNDS

G. L. Johnson and F. Svec, Jr.

Kenyon College
Gambier, Ohio

INTRODUCTION

The polyhalides and interhalogens have been known for some time.[1] Gay Lussac and Davy, in 1814, were the first to isolate an interhalogen when they prepared iodine chloride and iodine trichloride. Caventou and Peltier were the first to prepare a crystalline form of a polyhalide salt with the isolation of strychnine triiodide in 1819. Since then, many more polyhalides and interhalogens have been discovered.

The structures, thermodynamic stabilities, and mode of decomposition of several simple polyhalides and interhalogens have been extensively studied.[2-18] However, the chemical reactions of the polyhalides have not been extensively explored. Cremer and Duncan[11] have studied the reactions in the solid phase of dibromoiodide with ammonia. They concluded that the reaction product depended upon the decomposition pressure of the polyhalide. If the decomposition pressure of the IBr_2^- salt were less than 0.005 mm, the ammonia and the polyhalide would form an addition product, $RIBr_2 \cdot NH_3$, where R represents the cation. However, if the decomposition pressure of the polyhalide were greater than 0.005 mm, the reaction product would be nitrogen triiodide. To account for this, Cremer and Duncan postulated that at the lower pressures, the polyhalide reacts directly with ammonia whereas, at the higher decomposition pressures, the reaction of the ammonia occurs with the decomposition products. The reaction pathway proposed for the reaction of ammonia with the polyhalide at the higher decomposition pressures is[19]

$$3KIBr_2 \longrightarrow 3KBr + 3IBR$$
$$3IBr + NH_3 \longrightarrow 3HBr + NI_3$$
$$3HBr + NH_3 \longrightarrow 3NH_4^+ + Br^-$$

Also, a fluorination reaction has been noted for a polyhalide. Sharpe and Emeleus[20] found that potassium hexafluoroplatinate(IV), K_2PtF_6, is formed when potassium tetrafluorobromide, $KBrF_4$, is heated with metallic platinum. Since both BrF_3 and KF are inert towards platinum, the experimenters concluded that the polyhalide is responsible for the reaction.

The chemical properties of the interhalogens have been studied more fully than the chemical properties of the polyhalides. Their useful reactions depend upon the ability of the interhalogens to halogenate. In organic reactions, for example, it is possible to replace the chlorine atom in carbon tetrachloride with fluorine atoms through the use of bromine trifluoride, BrF_3.[21]

$$CCl_4 + BrF_3 \longrightarrow CCl_3F + BrF_2$$

$$CCl_3F + BrF_2 \longrightarrow CCl_2F_2 + BrF, \text{ etc.}$$

Similarly, salts containing the less electronegative halides are converted to fluoride salts by BrF_3.[20]

$$3KCl + BrF_3 \longrightarrow 3KF + Br + 3Cl$$

Many metals are attacked by interhalogens as demonstrated by the reaction between platinum and bromine pentafluoride.[22] A pre-equilibrium step is postulated.

$$BrF_5 + Br_2 \rightleftharpoons Br_2F_2 + BrF_3$$

$$5Br_2F_2 + Pt \longrightarrow (BrF_2)_2[PtF_6] + 4Br_2.$$

Recently, Chernyaev, Nibolaev, and Ippolitov[23] reported a reaction between potassium hexachloroplatinate(IV) and chlorine trifluoride. The reaction takes place at 500°C and yields potassium hexafluoroplatinate(IV) in surprisingly high yield.

$$K_2[PtCl_6] + 2ClF_3 \longrightarrow K_2[PtF_6] + 4Cl_2$$

Taube and Haim[24] showed that iodine chloride would react with an aqueous solution of iodopentamminecobalt(III) ion to oxidize the iodide and yield $[Co(NH_3)_5H_2O]^{+3}$ and iodine as the major products.

EXPERIMENTAL

Preparation of Starting Materials

All chemicals used in the experimental and preparative work except rubidium iodide were Baker reagent grade materials. The rubidium iodide was obtained from Alfa Inorganics, Inc.

The methods for the preparation of K_2PtCl_4[25] and K_2PtBr_4[26] have been described elsewhere.

Potassium hexachloroplatinate(IV), K_2PtCl_6, was prepared from metallic platinum by a modification of the method described by Kauffman and Teter.[27]

The sodium salt of the hexachloroplatinate(IV) ion, Na_2PtCl_6, was prepared in the same manner as the potassium salt, except that NaCl was added instead of KCl. Furthermore, due to solubility differences, the product was washed only with 95% ethanol.

The method Keller[28] used to synthesize $[Pt(NH_3)_4]Cl_2$ was adapted to prepare $[Pten_2]Cl_2$. The resulting chloride salt, dissolved in water, was titrated potentiometrically to an end point of 260 mv with a solution of $AgClO_4$ thus changing the anion from Cl^- to ClO_4^-. The AgCl was removed by filtration. The solution containing $[Pten_2](ClO_4)_2$ was reduced in volume and was added to a mixture containing 300 ml of 95% ethanol and 300 ml of ether. The crude product was recrystallized from an ethanol–water mixture.

The *trans*-dichlorobisethylenediamineplatinum(IV) ion was prepared from $[Pten_2](ClO_4)_2$. This was accomplished by bubbling chlorine gas through a 50-ml solution of 0.90 g of $[Pten_2](ClO_4)_2$ until the solution became noticeably turbid (approximately 45 min). After filtering, the volume of the solution was reduced to 10 ml. The crude *trans*-$[Pten_2Cl_2]$ $(ClO_4)_2$ was precipitated with ethanol and ether. The product was recrystallized from an ethanol–water mixture.

The rubidium and ammonium dibromoiodides[29] were prepared by placing an open beaker of liquid bromine and a weighing bottle containing a finely ground 5-g sample of the appropriate iodide in a desiccator over phosphorus pentoxide. The reactions which occur are represented by the equations

$$RbI + Br_2 \longrightarrow RbIBr_2$$

and

$$NH_4I + Br_2 \longrightarrow NH_4IBr_2$$

The progress of the reactions was followed by periodically noting the weight of the samples. Before weights were taken, however, the weighing bottles were placed in a desiccator of iodine to remove any adhering bromine. When the iodides had reacted with the correct amount of bromine, the samples were stoppered and stored in the desiccator containing iodine.

Analysis: Calculated for NH_4IBr_2: Br, 52.4%; I, 41.6%. Found: Br, 53.9%; I, 39.3%.†

---

† The dibromoiodide always gave too much bromine when analyzed. This may be due to either bromine adhering to the sample or to some decomposition of the IBr_2^- salt to the Br^- salt.

Reaction Techniques and Methods of Analyses

In order to determine the most favorable reaction conditions, the following parameters were studied: (a) the influence of reaction time upon the product; (b) the influence of added halide upon the product; and (c) the influence of the molar ratio of platinum(II) complex to IBr_2^- salt upon the product.

Most of the reactions between dibromoiodide and the platinum(II) complexes were carried out in the following manner: the weighed amounts of dibromoiodide and the platinum(II) complex were placed in separate 100 ml beakers and dissolved in a small amount of water, usually 10 ml. In studies involving extra halide ions, the appropriate alkali metal halide was dissolved in the same beaker that contained the platinum(II) complex. A Teflon-coated stirring bar was placed into one of the beakers. After stirring commenced, the contents of the other beaker were quickly added and the mixing continued. To remove the platinum complex for analysis a precipitating agent was used. If the reacting complex was an anionic species, a stoichiometric amount of $trans$-$[Pten_2Cl_2](ClO_4)_2$, dissolved in 10 ml of water, was added. On the other hand, if the reacting platinum(II) complex was cationic, the appropriate amount of $Na_2[PtCl_6]$, dissolved in 10 ml of water was added. The precipitating reactions may be represented by the following equations:

cationic platinum(II) species

$$[PtA_n]^{+2} + [PtCl_6]^{-2} \longrightarrow [PtA_n][PtCl_6]\downarrow$$

anionic platinum(II) species

$$[PtX_n]^{-2} + [Pten_2Cl_2]^{+2} \longrightarrow [Pten_2Cl_2][PtX_n]\downarrow.$$

The time recorded as "stopping time" for the reaction is the interval between the combination of the reactants and the addition of the precipitant. After the addition of the precipitating agent, the solution immediately (within 5 to 10 sec) became very cloudy due to the appearance of the precipitate. At this point, the solution was quickly filtered through a sintered glass filter and the precipitate washed first with 15 ml of water, then with 15 ml of either 95% ethanol or acetone, and finally with 15 ml of ether. The filter and its contents were placed in a vacuum desiccator over P_4O_{10} to dry.

An alternate method was used to shorten reaction time. In this method the precipitant and dibromoiodide salt were dissolved in 15 ml of water, and were added rapidly to a stirred 15 ml solution of the platinum(II) complex. Again, the solution would become very turbid in approximately 5 to 10 sec. The separation from the mother liquor and the washing and drying of the precipitate were all accomplished as previously noted. Stopping time for the reactions in which this technique was used will be noted in this paper as "O*".

To analyze the precipitate, a small weighed sample (usually between 0.0400 g and 0.0600 g) was placed into a weighed 100 ml beaker. Fifty ml of water and approximately 0.05 g of $N_2H_4 \cdot H_2SO_4$ were added. The solution was made basic by adding 2 ml of 1.8 M KOH and then heated while being stirred for 30 min. Under these conditions the complexes were rapidly reduced by hydrazine leaving a clear solution and platinum black. The samples were cooled. To ensure completion of the reduction, 0.02 g of $N_2H_4 \cdot H_2SO_4$ was added and the samples again heated and mixed for 10 min. After cooling the second time, the samples were filtered through a weighed, 2-ml sintered glass filter. The 100 ml beaker, filter, and Teflon stirring bar (whose weights were previously known) were placed in an oven, set at 70°, overnight to dry. The weight gained was used to calculate those percent platinum values marked with an asterisk (*).†

The clear solution which remained from the reduction was analyzed for halides by a potentiometric method adapted from that of Shiner and Smith.[30] In their method, Shiner and Smith used a commercial detergent, "Tergitol NPX," from Union Carbide to improve their accuracy. This detergent was tried but better results were obtained without it. Consequently, this step in their procedure was omitted. The first step in this procedure was to place the clear solution in a 250 ml beaker and add 50 ml of a buffer solution which was 1.2 M in both acetic acid and sodium acetate. This solution was titrated potentiometrically with $AgNO_3$ using a Sargent Model C constant rate buret, and a Leeds and Northrup Model 7401 potentiometer connected to a Model MR Sargent Automatic Recorder. The reference electrode was a saturated calomel in a 1.0 M KCl solution while the indicating electrode was a silver billet (Sargent Cat. No. S-30515). A salt bridge, which was one molar in KNO_3 and 4 % in agar-agar, was used to complete the circuit.

Correction factors were applied to the titrations. These correction factors were determined by titrating solutions of known concentrations which were made similar in proportions and amounts of those to be analyzed. None of the correction factors, however, would change the experimentally determined weight of halide by more than 2 %.

---

† This method for determining the platinum weight was not too successful. It was observed that after filtering, the mother liquor appeared slightly grey. Upon standing overnight, the solution cleared and a small quantity of platinum black was seen on the bottom of the beaker. Apparently, some of the platinum metal which was generated upon reduction was too finely divided to be retained by the filter, but would settle with time. Since the sample weight was about 0.0500 g, a loss of only a thousandth of a gram of platinum would cause a 2 % absolute error in its determination.

For a more accurate platinum weight, a carefully weighed sample of the precipitate (about a tenth of a gram) was placed in a weighed porcelain crucible. The crucible and its contents were heated for three hours in a muffle furnace set at 850°C. The weight gained was that of the metallic platinum. The percent platinum values obtained by this method are marked with a "+".

In parentheses for each percent platinum value is another value obtained by difference. The difference between the total sample weight and the weight due to the sum of the halide weights was assumed to be due entirely to platinum and ethylenediamine. Since there should be equimolar amounts of these two species,

$$\frac{195.09}{195.09 + 60.0} \times 100 \text{ or } 76.45\% \text{ of that weight is platinum.}$$

RESULTS AND DISCUSSION

The Reaction between $K_2[PtCl_4]$ and $NH_4(Rb)IBr_2$

Two products were obtained for the reaction of the iodine-dibromide ion and tetrachloroplatinate(II) ion. The products were the $[PtCl_4BrI]^{-2}$ and the $[PtCl_4Br_2]^{-2}$ ions. The product obtained depends upon the reaction conditions. The conditions are given below:

$[Pten_2Cl_2][PtCl_4BrI]$

g of $K_2[PtCl_4]$	0.0800 (0.193 mmoles)
g of KCl	0.200
g of $RbIBr_2$	0.0720 (0.193 mmoles)
g of $[Pten_2Cl_2](ClO_4)_2$	0.1000 (0.171 mmoles)
stopping time (in sec)	0*

Analysis:

	Calculated	Found
% Pt	41.9	39.5* (41.7)
% Cl	22.9	22.4
% Br	8.6	9.3
% I	13.7	13.7

$[Pten_2Cl_2][PtCl_4Br_2]$

g of $K_2[PtCl_4]$	0.0800 (0.193 mmoles)
g of KCl	2.0 g
g of NH_4IBr_2	0.4700 (1.541 mmoles)
g of $[Pten_2Cl_2](ClO_4)_2$	0.1000 (0.171 mmoles)
stopping time (in sec)	0*

Analysis:

Calculated	Found
% Pt 44.1	40.0* (42.7)
% Cl 24.1	23.7
% Br 18.1	19.0
% I 0.0	1.5

The Reaction between $K_2[PtBr_4]$ and NH_4IBr_2

Only one well-defined product was obtained, namely, $[PtBr_5I]^{-2}$. This product was obtained by using a $K_2[PtBr_4]$ to NH_4IBr_2 molar ratio of 1:1.1. As this ratio was increased, the amount of I in the product decreased and the amount of Br increased. However, the exact conditions for obtaining relatively pure $[PtBr_6]^{-2}$ have not as yet been found. The conditions which produce $[Pten_2Cl_2][PtBr_5I]$ are as follows:

g of $K_2[PtBr_4]$	0.0800 g (0.135 mmoles)
g of NH_4IBr_2	0.0450 g (0.148 mmoles)
g of $[Pten_2Cl_2](ClO_4)_2$	0.0800 g (0.137 mmoles)
stopping time (in sec)	0*

Analysis:

Calculated	Found
% Pt 35.2	35.3+(35.4)
% Cl 6.4	6.7
% Br 36.1	36.2
% I 11.4	10.8

The Reaction between $[Pten_2](ClO_4)_2$ and NH_4IBr_2

Again, only one well-defined product was obtained. The product was $[Pten_2BrI]^{+2}$. Neither an increase in the $[Pten_2](ClO_4)_2$ to NH_4IBr_2 molar ratio from 1:1.24 to 1:5, nor an increase in the stopping time from 3 to 30 sec affected the nature of the final product obtained in the reaction. The best conditions for obtaining $[Pten_2BrI][PtCl_6]$ are as follows:

g of $[Pten_2](ClO_4)_2$	0.0800 g (0.155 mmoles)
g of NH_4IBr_2	0.0586 g (0.192 mmoles)
g of $Na_2[PtCl_6]$	0.0704 g (0.155 mmoles)
stopping time (in sec)	3

Analysis:

Calculated	Found
% Pt 41.9	37.6* (42.2)
% Cl 22.9	21.4
% Br 8.6	9.7
% I 13.7	13.7

The reaction between the dibromoiodide ion and a platinum(II) complex results in the oxidation of the platinum(II) complex to an octahedral platinum(IV) complex. Furthermore, the rate of reaction is very fast and is complete within a few seconds.

The halide groups found in the product depend upon the reaction conditions. For instance, the product of the equimolar reaction between IBr_2^- and $K_2[PtCl_4]$ is $[PtCl_4BrI]^{-2}$, but if the IBr_2^- to $K_2[PtCl_4]$ ratio is 8:1 or greater, the product is $[PtCl_4Br_2]^{-2}$. This same general behavior may occur for the reactions between IBr_2^- and either $[PtBr_4]^{-2}$ or $[Pten_2]^{+2}$. There is some evidence to suggest that at a ratio of IBr_2^- to the platinum complex of 5 to 1, the amount of iodide ion in the complex is decreasing. However at this time the results are inconclusive. Furthermore, the product of either one of these reactions is influenced by the nature and amount of extra halide ions in the reaction mixture (see Tables 1 and 2).

Table 1
Variation of Extra Halide in Solution

Reaction conditions:
  0.0800 g (0.193 mmoles) of $K_2[PtCl_4]$
  0.0726 g (0.238 mmoles) of NH_4IBr_2
  0.1000 g (0.171 mmoles) of $[Pten_2Cl_2](ClO_4)_2$
  10 sec stopping time

Type and weight of added halide	Analysis Calculated for $[Pten_2Cl_2][PtCl_4BrI]$		
	% Cl, 22.9;	Br, 8.6;	I, 13.7
none	Found 18.1	14.4	13.2
1.0 g KCl	22.7	7.5	13.3
1.0 g KBr	10.5	29.4	10.3

Table 2
Variation of the Amount of KCl in Solution

Reaction conditions:
  0.0800 g (0.193 mmoles) of $K_2[PtCl_4]$
  0.0720 g (0.193 mmoles) of $RbIBr_2$
  0.1000 g (0.171 mmoles) of $[Pten_2Cl_2](ClO_4)_2$
  stopping time 0*

g of added KCl	Analysis: Calculated for $[Pten_2Cl_2][PtCl_4BrI]$		
	% Cl, 22.9;	Br, 8.6;	I, 13.7
0	Found 21.1	11.9	14.6
0.10 g	21.0	12.2	13.2
0.20 g	22.4	9.3	13.7
0.30 g	22.9	9.3	13.9
0.50 g	23.7	8.6	13.8

Ligand exchange reactions may account for the change in the product with variation of free halide concentration. Exchange might be expected to occur during the period of reaction. Exchange of ligands on platinum(IV) complexes is greatly accelerated when platinum(II) complexes are present. Dreyer[31] found the exchange of Cl^- for Br^- in $[PtCl_6]^{-2}$ to be substantially enhanced when $[PtCl_4]^{-2}$ was present. The exchange occurs through a bridged intermediate and may be represented as follows:

$$[PtCl_4]^{-2} + Br^- \longrightarrow [PtCl_4Br]^{-3}$$

$$[PtCl_6]^{-2} + [PtCl_4Br]^{-3} \rightleftharpoons [Cl_5-Pt\cdots Cl\cdots PtCl_4Br]^{-5}$$
$$[PtCl_4]^{-2} + Cl^- + [PtCl_5Br]^{-2}$$

During the reactions of platinum(II) complexes with IBr_2^-, there are quantities of both platinum(II) and platinum(IV) ions present. Further, Cremer and Duncan[13] report that bromide and iodide ions are the products when dibromoiodide is reduced. Bromide ion is expected to be a product in the formation of $[PtCl_4BrI]^{-2}$. Further, Dreyer shows that bromide ion exchanges more rapidly for a bound chloride than does the chloride ion for a bound bromide ion.

By adding chloride ion to the system, one should eventually statistically "overpower" the bromide (or iodide) exchange. Table 1 and Table 2 do indicate that some of the bromide ion in the reaction products is caused by ligand exchange. However, due to the continual coordination of one bromine and one iodine, even at high chloride ion concentrations, it appears that these two halides coordinated to the platinum(IV) during the oxidation of the platinum(II). At low ratios ($\sim 1:1$) of IBr_2^- to complex the halides added are specifically iodide and bromide and at high ratios (~ 8–$10:1$) of IBr_2^- to $[PtCl_4]^{-2}$, both halides are bromide.

Several species might be responsible for the oxidation of platinum(II) to platinum(IV). Cremer and Duncan[13] have stated that the hydrolysis products of IBr_2^- are I_2, IO_3^-, Br^-, and H_3O^+. Experiments were run in which iodine, sodium iodate, and potassium bromide were reacted with $K_2[PtCl_4]$ individually and in combination. The results are shown in Table 3.

Table 3
Reaction of I_2, $NaIO_3$, and KBr with $K_2[PtCl_4]$

Reactants ($+ K_2[PtCl_4]$)	Stopping time, sec	Product
$NaIO_3 + KBr$	10	$[PtCl_{4.28}]^{-2}$
$I_2 + KBr$	983	$[PtCl_{2.91}Br_{0.84}I_{0.49}]^{-2}$
$I_2 + KBr + KI$	10	$[PtCl_{1.92}I_{3.52}]^{-2}$
$I_2 + NaIO_3 + KBr$	10	$[PtCl_{3.99}Br_{0.03}]^{-2}$
$I_2 + NaIO_3 + KBr$	240	$[PtCl_{2.98}Br_{0.55}O_{0.04}]^{-2}$

None of the hydrolysis products give an oxidation reaction which is at the same time fast and specific in adding an iodide ion and a bromide ion. There are two remaining species that may be responsible for the reaction: dibromoiodide or its dissociation product iodine bromide ($IBr_2^- \rightleftharpoons IBr + Br^-$). Reactions between iodine bromide and $K_2[PtCl_4]$ were carried out at molar ratios of $1:1$ and $10:1$ (see Table 4).

Table 4
Reaction of IBr with $K_2[PtCl_4]$

$IBr:[PtCl_4]^{-2}$	Stopping time	Oxidation state of Pt in product	$Pt:Cl:Br:I$
1:1	0*	II	1:3.3:0:0.5
10:1	0*	IV	1:3.6:2.2:0.4

At low ratios of $IBr:[PtCl_4]^{-2}$, the $[PtCl_4]^{-2}$ was not oxidized to a platinum(IV) complex. At higher ratios of IBr to $[PtCl_4]^{-2}$, oxidation occurs and both iodide and bromide are present in the product. Therefore, the possibility that IBr is a contributing species in the reaction between IBr_2^- and $[PtCl_4]^{-2}$ must still be considered.

REFERENCES

1. E. H. Wiekenga, E. E. Havinga, and K. H. Boswijk, *Advances in Inorganic Chemistry and Radiochemistry*, Vol. 3, Academic Press Inc., New York, 1961, pp. 133–169.
2. W. F. Zelezny and N. C. Baenziger, *J. Am. Chem. Soc.* **74**: 6151 (1952).
3. R. W. G. Wyckoff, *J. Am. Chem. Soc.* **42**: 1100 (1920).
4. H. A. Tasman and K. H. Boswijk, *Acta Cryst.* **8**: 59 (1955).
5. S. Siegel, *Acta Cryst.* **10**: 380 (1957).
6. W. G. Sly and R. E. Marsh, *Acta Cryst.* **10**: 378 (1957).
7. R. C. L. Mooney, *Z. Krist.* **98**: 324 (1938) from E. H. Wiekenga, E. E. Havinga and K. H. Boswijk, *Advances in Inorganic Chemistry and Radiochemistry*, Vol. 3, Academic Press Inc., New York, 1961, p. 150.
8. J. Magnuson, *J. Chem. Phys.* **26**: 233 (1957).
9. R. C. Lord, M. A. Lynch, W. C. Schumb, and E. J. Slowinski, *J. Am. Chem. Soc.* **72**: 522 (1950).
10. H. W. Cremer and D. R. Duncan, *J. Chem. Soc.* 2243 (1931).
11. H. W. Cremer and D. R. Duncan, *J. Chem. Soc.* 182 (1933).
12. F. Ephraim, *Ber.* **50**: 1069 (1917).
13. H. W. Cremer and D. R. Duncan, *J. Chem. Soc.* 2031 (1932).
14. N. V. Sidgwick, *The Chemical Elements and Their Compounds*, Vol. *II*, Clarendon Press, London, 1952, p. 1196.
15. J. W. Mellor, *Comprehensive Treatise on Chemistry* Supp II, Part I, Longmans, Green, New York, 1956, pp. 369, 482, 488, 709, 744, 837.

16. N. S. Grace, *J. Phys. Chem.* **37**: 347 (1933).
17. Werner, *Nevere Anschauungen*, Ed. 3, 1913, p. 110, from Sidgwick, *The Chemical Elements and Their Compounds*, *Vol. II*, p. 1191, Clarendon Press, London, 1952.
18. R. L. Scott, *J. Am. Chem. Soc.* **75**: 1550 (1953).
19. H. W. Cremer and D. R. Duncan, *J. Chem. Soc.* 2750 (1930).
20. A. G. Sharpe and H. J. Emeleus, *J. Chem. Soc.* 2135 (1948).
21. A. A. Banks, H. J. Emeleus, R. N. Haszeldine, and V. Kerrigan, *J. Chem. Soc.* 2188 (1948).
22. I. I. Chernyaev, N. S. Nikolaev, and E. G. Ippolitov, *Dok. Akad. Nauk, S.S.S.R.* **130**: 1041 (1960).
23. *Ibid.* **132**: 378 (1960).
24. A. Haim and H. Taube, *J. Am. Chem. Soc.* **85**: 3108 (1963).
25. R. N. Keller, *Inorganic Synthesis* **2**: 247 (1946). G. B. Kauffman and D. O. Cowan, *Inorganic Synthesis* **7**: 240 (1963).
26. C. M. Davidson and R. F. Jameson, *Trans. Farad. Soc.* **61**: 133 (1965).
27. G. B. Kauffman and L. A. Teter, *Inorganic Synthesis* **7**: 232 (1963).
28. R. N. Keller, *Inorganic Synthesis* **2**: 250 (1946).
29. H. W. Cremer and D. R. Duncan, *J. Chem. Soc.* 1859 (1931).
30. V. J. Shiner and M. L. Smith, *Anal. Chem.* **28**: 1043 (1956).
31. R. Dreyer, *Physik. Chem.* **29**: 347 (1961).

GAS CHROMATOGRAPHIC AND RELATED STUDIES OF METAL COMPLEXES

R. E. Sievers

Aerospace Research Laboratories, ARC
Wright–Patterson Air Force Base, Ohio
and
Chemisches Institut der Universität
Tübingen, Germany

INTRODUCTION AND GENERAL SURVEY

In the few years since its discovery, gas chromatography has revolutionized the conduct of research in organic chemistry and biochemistry. Its application to problems in inorganic chemistry has been slower in coming, but it is now clear that gas chromatography has much to offer to the inorganic chemist as well. For a more detailed discussion, bibliography, and historical development of inorganic studies, the reader is referred to the book entitled *Gas Chromatography of Metal Chelates*. Many more recent papers are listed in Ref. 2 and 3.

The properties that make gas–liquid chromatography so attractive are extraordinary sensitivity, the speed and ease of separating mixtures of closely related compounds, and the easy rapid measurement of thermodynamic data. The great sensitivity of electron capture detectors has made it possible to use gas chromatography for ultratrace metal analysis.[1,2] The lower limit of detectability for some metals is of the order of 10^{-13}g. Consequently, for some elements the gas chromatographic method is more sensitive than atomic absorption, neutron activation, or any other known technique. The most significant areas in which gas chromatography is useful are studies of stereochemistry, ligand exchange, isomerization, metal-ligand stoichiometry, interactions of complexes with weak donors, separation of geometrical and optical isomers, and ultratrace metal analysis.[1]

For most purposes, but not all, the metal complex must exhibit a vapor pressure of the order of 0.1 to 1 mm in order to have gas-phase migration through the column at a reasonable rate. This means that the complexes must be unusually volatile, and thermally and solvolytically stable in the chromatographic column. Several classes of compounds have been studied, e.g., β-diketonates, carbonyls, alkyls, cyclopentadienyls, fluorides, chlorides, bromides, iodides, etc. The β-diketonates have been the most thoroughly examined. The ligands of principal interest are:

$$\begin{array}{c} O \quad\quad\ O^- \\ \| \quad\quad\quad | \\ R^1-C-CH=C-R^2 \end{array}$$

CH_3-	$-CH_3$	acac
CF_3-	$-CH_3$	tfa
CF_3-	$-CF_3$	hfa
$C(CH_3)_3-$	$-C(CH_3)_3$	thd
$CF_3CF_2CF_2-$	$-C(CH_3)_3$	fod
$CF_3CF_2CF_2-$	$-CF_3$	dfhd

The classical work of Morgan and Moss[4] and others showed that certain of the acetylacetonates are volatile and can be sublimed. Unfortunately, most of the acetylacetonates are not sufficiently volatile and thermally stable to be chromatographed in the gas phase. Noteworthy exceptions are the beryllium, aluminium, scandium, and chromium complexes. However, it is now well established that the volatility of complexes and the ease with which they can be eluted is greatly increased by substitution of fluorine for hydrogen in the "ligand shell." For example, the vapor pressures of various chromium(III) chelates at 125° are: $Cr(acac)_3$, 0.01 mm; $Cr(fod)_3$, 0.6 mm; $Cr(tfa)_3$, 0.7 mm; and $Cr(hfa)_3$, 50 mm.[5] In practical terms, it is more useful to consider what temperatures will be required for the compounds to exhibit a vapor pressure of 1 mm. The temperatures for the four compounds are, 179, 133, 128, and 67°, respectively.[5] With nonpolar liquid stationary phases the ease with which the complexes can be eluted roughly parallels their saturation vapor pressures. While column temperatures between 150 and 220° are required for the acetylacetonates (often with extensive sample decomposition), much lower column temperatures (100 to 150°) can be used for the trifluoroacetylacetonates and still lower (30 to 80°) for the more extensively fluorine-substituted hexafluoroacetylacetonates.[6,7] This discovery has reduced problems with sample decomposition and has broadened greatly the applicability of gas chromatography to trace metal analysis. It has also led to the development of an empirical rule that enables one to predict the identity of unknown compounds from a knowledge of their chromatographic retention behavior: *For complexes of a given metal ion, increasing substitution of fluorine for hydrogen reduces the retention time.* We have observed more than fifty examples of this phenomenon. Thus, it was possible to predict the identities of the peaks shown in Fig. 1 with a rather high level of confidence before confirmation by other independent techniques. The complex with 18 fluorines is eluted more readily than that

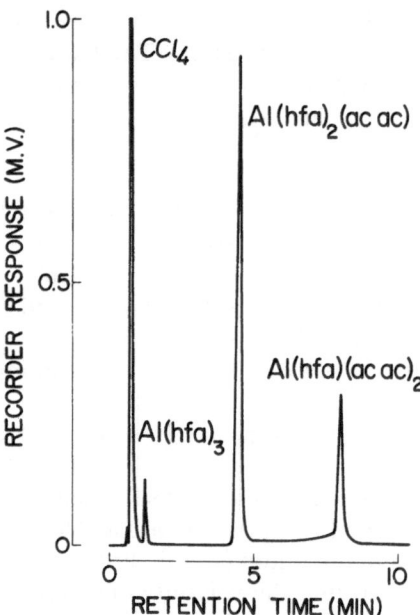

Fig. 1. Gas chromatography of mixed-ligand complexes.[1] Glass column: 4 ft × 4 mm i.d., packed with 5% D.C. Silicone Grease on Chromosorb W. Col. temp. programmed from 65 to 135° at 7.9° min^{-1}. Sample: 0.5 μl of a mixture of the complexes in CCl_4.

with 12 fluorines, which in turn has a shorter retention time than that with 6 fluorines. Unfluorinated Al(acac)$_3$ has a much longer retention time yet.

Figure 2 shows a summary of some of the many gas chromatographic studies that have been made on the various chelates.[1,3,6,8–12] The complexes with three asterisks can be eluted without apparent evidence of decomposition. The complexes denoted with two asterisks produce chromatographic peaks, but with these compounds there is evidence of some decomposition. A single asterisk marks complexes known to be volatile at temperatures at which their chromatography should be possible, but their properties have not yet been thoroughly examined. In the experiments of Genty et al.,[13] radioactive tracers were employed to determine whether any part of the samples was irreversibly adsorbed or decomposed in the injection port or chromatographic column. This is experimentally more difficult to perform, but it is more definitive than visual examination of the injection port liner for sample decomposition. Sample adsorption and instability become most important when gas chromatography is used for ultratrace metal analysis. In other applications some sample loss can occasionally be tolerated.

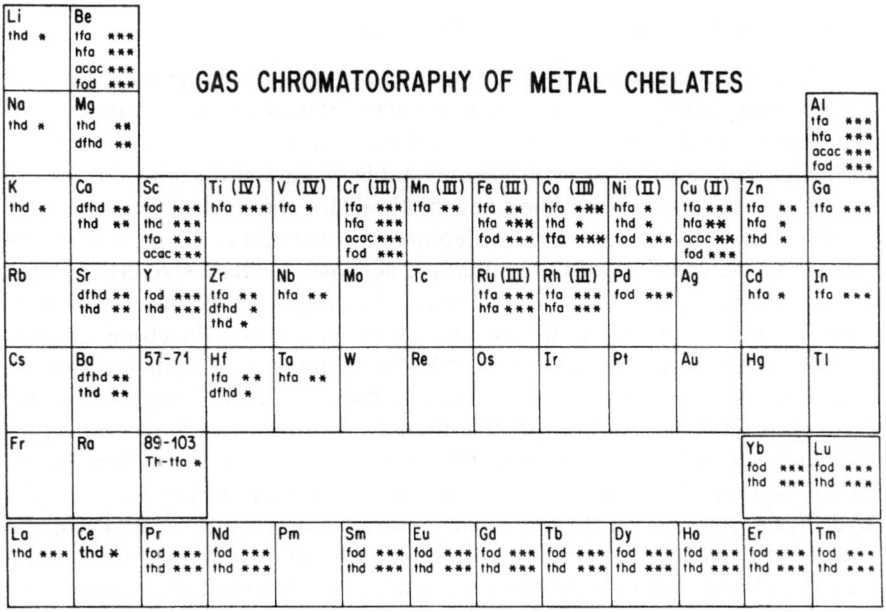

GAS CHROMATOGRAPHY OF METAL CHELATES

Li	Be	Sc	Ti (IV)	V (IV)	Cr (III)	Mn (III)	Fe (III)	Co (III)	Ni (II)	Cu (II)	Zn	Ga / Al
thd *	tfa ***, hfa ***, acac ***, fod ***											Al: tfa ***, hfa ***, acac ***, fod ***
Na thd *	**Mg** thd **, dfhd **											
K thd *	**Ca** dfhd **, thd **	**Sc** fod ***, thd ***, tfa ***, acac ***	**Ti (IV)** hfa ***	**V (IV)** tfa *	**Cr (III)** tfa ***, hfa ***, acac ***, fod ***	**Mn (III)** tfa **	**Fe (III)** tfa **, hfa ***, fod ***	**Co (III)** hfa ***, thd *, tfa ***	**Ni (II)** hfa *, thd *, fod ***	**Cu (II)** tfa ***, hfa **, acac **, fod ***	**Zn** tfa **, hfa *, thd *	**Ga** tfa ***
Rb	**Sr** dfhd **, thd **	**Y** fod ***, thd ***	**Zr** tfa **, dfhd *, thd *	**Nb** hfa **	**Mo**	**Tc**	**Ru (III)** tfa ***, hfa ***	**Rh (III)** tfa ***, hfa ***	**Pd** fod ***	**Ag**	**Cd** hfa *	**In** tfa ***
Cs	**Ba** dfhd **, thd **	**57–71**	**Hf** tfa **, dfhd *	**Ta** hfa **	**W**	**Re**	**Os**	**Ir**	**Pt**	**Au**	**Hg**	**Tl**
Fr	**Ra**	**89–103** Th-tfa *									**Yb** fod ***, thd ***	**Lu** fod ***, thd ***

La	Ce	Pr	Nd	Pm	Sm	Eu	Gd	Tb	Dy	Ho	Er	Tm
thd ***	thd *	fod ***, thd ***	fod ***, thd ***		fod ***, thd ***	fod ***, thd ***	fod ***, thd ***	fod ***, thd ***	fod ***, thd ***	fod ***, thd ***	fod ***, thd ***	fod ***, thd ***

Fig. 2. Gas chromatography of metal chelates.

All of the ligands shown in Fig. 2 are oxygen donors. Very recently, we have synthesized complexes with sulfur donors of the type shown in Fig. 3.[16] In some cases they are more volatile than the corresponding β-diketonates. For example, the complex in which M is nickel, X is oxygen, and y is 2 is more volatile than any other known nickel(II) chelate. It sublimes *in vacuo* at room temperature and is presently being examined by gas chromatography. The corresponding green nickel(II) hexafluoroacetylacetonate occurs as a dihydrate; the coordinated water may account for the lower volatility relative to the sulfur containing complex.

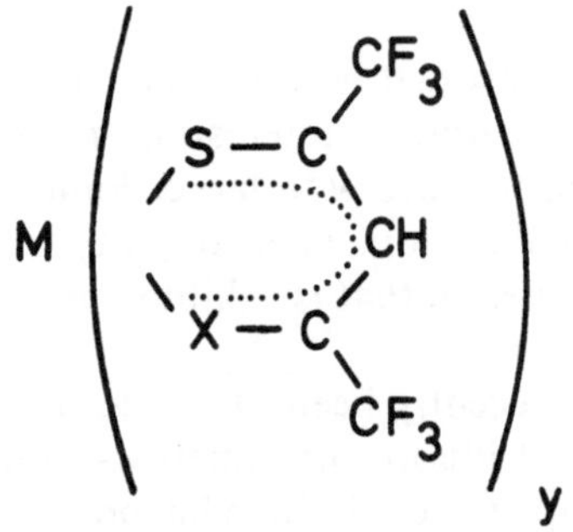

Fig. 3. New volatile complexes with sulfur donors.[16]

VOLATILE LANTHANIDE CHELATES

It will be noted that Fig. 2 shows examples of volatile complexes of almost every class of metals. Of particular interest are the complexes of the lanthanides. For many years workers have been searching for complexes of the lanthanides that are volatile and stable.[14] Part of the stimulus for the search was the desire to develop new methods for the separation of the rare earths based upon hoped-for differences in volatility. Early statements regarding the volatility of the lanthanide acetylacetonates were later shown to be incorrect.[14] In 1940 at the University of Illinois, Brimm found that lanthanide complexes of dibenzoylmethane and benzoylacetone are decomposed by traces of moisture with the formation of hydroxy species, which are nonvolatile.[15] Most of the tris β-diketonates of the lanthanides have coordination numbers greater than six and occur as hydrates. Other evidence suggested that the presence of water in the coordination sphere of the lanthanide chelate contributes to thermal instability due to self-hydrolysis of $Ln(acac)_3(H_2O)_{1-3}$ at elevated temperature.[17] In 1960, the search for volatile lanthanide compounds was again intensified, with the added motivation that if suitable compounds could be found, their gas chromatography and the trends in volatility exhibited might prove interesting. In the early stages of this research, it was reasoned that the nature of the "ligand shell" should greatly influence such physical properties of the complexes as volatility.[18] The striking effect of incorporating fluorine has already been mentioned above. A more subtle but equally important facet of this concept is the control of the number and type of coordinated groups by careful selection of the substituents attached to the donors. Therefore, one might conclude that the most promising approach to obtain anhydrous, stable, volatile tris complexes would be to choose ligands that are sufficiently bulky to decrease the likelihood that hydrates would be formed. An extensive investigation of the lanthanide complexes of thd revealed that they can be isolated as anhydrous tris chelates with unprecedented volatility and thermal stability.[9-11]

Examination of the chromatographic data for complexes of $Ln(thd)_3$ and $Ln(fod)_3$ revealed an extremely interesting trend depicted in Fig. 4.[9-11] The complexes of the lanthanides with larger ionic radii were eluted much later than those with smaller radii. This volatility trend was later shown to be a quite general phenomenon that is exhibited by other classes of chelates as well.[10,12,19-23]

This trend has very recently been observed by Shigematsu et al.[20] for tris complexes of 1,1,1,-trifluoro-5,5-dimethyl-2,4-hexanedione. Thermogravimetric data for complexes of all three ligands and gas chromatographic data with different liquid phases indicate that the phenomenon does not

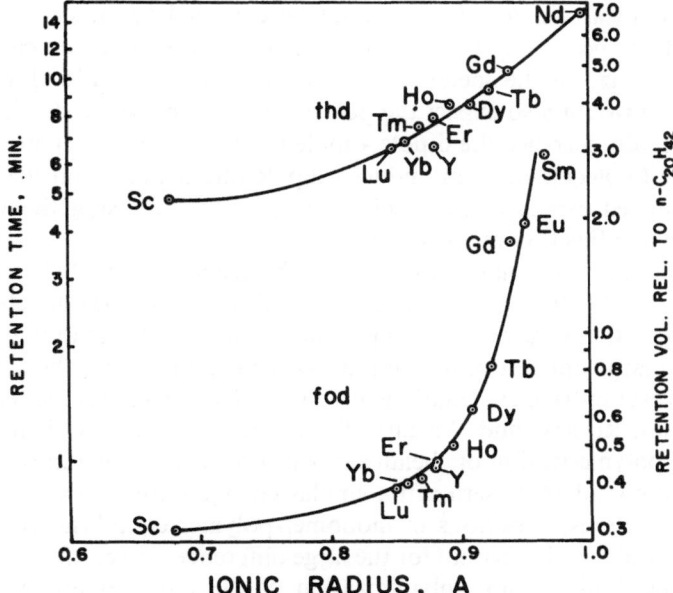

Fig. 4. Gas chromatographic retention data as a function of the ionic radius of the metal.[11] Column temp., 171°. Liquid phase, polydimethylsiloxane. Sample: Ln(thd)₃ or Ln(fod)₃.

arise from any specific solvation effects in a particular liquid phase, but rather reflects actual changes in the volatility of the complexes.[11,19,20]

Richardson[21] has synthesized other interesting types of volatile lanthanide complexes of the form Ln(dfhd)₃(2,2-dimethoxypropane) and NH₄Ln-(dfhd)₄. Preliminary thermogravimetric data indicate that these tetrakis and tris-adduct complexes exhibit the same volatility trend as the simple tris complexes. The tetrakis complexes probably undergo sublimation as ion pairs. Thermogravimetric data[21] for a series of erbium(III) chelates indicates the following order of increasing volatility: $Er(thd)_3$ < NH_4Er-$(hfa)_4$ < $Er(fod)_3$ < $Er(dfhd)_3(H_2O)_2$ ≅ $NH_4Er(dfhd)_4$ < $Er(hfa)_3(2,2$-dimethoxypropane). This order must be considered only approximate, however, because some of the complexes undergo slight thermal decomposition.

Several possible explanations can be offered to account for the ubiquitous correlation between the volatility of the complexes and the size of the metal ion. In the case of tris complexes with unsymmetrical ligands such as $Ln(fod)_3$, the molecules have permanent molecular dipoles. Assuming that the size of the molecule decreases with the radius of the metal ion, the size of the molecular dipole should also decrease. Furthermore, all β-diketonate chelates, irrespective of the symmetry of the ligand, have small permanent local dipoles. These would be expected to either decrease in magnitude or become more effectively shielded by a more compact ligand shell as the size of the metal ion is decreased. As the molecules become smaller one would also expect the polarizability to be reduced. All of these factors should decrease the dipole–dipole (molecular or local) interactions, the dipole (molecule or local)-induced dipole interactions, and the induced dipole-induced dipole interactions, consequently increasing the volatility as the ionic radius becomes smaller.[9,10]

Other possible explanations have also been considered. Mass differences do not appear to affect the volatilities noticeably. In Fig. 4, the gas chromatographic retention behavior of $Y(fod)_3$ is almost identical with that of $Er(fod)_3$ in spite of large differences in mass. Because yttrium(III) has the same ionic radius as erbium(III), explanations involving size differences are reinforced. Muetterties, Roesky, and Wright[24] found that there was a change in the extent of polymerization of chelates of γ-isopropyltropolone in progressing through the lanthanide series. If a similar change were to occur in the fod or thd complexes, variations in monomer-polymer equilibria with change in radius could easily account for the large differences in volatility. However, there is no evidence for polymerization of the thd chelates in solution. Molecular weight data obtained in a variety of nonpolar solvents shows that the complexes are monomeric.

Sicre et al.,[23] have made a detailed study of vapor pressures of the $Ln(thd)_3$ complexes as a function of temperature. The Clausius-Clapeyron plots in Figure 5 show that significant differences exist in the vapor pressures of the complexes at a given temperature. For example, at 200° vapor pressures vary from about 0.2 mm for the lanthanum(III) complex up to 5 mm for the ytterbium(III) complex. The plots reflect the effects of the lanthanide contraction with remarkable fidelity, the elements falling in almost perfect order. It is interesting to compare the volatility of the complexes with more familiar organic compounds such as saturated hydrocarbons. If we again select an arbitrary vapor pressure of 1 mm., we find that the lutetium(III) complex exhibits this pressure at about 170°, while the corresponding temperature for n-tetracosane is 184°. This is quite surprising when one compares the empirical formula of the complex, $Lu(C_{33}H_{57}O_6)$, with that of tetracosane, $C_{24}H_{50}$. The extremely low volatility of virtually all previously known lanthanide compounds has often in the past been ascribed to large ionic contributions in bonding. That most of the $Ln(thd)_3$ complexes exhibit higher vapor pressures than a saturated hydrocarbon with a much lower carbon number must be considered remarkable. Even considering shape differences and partial shielding of the six metal–oxygen bonds, the vapor pressures are higher than expected.

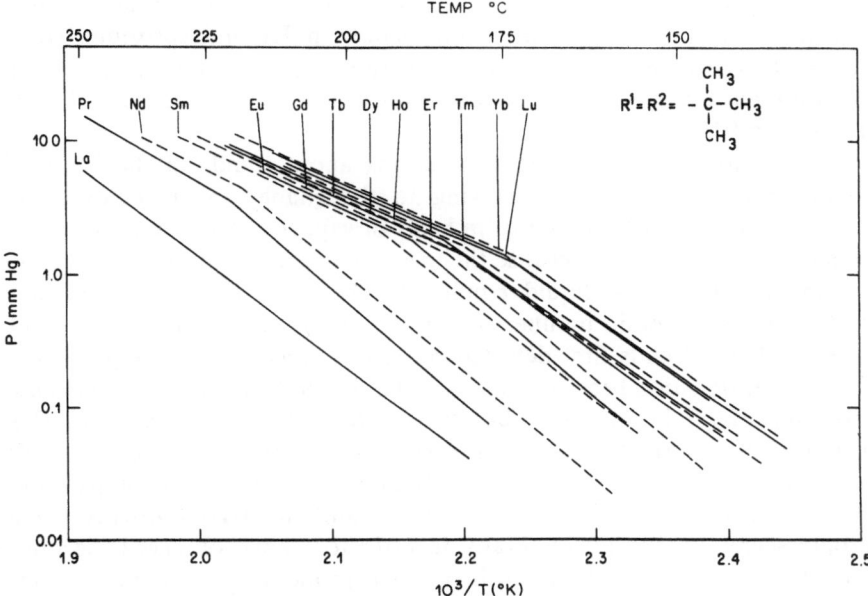

Fig. 5. Clausius–Clapeyron plots for Ln(thd)₃ complexes.[23] For clarity complexes of metals with even atomic number are depicted with dashed lines; no other significance is intended.

The thermodynamics of vaporization are worth considering in more detail for the thd complexes because an excellent opportunity is offered to examine the effects of systematically changing the size of very closely related metal ions. It will be noted in Fig. 5 that all of the Clausius–Clapeyron plots (excepting La) show sharp discontinuities at the experimentally observed[9] melting points of the complexes, a consequence of large heats of fusion (11 to 19 kcal/mole). Analysis of the data in the two segments of each plot yielded some extremely interesting trends in the heats of vaporization and heats of sublimation. The melting point of the lanthanum(III) complex was too high to permit ready measurement of the heat of vaporization, so the plot for this compound represents only solid–vapor equilibria.

The heats of sublimation appear to fall into two main groups; the values for the elements from lanthanum(III) through gadolinium(III) lie between 34.3 and 39.5 kcal/mole, while those for dysprosium(III) through lutetium(III) are almost constant at 31.4 to 32.1 kcal/mole. Terbium(III) is intermediate at 33.8 kcal/mole. The heats of fusion show a similar pattern with a general increase from 13.4 to 18.6 kcal/mole for praseodymium(III) through gadolinium(III), while for dysprosium(III) through lutetium(III) they are essentially constant at 11 to 12 kcal/mole. Essentially, the same pattern is exhibited in plots of the entropy of fusion *vs.* atomic number. All of this suggested a change in crystal structure in the vicinity of terbium(III) or dysprosium(III).[23] Recent X-ray studies have shown that these complexes do, in fact, exist in two different crystallographic modifications, with the change occurring at dysprosium(III).[25]

Attempts to draw further conclusions about trends in the heats of sublimation were not fruitful owing to the change in crystallographic structure. By contrast, the trends in heats of vaporization were expected to be more straightforward because the measurements are all made above the melting points on gas–liquid equilibria. The heats of vaporization are plotted against atomic number in Fig. 6. The gross features of the plot agree with the trends first detected by gas chromatography (Fig. 4); the heats of vaporization become smaller as the size of the metal ion diminishes. However, Fig. 6 shows a small but distinct discontinuity at gadolinum(III). The heat of vaporization for the proseodymium(III) complex is 26.1 kcal/mole, falling to 23.7 for neodymium(III), then to 21.2 for samarium(III) and 20.9 for europium(III), but rising to 21.5 for gadolinium(III) before becoming progressively smaller again. Many investigators have observed discontinuities or so-called "gadolinium breaks" in properties of lanthanide complexes as a function of atomic number.[25,27] For example, discontinuities

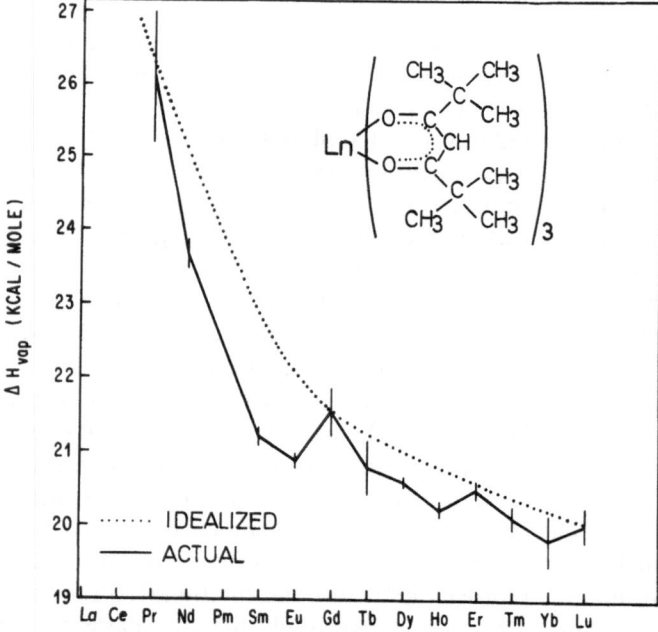

Fig. 6. Heats of vaporization of Ln(thd)$_3$ complexes *vs.* atomic number.[23]

have been found in formation constants, heats of hydration, unit cell volumes, separation factors, *etc.* Unfortunately, the properties examined are so complicated that it was often impossible to determine the exact cause of the discontinuities with confidence. As an illustration, the polyamino-carboxylic acid complexes of the lanthanides show irregularities that can be caused in a number of ways. Possible changes in coordination number, hydration number, and the dentate number of the multidentate ligand can all complicate interpretation of the data. Multidentate ligands such as ethylenediaminetetraacetic acid are able to function with all or only part of the potential donors bonded to the metal, depending on the precise stereo-chemical factors operative. In contrast, the $Ln(thd)_3$ complexes would appear to provide a much simpler model system.

Stavely *et al.*[27] found a relationship very similar to that in Fig. 6 when they plotted the heats of formation of dipicolinate complexes against atomic number. All of the complexes are more stable relative to those of lanthanum-(III), gadolinium(III), and lutetium(III) by amounts of the order of a few tenths of a kcal/mole. They attributed the extra stability to $4f$ electron crystal field stabilization effects on a scale between one and two orders of magnitude smaller than in the $3d$ transition series. Since no stabilization can occur for the unfilled lanthanum(III), the half filled gadolinium(III), and the completely filled lutetium(III), the heats of formation of these complexes represented maxima in the plot. It is possible that the irregularity discovered in the present study also arises from crystal field effects. The focal point of local dipoles responsible for interactions in the melt lies in the Ln-O-C moieties. Any factor which acts to decrease the local dipoles, e.g., to decrease the bond lengths or magnitudes of charge separation, will produce a concomitant reduction in intermolecular forces. As mentioned earlier, the heat of vaporization of the lanthanum(III) complex is not available but it is assumed to be at least as large as that of praseodymium(III), and probably higher. In an idealized model, if one draws a smooth curve through lanthanum(III), gadolinium(III), and lutetium(III) the actual values for the remaining complexes fall below the "unstabilized" values. The effect is most apparent in the section from samarium(III) to europium(III) to gado-linium(III).

One must also consider a different possibility, namely, the occurrence of a discontinuous change in the structure or ordering of the melt. Some support for this explanation is found by comparing entropy change trends. If differences in the structure of the liquid melt exist, the question arises as to whether this is an accidental function of packing or reflects a basic change in the molecular structure, e.g., from octahedral to distorted octahedral or trigonal prismatic. The infrared spectra of the complexes dissolved in CCl_4 are all almost identical, so if a fundamental change in molecular structure occurs, it was not detected. The two explanations offered for the discontinuity are not mutually exclusive and may be interrelated.

The differences in volatility of the Ln(thd)$_3$ chelates have made it possible to separate mixtures of the complexes by fractional distillation.[28] This has been accomplished by using codistilling agents, which act as solvents for the chelates and cause concomitant alterations in the partial pressures exhibited by the complexes.

Other classes of metal chelates have also shown size-related trends in volatility. Schwarberg et al.,[12] have gas chromatographic and thermo-gravimetric data for the complexes of the alkaline earths that show the volatility and ease of elution increasing in the order: Ba(II) < Sr(II) < Ca(II) < Mg(II). Eisentraut[22] has demonstrated that the volatility of the thd complexes of the alkali metals also increases with decreasing ionic radius: K(I) < Na(I) < Li(I). The trifluoroacetylacetonates of aluminum(III), gallium(III), and indium(III) show the same effect. In the light of the continually growing mass of evidence, a second general rule concerning gas chromatographic retention behavior can be advanced: *For a given class of metal ions, the ease of elution of chelates with a common ligand is an inverse function of the ionic radius.*

STEREOCHEMICAL STUDIES

It is in studies of stereochemistry that gas chromatography may be most useful to the coordination chemist. An illustration of the separation of geometrical isomers of Cr(tfa)$_3$ is shown in Fig. 7.[6] The technique is especially effective with mixtures of isomers possessing only subtle differences in structure. The coordination sphere is the same for both isomers; only R^1 and R^2 are different. In the *cis* (or facial) isomer the CF$_3$ groups are mutually adjacent and lie above the upper front face to the octahedron.[29] In the *trans* (or meridianal) isomer one of the CF$_3$ groups is interchanged with a CH$_3$. *Cis-trans* isomers of Rh(tfa)$_3$ have also been separated by gas chromatography,[20] and in both this and the chromium case the somewhat less polar *trans* isomer is eluted before the *cis*. The integrated peak areas showed 79.8 % *trans*-Rh(tfa)$_3$ and 20.2 % *cis*-Rh(tfa)$_3$, in good agreement with the relative yields determined independently by Fay and Piper.[29] Tanikawa et al.,[31] have reported that they were able to achieve a partial separation of *cis-trans*-Ga(tfa)$_3$ at a column temperature of 140°. The appearance of the peaks was very poor, however, and is reminiscent of what one obtains when the sample undergoes partial decomposition in the column. Nuclear magnetic resonance measurements[32] have shown that Ga(tfa)$_3$ undergoes rapid *cis-trans* isomerization in solution at 96°, so it seems unlikely that any significant separation could be effected in the column operated at 140°.

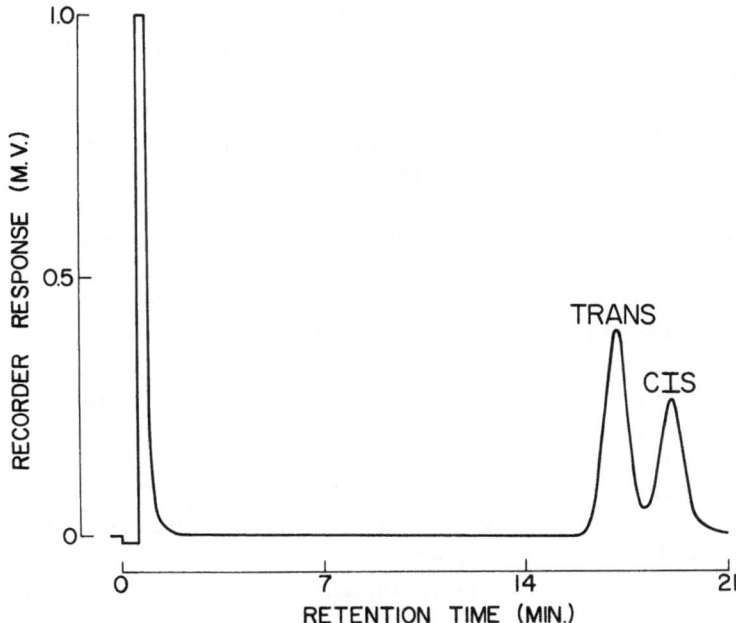

Fig. 7. Separation of geometrical isomers of Cr(tfa)$_3$ by gas chromatography.[6]

Figure 1 demonstrates that mixed-ligand complexes can easily be separated and analyzed by gas chromatography. In this instance both of the ligands are symmetrical. If one or both are unsymmetrical, several geometrical isomers exist.[33] For example, if mixtures of hexafluoroacetylacetone and trifluoroacetylacetone are mixed with chromium(III) nitrate nonahydrate in ethanol, seven compounds are formed with the formulae Cr(tfa)$_x$(hfa)$_{3-x}$. Under various conditions, all seven compounds can be separated by gas chromatography.[1] The peak with the shortest retention time is caused by Cr(hfa)$_3$, the second peak by Cr(hfa)$_2$(tfa), the third through fifth peaks by the three geometrical isomers (*trans-cis*, *cis-cis* and *cis-trans*) of Cr(tfa)$_2$(hfa), and the sixth and seventh by the *trans* and *cis* isomers of Cr(tfa)$_3$. The order of elution is in complete accord with the first rule stated above. Data of this type with even more complicated ligand systems should prove extremely useful in studies of the kinetics and equilibria of isomerization, ligand substitution, and ligand-exchange reactions.

The separation of optical isomers presents an entirely different problem. By definition, *dextro* and *levo* isomers have identical dipole moments, vapor pressures, *etc.* Therefore, there can be no separation unless an asymmetric environment exists within the column. In gas–liquid chromatography this condition is fulfilled when an optically active liquid is employed as the stationary phase. Separation is achieved by stronger stereospecific solvation of one isomer relative to the other by the optically active liquid. In gas–solid chromatography the solid adsorbent must be optically active, and in this instance stereospecific adsorption is the key to successful separation. Sievers, Moshier, and Morris[7] achieved a partial separation of *dextro-* and *levo-* Cr(hfa)$_3$ by gas–solid chromatography using a column filled with powdered quartz, prepared by pulverizing a large single crystal of *dextro* quartz. Early attempts using gas–liquid chromatography were unsuccessful,[6] but the recent work of Feibush and Gil-Av[34] in separating optical isomers of various organic compounds proves clearly what a powerful tool this can be. It has been proposed that it should be possible to determine absolute configurations of closely related optically active compounds based solely on gas chromatographic retention measurements.[18,34,35] This requires only a knowledge of the absolute configuration of a reference isomeric pair and that the mode of stereospecific solvation (or adsorption) is not significantly different for the unknown enantiomers. The validity of this hypothesis was recently confirmed experimentally in gas chromatographic retention studies of derivatives of optically active amines with known absolute configurations.[34] In each case, the isomer with one configuration emerged from the column before its antipode.

Since the time of Werner, chemists have been fascinated by optical isomers of octahedral chelates with bidentate ligands and have puzzled over the possible mechanisms by which one isomer undergoes optical inversion to form its mirror image. Several possible mechanisms have been advanced, some requiring the rupture of one or more bonds plus bond reorganization, and others involving only internal reorganization of the bonds without bond rupture.[14,32,33,36–43] In the Thomas mechanism, the bidentate ligand is completely dissociated, and, following reorganization of the remaining bonded ligands, the dissociated ligand reattaches to regenerate the same or the opposite isomer.[43] In the Werner mechanism one bond to one of the bidentate ligands must rupture, followed by reorganization of the remaining bonds to form a trigonal bipyramid; and then the ruptured ligand again reattaches, followed again by reorganization of the remaining bonds.[42] Other studies have been performed in which the solvent is thought to play a role in optical inversion. Some mechanisms that do not require bond rupture, but rather only bond reorganization are depicted in Fig. 8.[36–39] The case is selected in which the ligands are unsymmetrical so as to also show whether geometrical isomerization accompanies optical inversion. The "twist" mechanisms have been suggested by some authors as possible modes of racemization. For some β-diketonates and other chelates inferential evidence has suggested that isomerizations occur principally by bond rupture plus bond reorganization rather than by bond reorganization without bond

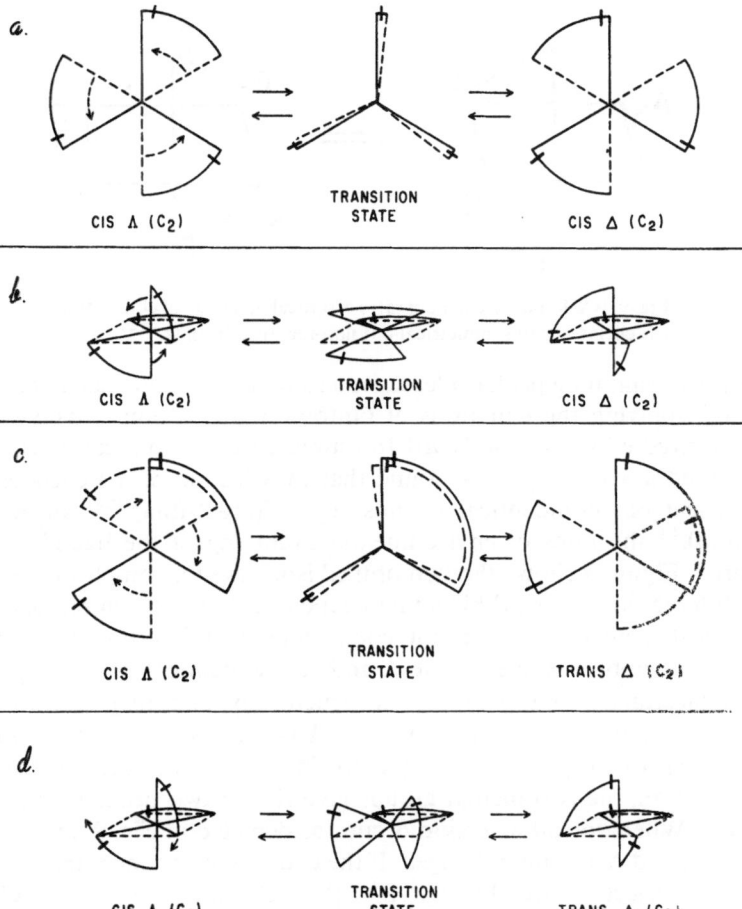

Fig. 8. (a) The Bailar twist about the C_3 axis of a tris chelate as viewed along the C_3 axis. (b) The rigid-ring analog of the Bailar twist about the C_3 axis. (c) The Bailar twist about an imaginary C_3 axis. (d) The Ray–Dutt rigid-ring analog of the Bailar twist about an imaginary C_3 axis.[39]

rupture. However, in most cases there is no truly definitive evidence that allows a clear choice between the Werner bond rupture mechanism and the various "twist" mechanisms. The stereochemical rigidity of six-coordinate complexes, though greater than that of five- or seven-coordinate species, should not be inordinately great. For a more detailed discussion of this topic the reader should refer to the recent interesting paper by Muetterties.[38]

Fay and Piper[32] used NMR to measure the rates of geometrical isomerization of octahedral chelates with unsymmetrical ligands, such as

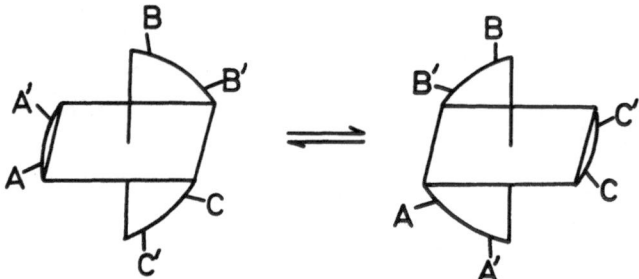

Fig. 9. NMR measurement of rates of optical inversion of octahedral tris chelates with symmetrical but non-identical ligands.[40,41]

Al(tfa)$_3$, by examining peak coalescence as a function of increasing temperature and applying the equations of Gutowsky and Holm.[44] This elegant work inspired efforts to use NMR to estimate rates of optical inversion of tris octahedral chelates. It was found that by selection of tris chelates with symmetrical but nonidentical ligands, e.g., Al(acac)$_2$(hfa)[40,45] or Al(acac) (hfa) (thd),[41] the rates of environmental averaging of the ligands can be measured. Figure 9 shows the two optical isomers of a complex containing three different symmetrical bidentate ligands. It should be emphasized that geometrical isomerization cannot complicate data interpretation in this system; only optical isomers exist. A and A' are identical groups, e.g., CH$_3$ in acetylacetonate, but they are in structurally and magnetically nonequivalent positions. In each isomer one A is *trans* to a B group while the other is *trans* to a C group. Consequently, in the absence of rapid structural reorganization, the two methyl groups give rise to two peaks in the NMR spectrum. When optical inversion occurs, as depicted in Fig. 9, the environments of A and A' are interchanged. If the end groups on the remaining two ligands are selected carefully, e.g., B = B' = CF$_3$ and C = C' = C(CH$_3$)$_3$, they too will each appear as two peaks in the fluorine and proton spectra.

As the temperature is raised each pair of peaks coalesces, and the rate of environmental averaging for each group can be calculated and related to the rate of optical inversion. Two important features of this technique should be stressed. First, the measurements are made on racemic mixtures at equilibrium; no resolution of isomers is required. Second, measurements can be made on compounds which racemize much too fast to allow study by classical techniques.

METAL ANALYSIS BY GAS CHROMATOGRAPHY

More studies have been conducted in this area than in any other. In the limited space available it is not possible to discuss this topic in detail but most of the work is discussed in the references cited.[1–3,6,8–13,18,30,31,]

[46-55] The principle is extremely simple; the sample to be analyzed is converted by solvent extraction or by simple homogeneous or heterogeneous reactions to volatile metal complexes which are then simultaneously separated and analyzed by gas chromatography. Although several ligands have been examined, trifluoroacetylacetone is used in quantitative analysis more than any other, owing principally to the easy formation and chromatography of the chelates, and the extraordinary sensitivity with which they can be detected by electron capture detectors.

Quantitative gas chromatographic schemes now exist for the determination of beryllium in blood, urine, and tissue,[2,46] chromium in serum,[47] aluminum in uranium,[13] aluminum, gallium, and indium, in aqueous solutions,[48,50] iron in ore,[3] chromium in steel,[52] titanium in bauxite,[53] aluminum, iron, and copper in alloys,[49,51] uranium, tungsten and molybdenum in alloys and ores,[54] and the list continues to grow rapidly. In the ultratrace analysis of beryllium the lower limit of detectability is *ca.* 10^{-13} g.[2] The gas chromatographic determination of beryllium can be accomplished using only a single drop of blood (50 μl) or other biological fluid or tissue, making the technique extremely useful in biological studies.[46] A modification of this method is now in regular use in studies of pollution control.

STUDIES OF NONVOLATILE COMPLEXES BY GAS CHROMATOGRAPHY

Essentially all of the studies discussed above deal with volatile metal complexes that are carried in the gas stream through the column. Since the vast majority of metal coordination compounds are not volatile, it should be emphasized that gas chromatography can still yield valuable information in investigations of these as well. One must only reverse the operational procedure and use the metal complex as the stationary phase. Then it is possible to study interactions with various volatile weak-donor ligands that are passed through the column in the gas phase. In the early days of gas chromatography, Bradford *et al.*,[56] and Bayer[57] demonstrated that if the stationary liquid phase contained an appropriate metal ion, the elution of organic compounds with certain functional groups was slower than in the absence of the metal. When silver nitrate was dissolved in the liquid phase, the elution of olefins was significantly retarded. When nickel caproate was present in the liquid, esters of amino acids showed markedly longer retention times. In both cases, this can be attributed to complex formation. Cartoni, Lowrie, Phillips, and Venanzi[58] employed N-dodecylsalicyldimines of of nickel, palladium, platinum, and copper, and methyl-*n*-octylglyoximes of nickel, palladium, and platinum as column liquids. They selected complexes that are coordinatively unsaturated, in hopes of detecting complex formation in the vacant fifth and sixth coordination positions, and found that amines, alcohols, and ketones were preferentially retarded.

Recently, we have studied the interaction of various organic donors with several tris lanthanide chelates of 3-trifluoroacetyl-d-camphorate.[59] The corrected relative retention volumes, α, of potential ligands were determined for liquid phases containing the complexes of lanthanum(III), praseodymium(III), samarium(III), terbium(III), erbium(III), yttrium(III), and lutetium(III) dissolved in squalane. As expected, no differences in interaction were seen for such poor donors as 3-methylcyclohexene or di-n-propyl ether, but with tetrahydrofuran there were large differences in the α values. The complexes containing ions with smaller radii apparently interact with tetrahydrofuran to a greater extent than those with larger radii. When α was plotted against the inverse of the ionic radius a regular increase was observed. Similar behavior was found for esters, ketones, and alcohols. The interpretation of this data is complicated by the fact that these chelates, unlike the monomeric Ln(thd)$_3$, tend to dimerize in solution. Consequently, if it is assumed that the monomer interacts more readily with the ligand carried in the gas phase than the dimer, the position of monomerdimer equilibria can contribute significantly to the differences in α.

Gas chromatographic retention data in some instances can be used in the determination of equilibrium constants, free energies, heats and entropies for the interactions between column liquids containing metal complexes and a great variety of volatile ligands that may be passed through the column.[1,60,61] Stability constants have been determined for sixteen silver-olefin complexes, and it was demonstrated that steric as well as electronic factors affect the stability of the complexes.[60,61] The presence of methyl substituents at the double bond reduces the stability; and, as the size of the substituent is increased the stability is further decreased. The constants obtained by gas chromatography agreed very well with those measured by a much more laborious classical method. This technique merits careful consideration for many types of problems. In future years it may become one of the most important tools available to the coordination chemist for the study of weak-donor interactions.

ACKNOWLEDGMENTS

To Prof. John C. Bailar, Jr., who has been my *Doktorvater* in the fullest sense, will always go my deepest gratitude. I am also grateful to my co-workers, post-docs, and students who have contributed to this research. Finally, my thanks are due to Prof. Ernst Bayer and the other members of the staff of Tübingen Universität for many stimulating discussions and the cooperation and hospitality extended me during 1968–69.

REFERENCES

1. R. W. Moshier and R. E. Sievers *Gas Chromatography of Metal Chelates*, Pergamon Press, Oxford, 1965, and references therein. See also Chapter 8 by R. S. Juvet, Jr. and F. Zado, of *Advances in Chromatography*, Vol. 1, Marcel Dekker, New York, 1966.
2. W. D. Ross and R. E. Sievers, *Talanta* 15: 87 (1968).
3. R. E. Sievers, J. W. Connolly, and W. D. Ross, *J. Gas Chromatography* 5: 241 (1967).
4. G. T. Morgan and H. W. Moss, *J. Chem. Soc.* 105: 189 (1914), *et seq.*, and references therein.
5. R. A. Shulstad and G. I. John, as reported in: R. A. Shulstad, Masters Thesis, Air Force Institute of Technology, June, 1968; W. R. Wolf, G. L. Brown, R. E. Sievers, unpublished data.
6. R. E. Sievers, B. W. Ponder, M. L. Morris, and R. W. Moshier, *Inorg. Chem.* 2: 693 (1963).
7. R. E. Sievers, R. W. Moshier, and M. L. Morris, *Inorg. Chem.* 1: 966 (1962).
8. H. Veening, W. E. Bachman, and D. M. Wilkinson, *J. Gas Chromatography* 5: 248 (1967); H. Veening and J. F. K. Huber, *ibid.* 6: 326 (1968).
9. K. J. Eisentraut and R. E. Sievers, *J. Am. Chem. Soc.* 87: 5254 (1965).
10. C. S. Springer, Jr., D. W. Meek, and R. E. Sievers, *Inorg. Chem.* 6: 1105 (1967).
11. R. E. Sievers, K. J. Eisentraut, C. S. Springer, Jr., and D. W. Meek, *Advances in Chemistry* 71: 141 (1967).
12. R. E. Sievers, J. W. Connolly, A. S. Hilton, M. F. Richardson, J. E. Schwarberg, and W. D. Ross, 155th Amer. Chem. Soc. Mtg., San Francisco, Calif., April 1968; J. E. Schwarberg, R. W. Moshier, and R. E. Sievers, to be published.
13. C. Genty, C. Houin, and R. Schott, 7th International Symposium on Gas Chromatography, Copenhagen, June 1968.
14. J. C. Bailar, Jr. and D. H. Busch, *Chemistry of the Coordination Compounds*, Reinhold, New York, 1956, and reference therein.
15. E. O. Brimm, Ph.D. Thesis, University of Illinois, 1940.
16. E. Bayer, R. Sundermeier, and R. E. Sievers, unpublished data.
17. G. W. Pope, J. F. Steinbach, and W. F. Wagner, *J. Inorg. Nucl. Chem.* 20: 304 (1961).
18. R. E. Sievers, B. W. Ponder, and R. W. Moshier, 141st Amer. Chem. Soc. Mtg., Washington, March 1962; Anon., *Chem. Eng. News* 40: 50 (April 2, 1962).
19. K. J. Eisentraut and R. E. Sievers, *J. Inorg. Nucl. Chem.*, 29: 1931 (1967); J. E. Schwarberg, D. R. Gere, R. E. Sievers, and K. J. Eisentraut, *Inorg. Chem.* 6: 1933 (1967), and references therein.
20. T. Shigematsu, M. Matsui, and K. Utsunomiya, *Bull. Chem. Soc. Japan* 41: 763 (1968).
21. M. F. Richardson and R. E. Sievers, unpublished data; N. J. Rose and others have observed similar volatile tetrakis complexes (e.g., see Ref. 1, p. 36).
22. R. E. Sievers and K. J. Eisentraut, 156 Am. Chem. Soc. Mtg., Atlantic City, Sept. 1968.
23. J. E. Sicre, J. T. Dubois, K. J. Eisentraut, and R. E. Sievers, Proc. 10th Intl. Conf. Coord. Chem., Chemical Soc. Japan, Tokyo, 1967, p. 165; *J. Am. Chem. Soc.*, in press.
24. E. L. Muetterties, H. Roesky, and C. M. Wright, *J. Am. Chem. Soc.* 88: 4856 (1968).
25. E. A. Boudreaux, private communication, Nov. 1967; A. Mode and G. Smith, private communication, Aug. 1968.
26. T. Moeller, D. F. Martin, L. C. Thompson, R. Ferrús, G. R. Feistel, and W. J. Randall, *Chem. Revs.* 65: 1 (1965), and references therein.
27. L. A. K. Staveley, D. R. Markham, and M. R. Jones, *J. Inorg. Nucl. Chem.* 30: 231 (1968), and references therein.
28. C. W. Harris, R. E. Sievers, and K. J. Eisentraut, patents pending.
29. R. C. Fay and T. S. Piper, *J. Am. Chem. Soc.* 85: 500 (1963).
30. W. D. Ross, R. E. Sievers, and G. Wheeler, Jr., *Anal. Chem.* 37: 598 (1965).
31. K. Tanikawa, K. Hirano, and K. Arakawa, *Chem. Pharm. Bull.*, 15: 915 (1967).

32. R. C. Fay and T. S. Piper, *Inorg. Chem.* **3**: 348 (1964).
33. R. A. Palmer, R. C. Fay, and T. S. Piper, *Inorg. Chem.* **3**: 875 (1964).
34. B. Feibush and E. Gil-Av, *Advances in Chromatography 1967*, (ed. A. Zlatkis) Preston Tech. Abstracts Co., Evanston, Ill., p. 100, and references therein.
35. R. E. Sievers, Gas Chromatography 1966, (ed. A. B. Littlewood) Inst. of Petroleum, Elsevier, Amsterdam, 1966, p. 292.
36. J. C. Bailar, Jr., *J. Inorg. Nucl. Chem.* **8**: 165 (1958), Independent proposals by W. G. Gehman, Ph.D. Thesis, Pennsylvania State University, 1954 and L. Seiden, Ph.D. Thesis, Northwestern University, 1957.
37. P. Rây and N. K. Dutt, *J. Indian Chem. Soc.* **20**: 81 (1943).
38. E. L. Muetterties, *J. Am. Chem. Soc.* **90**: 5097 (1968). Construction of simple three-dimensional models that greatly facilitate visualization of various inversion mechanisms is described in J. A. Broomhead, M. Dwyer, and A. Meller, *J. Chem. Educ.* **45**: 717 (1968).
39. C. S. Springer, Jr. and R. E. Sievers, *Inorg. Chem.* **6**: 852 (1967).
40. J. J. Fortman and R. E. Sievers, *ibid.* **6**: 2022 (1967), and references therein.
41. J. J. Fortman and R. E. Sievers, 155th Amer. Chem. Soc. Mtg., San Francisco, Calif., April 1968.
42. A. Werner, *Ber.* **45**: 3061 (1912).
43. A. A. Thomas, *J. Chem. Soc.* **121**: 196 (1922).
44. H. S. Gutowsky and C. H. Holm, *J. Chem. Phys.* **25**: 1228 (1956).
45. R. G. Linck and R. E. Sievers, *Inorg. Chem.* **5**: 806 (1966).
46. M. L. Taylor, E. L. Arnold, and R. E. Sievers, *Anal. Letters* **1**: 735 (1968).
47. J. Savory, P. Musak, N. O. Roszel and F. W. Sunderman, Jr., Federation Proceedings (Biochem.) Abstracts 52nd Meeting, Atlantic City, N.J., April 1968, p. 777.
48. J. E. Schwarberg, R. W. Moshier, and J. H. Walsh, *Talanta* **11**: 1213 (1964).
49. R. W. Moshier and J. E. Schwarberg, *Talanta* **13**: 445 (1966).
50. G. P. Morie and T. R. Sweet, *Anal. Chem.* **37**: 1552 (1965).
51. G. P. Morie and T. R. Sweet, *Anal. Chim. Acta* **34**: 314 (1966).
52. W. D. Ross and R. E. Sievers, 156 Am. Chem. Soc. Mtg., Atlantic City, N.J., Sept. 1968.
53. R. E. Sievers, G. Wheeler, Jr., and W. D. Ross, *Anal. Chem.* **38**: 306 (1966).
54. R. S. Juvet, Jr. and R. L. Fischer, *ibid.* **38**: 1860 (1966), and references therein.
55. W. G. Scribner, M. T. Borchers, and W. J. Treat, *ibid.* **38**: 1779 (1966), and references therein.
56. B. W. Bradford, D. Harvey, and D. E. Chalkley, *J. Inst. Petrol.* **41**: 80 (1955).
57. E. Bayer, "Gas Chromatography 1958" (ed. D. H. Desty) Butterworths, London, 1958, p. 333.
58. G. P. Cartoni, R. S. Lowrie, C. S. G. Phillips, and L. M. Vananzi, *Gas Chromatography 1960* (ed. R. P. W. Scott) Butterworths, London, 1960, p. 273; see also G. P. Cartoni, A. Liberti, and R. Palombari, *J. Chromatography* **20**: 278 (1965).
59. B. Feibush, C. S. Springer, Jr., and R. E. Sievers, unpublished data.
60. C. S. G. Phillips and P. L. Timms, *Anal. Chem.* **35**: 505 (1963), and references therein; M. A. Muks and F. T. Weiss, *J. Am. Chem. Soc.* **84**: 4697 (1962).
61. E. Gil-Av and J. Herling, *J. Phys. Chem.* **66**: 1208 (1962).

LIQUID–LIQUID EXTRACTION OF THE RARE EARTHS AND RELATED ELEMENTS*

D. F. Peppard and G. W. Mason

Argonne National Laboratory
Argonne, Illinois

Our group's current connection with the field of coordination is somewhat tenuous, as attested by almost any one of our publications, the discussion section being liberally sprinkled with "it appears that the coordination number is six," "probably the extracted entity contains coordinated extractant dimers," and similar expressions which confess to factual ignorance but reaffirm an abiding faith in the principles of coordination chemistry. This faith in the principles of coordination chemistry, coupled with a long-held interest in the rare earths, has led to a research program encompassing portions of the fields of interest of two outstanding University of Illinois research chemists, Prof. John C. Bailar, Jr. and Prof. B. S. Hopkins.

In our study of the extraction of the rare earths by acidic phosphorus-based extractants, we have assumed that in the extracted entity the metal is bound, in an undetermined fashion, to coordinated anions of the extractant. The extraction of the trivalent rare earth might then be represented:

$$M_A^{+3} + 3HY_O \rightleftharpoons MY_{3O} + 3H_A^+ \tag{1}$$

where HY represents the monoprotic phosphorus-based extractant (here assumed to be monomeric) dissolved in a carrier diluent sensibly immiscible with water and the subscripts A and O refer to the mutually equilibrated aqueous and organic phases. In most of the studies reported here, M^{+3} is a radioactive element in trace-level concentration, so the extraction of M^{+3} does not produce a measurable change in the concentration of extractant HY or in the concentration of H^+.

*Work performed under the auspices of the U.S. Atomic Energy Commission.

Assuming the applicability of the mass law expression:

$$\frac{[MY_3]_O[H^+]_A^3}{[M^{+3}]_A[HY]_O^3} = k \tag{2}$$

and assuming that MY_3 and M^{+3} are the only significant forms in which the rare earth appears in the mutually equilibrated organic and aqueous phases, respectively, the distribution ratio K for the specific rare earth may be expressed as

$$K = k[HY]_O^3/[H^+]_A^3 \tag{3}$$

since, under these conditions, $K = [MY_3]_O/[M^{+3}]_A$.

In our early studies involving a solution of di 2-ethyl hexyl phosphoric acid, $(C_8H_{17}O)_2PO(OH)$, symbolized as HDEHP, in benzene or toluene as the organic phase, the data fit equation (3) very well. So we were erroneously led to assume the validity of equation (1).[1]

Since, on the basis of fragmentary Z-data, it appeared that a plot of log K vs. Z approximated a straight line indicating, on the average, an r value of about 2.5, where $r = K_{Z+1}/K_Z$, the system was investigated for possible applicability to the mutual separation of gross rare earths. See Fig. 1[2] for this and analogous systems. The extractants referred to in Fig. 1 are, respectively: HEH[ΦP], 2-ethyl hexyl hydrogen phenyl phosphonate, $(C_8H_{17}O)(C_6H_5)PO(OH)$; TBP, tri n-butyl phosphate, $(C_4H_9O)_3PO$; HDHoEP, di hexoxyethyl phosphate, $(n$-$C_6H_{13}OC_2H_4O)_2PO(OH)$; and H_2MEHP, mono 2-ethyl hexyl phosphoric acid, $(C_8H_{17}O)PO(OH)_2$.

The preliminary results were promising, so an 11-stage, center-feed, countercurrent system, using separatory funnels as discrete stages, was set up—the feed being a mixture of rare earths, closely approximating the composition found in monazite, introduced as chlorides.

But reaching steady-state operation in a system of funnels, with some of the rare earths splitting nearly equally between opposing phases, requires an exhaustingly large number of contacts per funnel. Of course we were anxious for data, so the products from the two ends of the system were analyzed—long before the system had reached a steady state—and found to demonstrate an excitingly sharp separation of the rare earths into two groups.

Unfortunately, as the system came closer to a steady state, the organic phase in the central stages became disturbingly more viscous. Finally, after mixing, the two phases in the central stage failed to disengage and the entire funnel contents set to a thin gel. This effectively stopped the operation.

In an investigation involving a single gross rare earth, it was found that gellation occurred when the ratio of the concentration of M in the organic phase to the concentration (in equivalents) of the extractant initially present exceeded 1 : 6, whereas from equation (1) saturation (exhaustion of extractant)

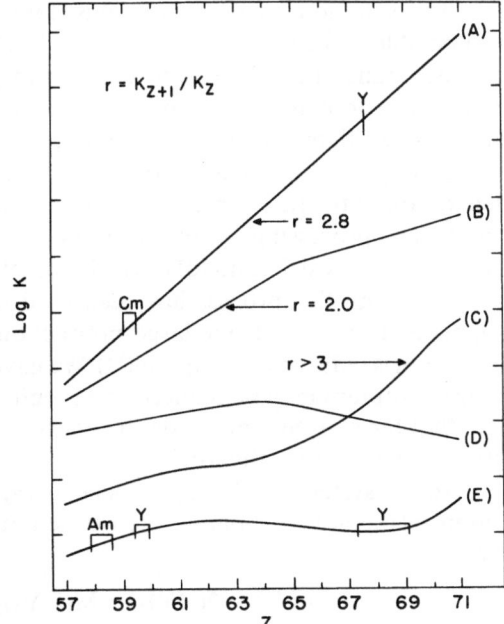

Fig. 1. Variation of log K with Z for lanthanides(III) in selected systems. (Smoothed curves. Log K-values at Z-axis, respectively, for curves A to E, are -7, -4, -2, -5, and 1.)[2] (A) 1.0 F HEH[ΦP] (in toluene) *vs.* 2.0 M HCl; (B) Undiluted TBP *vs.* 15.6 M HNO$_3$, Hesford, Jackson, and McKay; (C) 0.0375 F HDHoEP (in benzene) *vs.* 0.25 M HClO$_4$; (D) 0.71 F TBP (in CCl$_4$) *vs.* 1.96 M HNO$_3$; (E) 0.10 F H$_2$MEHP (in toluene) *vs.* 0.25 M HCl.

should occur at $1:3$. Attempted rationalization of the factor of 2 suggested a dimeric state for the extractant, so that equation (1) would be replaced by:

$$M^{+3} + 3(HY)_2 \rightleftharpoons M(HY_2)_3 + 3H^+ \qquad (4)$$

the subscripts A and O being understood. The argument was completed by noting that 1 mole of M^{+3} would saturate 3 *moles* of extractant, or 6 *equivalents* of it, according to equation (4). Presumably, then, as more M^{+3} was added the protons of the extracted entity would be displaced through bonding of Y to another metal cation, resulting in the formation of 3-dimensional polymers and leading to gellation.

This explanation seems reasonable, now that HDEHP has been established to be dimeric in dry benzene, as shown by both cryoscopic and isopiestic studies.[3–5]

After publishing what we considered the startling fact that a (AlkO)$_2$-PO(OH), where Alk is 2-ethyl hexyl, is an extremely stable dimer in benzene, we discovered that Kosolapoff and Powell[6] had, nearly seven years earlier, shown (Alk)$_2$PO(OH), where Alk is *n*-butyl or *n*-hexyl, to be dimeric in naphthalene. It is true that the existence of di-alkyl phosphinic acids as dimers doesn't prove that di-alkyl phosphoric acids are dimers, but it would certainly have provided a valuable hint.

The gellation phenomenon constitutes an embarrassment but not an insuperable difficulty in applying the "HDEHP in carrier diluent vs. an aqueous mineral acid" system to the individual mutual separation of rare earths. The process has been patented and is presently in use in at least three large rare earth production facilities in this country. It has proved applicable in four ways: (a) the isolation of rare earths and yttrium as a group; (b) the separation of the rare earths and yttrium into two groups to be used as feed for further processing by this process or another; (c) the isolation of an individual rare earth; and (d) the isolation of yttrium.

Although this process has been used to isolate europium it has been supplanted in at least one large production facility by a process in which europium is reduced to europium(II). We have found that even in the 10^{-10} M range europium may be reduced to europium(II). Advantage is taken of the fact that for a given lanthanide the distribution ratio K for the M(III) is far greater than that for the M(II).

In the system employing toluene as carrier diluent and chloride as the anion in the aqueous phase, the extraction of europium(II) is represented by[7]:

$$M^{+2} + 3(HY)_2 \rightleftharpoons M(HY_2)_2(HY)_2 + 2H^+ \qquad (5)$$

So, from equations (4) and (5) it is apparent that the K values for europium(III) and europium(II) are respectively represented, in terms of extractant concentration in the organic phase and hydrogen ion concentration in the opposing aqueous phase, by

$$Eu(III): K_3 = k_3[(HY)_2]^3/[H^+]^3 \qquad (6)$$

$$Eu(II): K_2 = k_2[(HY)_2]^3/[H^+]^2 \qquad (7)$$

From equations (5) and (7) it is evident that the extractant dependency of the K for a cation is not a valid indicator of the numerical value of the charge on the cation.

From a combination of (6) and (7), β, the ratio K_3/K_2, is expressed as

$$\beta = (k_3/k_2)/[H^+] \qquad (8)$$

Specifically, under one set of operating conditions the ratio of the K for europium(III) to that for europium(II) is greater than 10^5. Consequently, all of the trivalent rare earths and yttrium report in high yield in the organic phase while europium(II) reports principally to the aqueous phase.[7]

In general, in systems employing a mono-acidic phosphorus-based extractant and which are represented by equation (4), yttrium ($Z = 39$) reports in the Ho ($Z = 67$)–Er ($Z = 68$) region and Sc ($Z = 21$) extracts with a K far greater than that for any rare earth. Consequently, separation of scandium from yttrium and all rare earths is readily accomplished. Since the K is so large the real problem lies in returning scandium to an aqueous phase.

In the isolation of cerium, advantage is taken of the tetravalent state. In the system 0.15 F HDEHP (*n*-heptane) *vs.* 10 M HNO_3, the ratio of the respective K values for cerium(IV) and cerium(III) is greater than 10^6, the former being greater than 10^2 and the latter less than 10^{-4}.[8] Under the same conditions the respective K values for berkelium(IV) and berkelium(III) are greater than 10^3 and less than 5×10^{-3}, the β for berkelium in the two valence states being greater than 10^5.[9]

From the data for europium, cerium, and berkelium it may be assumed, as a working hypothesis, that for a given lanthanide or actinide which may exist in two or more of the II, III, and IV valence states the K values will be ordered (IV) > (III) > (II). In the instances quoted the "greater than" represents factors exceeding 10^5.

These valence state effects have been used to advantage in devising a procedure for the separation of neptunium, Np ($Z = 93$) from fission-product rare earths and yttrium and from uranium, U ($Z = 92$) and from plutonium, Pu ($Z = 94$); americium, Am ($Z = 95$); and curium, Cm ($Z = 96$).

In this system, a di-acidic extractant, mono 2-ethyl hexyl phosphoric acid, $(C_8H_{17}O)PO(OH)_2$, H_2MEHP is employed. The carrier diluent is toluene and the aqueous phase 12 M or 6 M HCl.[10] In the system, 0.48 F H_2MEHP (toluene) *vs.* HCl (6–12 F range) the K for neptunium(IV) exceeds that for neptunium(VI) by a factor greater than 10^5. [It should be noted that both neptunium(VI) and uranium(VI) are oxygenated ions formulated as MO_2^{+2}.] See Figs. 2[10] and 3.[11] The K for neptunium(IV) exceeds that for any M(III) or M(VI) lanthanide or actinide by such a large factor that its isolation in the organic phase as a pure product is readily accomplished. But as in the isolation of scandium, the difficulty lies in returning it to an aqueous phase.

A simple way to accomplish this return is to utilize the antisynergistic (or inhibiting) effect of tri butyl phosphate, $(n\text{-}C_4H_9O)_3PO$, TBP, upon the extraction of M(IV) and M(III) actinides and lanthanides,[12] under operating conditions (H_2SO_4 as aqueous phase) 1.8 F TBP depressing the K for neptunium(IV) by a factor greater than 10^4, as shown in Fig. 4.[10]

The system has been used to effect the isolation and final purification of gram quantities of Np^{237}. Interestingly, the only difficulty encountered is in maintaining uranium in the uranium(VI) state, since in this system uranium(IV) closely resembles neptunium(IV) in its extraction behavior. So after two to three aqueous HCl scrubs to remove contaminants from the neptunium product in the organic phase an oxidation cycle must be inserted to oxidize uranium(IV) to uranium(VI). Under the oxidizing conditions used, neptunium(IV) is oxidized to neptunium(VI). But through the action of hydroquinone, following the oxidation step, neptunium(VI) is rapidly reduced to

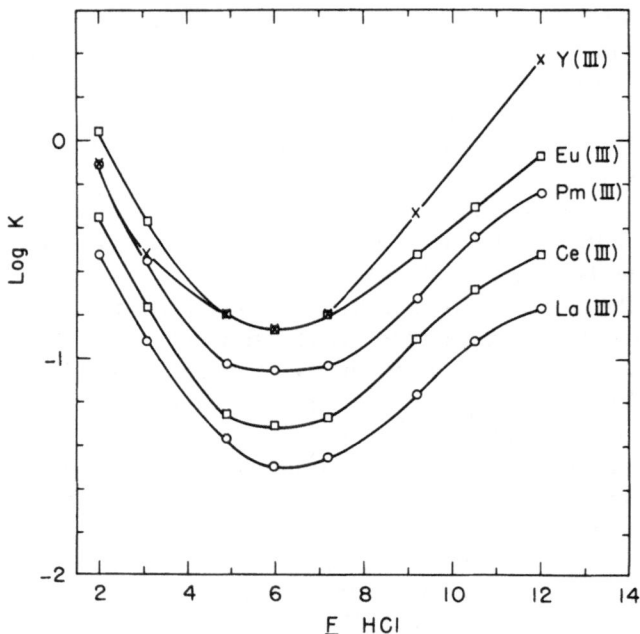

Fig. 2. Extraction of selected lanthanide (and yttrium) cations into
0.48 F H_2MEHP (toluene) as a function of HCl concentration.[10]

neptunium(IV), whereas uranium(VI) is very slowly reduced to uranium(IV).
Consequently, successful application of the process is dependent upon the
differing rates of approach to oxidation–reduction equilibrium for neptunium
and uranium.

Although the H_2MEHP (toluene) *vs.* 6–12 M HCl system has been
employed successfully in a production facility for gram quantities of neptunium, it has not been studied in a basic way. Whether the extracted M(III),
M(IV) and M(VI) entities contain anion of the mineral acid employed has not
been determined.

However, in the lower range of acidity the extractant is shown to extract
by a "large aggregate-proton displacement" mechanism[13] the extraction for
M^{+3} being represented as

$$M^{+3} + (H_2Y)_x \;\rightleftharpoons\; M(H_{2x-3}Y_x) + 3H^+ \tag{9}$$

where H_2Y represents H_2MEHP which has been shown to be highly polymeric.[14]

The extraction of UO_2^{+2} is similarly represented:

$$UO_2^{+2} + (H_2Y)_x \;\rightleftharpoons\; UO_2(H_{2x-2}Y_x) + 2H^+ \tag{10}$$

Whether Th^{+4} would also display a first-power extractant dependency could not be determined since the K values were too high to be measured with the required precision.

The fact that equation (4) represents the extraction of lanthanides(III) and actinides(III) suggests the use of such a system in the determination of stability constants of chloride and nitrate complexes of these elements in the aqueous phase. Assuming that M^{+3} and MX^{+2} are the only species of significance present in the equilibrated aqueous phase containing 1 M $(HClO_4 + HX)$, the K for M(III) may be represented as

$$K = [M(HY_2)_3]/\{[M^{+3}] + [MX^{+2}]\} \tag{11}$$

from which it follows from equation (4) that

$$K = k[(HY)_2]^3/\{[H^+]^3(1 + k_c[X^-])\} \tag{12}$$

where X^- is Cl^- or NO_3^- and k_c is the stability constant of MX^{+2}. For $[X^-] = 0$, this reduces to K_0, the K for M(III) from 1 M $HClO_4$.

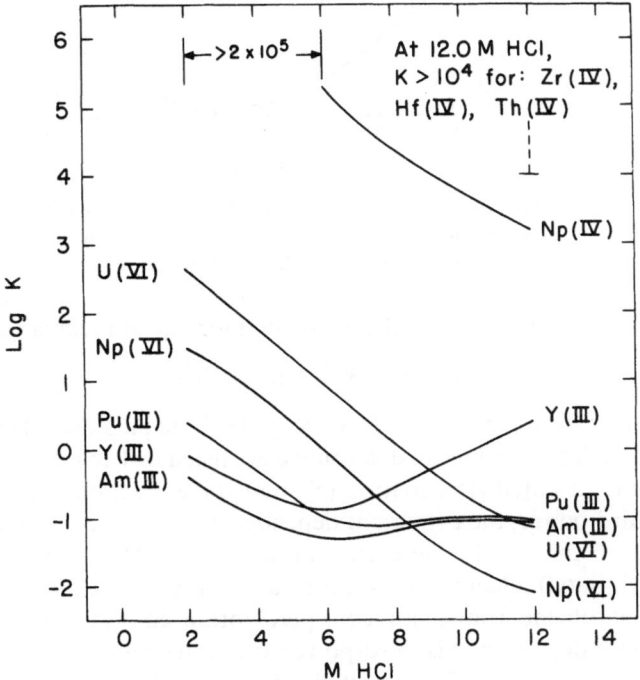

Fig. 3. Variation of log K with concentration of HCl for selected M(III), M(IV), M(VI) in the system: 0.48 F H_2MEHP (in toluene) vs. HCl.[10,11]

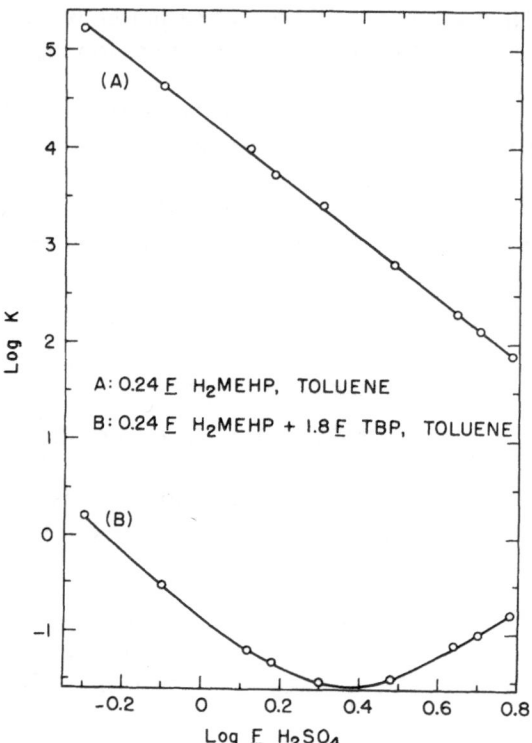

Fig. 4. Extraction of Np(IV) into (A) 0.24 F H_2MEHP
(toluene) and (B) 0.24 F H_2MEHP + 1.8 F TBP (toluene),
as a function of H_2SO_4 concentration.[10]

Dividing equation (11) by the expression for K_0 and rearranging terms:

$$(1/K) = (1/K_0) + (k_c/K_0)[X^-] \tag{13}$$

Consequently, from a plot of $1/K$ vs. $[X^-]$ the slope plus the intercept serve to evaluate k_c. The k_c values so determined are listed in Table 1.[15]

It has been noted [in equation (5)] that the extractant dependency for Eu(II) is third power, the extractant dependency being one unit greater than the charge on the ion. In the extraction of thorium(IV) by di[para(1,1,3,3-tetramethyl butyl) phenyl] phosphoric acid, $(p\text{-}C_8H_{17}\cdot C_6H_4O)_2PO(OH)$, HDOΦP, in toluene from an aqueous perchlorate, chloride, or nitrate phase the extractant dependency is third-power, the value being one unit less than the charge on the ion. The expected inverse fourth-power hydrogen ion dependency is found. So the extraction by HDOΦP may be represented:

$$Th^{+4} + 3(HY)_2 \rightleftharpoons Th(HY_2)_2(Y)_2 + 4H^+ \tag{14}$$

Table 1

Stability Constants for MX^{+2}, $\mu = 1.00$, $22 \pm 1°C^{15}$

M(III)	k_c for MX^{+2}	
	$X^- = Cl^-$	$X^- = NO_3^-$
La	0.9 ± 0.3	1.3 ± 0.3
Ce	0.9 ± 0.3	1.3 ± 0.3
Pr	0.9 ± 0.3	1.7 ± 0.3
Eu	0.9 ± 0.3	2.0 ± 0.3
Tm	0.8 ± 0.3	0.7 ± 0.2
Yb	0.6 ± 0.2	0.6 ± 0.2
Lu	0.4 ± 0.2	0.6 ± 0.2
Am	0.9 ± 0.2	1.8 ± 0.3

where the formulation of the extracted entity is in no sense meant to indicate any particular type of bonding, since it is obvious that several other formulations, including $ThY_4(HY)_2$, are also consistent with the data.[16]

Careful studies with gross thorium from perchlorate, chloride and nitrate solutions proved the absence of the aqueous phase anion from the HDOΦP extract.

However, in the HDEHP (toluene) $vs.$ HX system it was found, in gross experiments, that nitrate was extracted, the ratio of nitrate to Th in the extractant phase approximating unity. This is consistent with the observation that tracer level Th^{+4} is extracted with a far higher K from a nitrate phase than from a chloride or perchlorate phase.

Hydrogen ion, HDEHP and nitrate dependency studies showed the extraction to be represented as

$$Th^{+4} + NO_3^- + 3(HY)_2 \rightleftharpoons Th(NO_3)(HY_2)_3 + 3H^+ \qquad (15)$$

This was the first reported instance of an "extracted mixed complex" containing inorganic anion and extractant anions.[16] From Fig. 5,[16] the substitution of NO_3^- for ClO_4^- in the aqueous phase is seen to depress the K for thorium(IV) in the HDOΦP system but enhance it in the HDEHP system. These changes indicate inhibition by aqueous phase complexing in the first instance and enhancement through formation of a "mixed complex" in the second.

Other instances have since been reported of $Th(NO_3)(HY_2)_3$, $Th(NO_3)_2$-$(HY_2)_2(HY)_2$ and $Th(Cl)(HY_2)_3$ as extracted entities where HY is 2-ethylhexyl hydrogen 2-ethylhexyl phosphonate, $(C_8H_{17}O)$ $(C_8H_{17})POOH$, HEH-[EHP].[17]

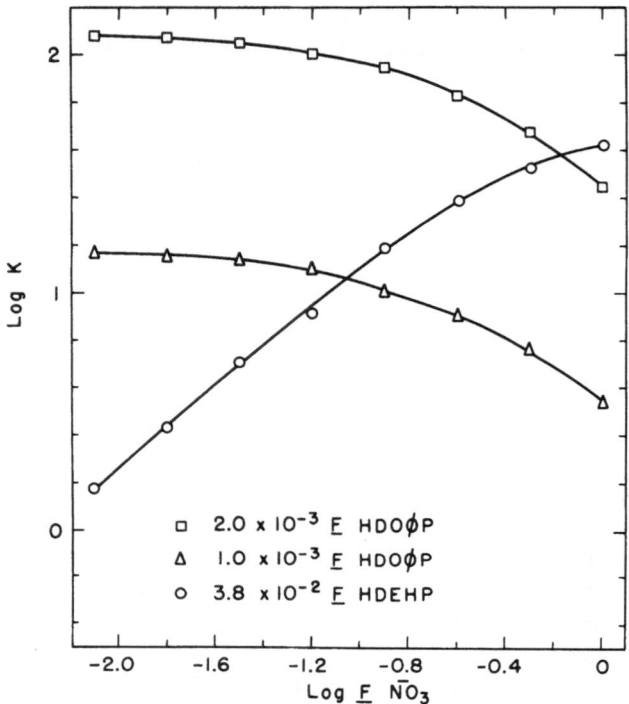

Fig. 5. Effect of substitution of NO_3^- for ClO_4^- upon the extraction of Th(IV) into HDOΦP (toluene) and HDEHP (toluene) from aqueous $HClO_4 + HNO_3$, $\mu = 1.0$, $[H^+] = 1.0$.[16]

It should be noted that in each of the systems reported the extractant dependency for Th^{+4} is third power, the extraction of the "dinitrate" being represented[16] as

$$Th^{+4} + 2NO_3^- + 3(HY)_2 \rightleftharpoons Th(NO_3)_2(HY_2)_2(HY)_2 + 2H^+$$

(Throughout this discussion the total number of Y and H groups, and of nitrate or chloride groups, is obtained from considerations of stoichiometry. No knowledge of the mode of attachment is implied.)

Liquid–liquid extraction has proved a practical technique for the isolation of rare earths and yttrium as a group and for the isolation of individual members of this group, the separation being accomplished especially easily in instances in which a valence other than $+3$ might be employed. It has proved useful in the determination of stability constants of lanthanide and actinide chloride and nitrate complexes. And it has demonstrated the existence of mixed extracted entities such as $Th(NO_3)(HY_2)_3$.

In the field of engineering research, it has been applied in such a way that the usual $X–Y$ diagram approach may be replaced by an algebraic treatment.

The system, HDEHP (toluene), HNO_3, and tracer-level yttrium and promethium behaves ideally within the limits of measurement, the extraction of the two M^{+3} elements being represented by equation (4). By employing HDEHP (toluene) and aqueous HNO_3 which have been mutually pre-equilibrated as extractant and scrub phases respectively and either pre-equilibrated extractant phase or HNO_3 phase in the feed makeup, a counter-current extraction assembly may be operated without any gross transfers occurring. Basically, only the transfer of Y^{+3}, Pm^{+3} and H^+ ions need be considered; and since the Y^{+3} and Pm^{+3} are in tracer-level concentrations the transfer of M^{+3} represented by equation (4) does not measurably affect the concentration of $(HY)_2$ in the organic phase or the concentration of H^+ in the aqueous phase.

Consequently, the behavior of the extraction assembly may be analyzed in terms of

$$K_Y = k_Y[(HY)_2]_0^3/[H^+]_A^3 \qquad (17)$$

$$K_{Pm} = k_{Pm}[(HY)_2]_0^3/[H^+]_A^3 \qquad (18)$$

without the need for stage-to-stage corrections for gross transfers. Such a study has been reported in which the extraction assembly was a perforated plate pulse column.[18] Perhaps the most obvious application is in evaluation of parameters determining the number of theoretical stages in a column and in determining the rate of approach to steady state operation.

Finally, in the "interesting observations" category, we have noted that the 15 members of the rare earths in the trivalent state and in tracer-level concentration display a tetrad effect in a variety of the liquid–liquid extraction systems we have investigated. This tetrad effect is evident in a plot of log K vs. Z, the plot consisting of four separate smooth curves, each encompassing four points, the two central curves meeting to form a cusp at gadolinium ($Z = 64$, the middle element of the 15) and the others, extended, intersecting to form cusps respectively between $Z = 60$ and 61 and between $Z = 67$ and 68.*

In each of the systems found to display the tetrad effect, each of the separate curves is concave downward. Sometimes one or more of the curves is nearly straight, but no instance of upward concavity has yet been found. The plot has maxima in some systems but not in all. In a number of systems, the plot has a minimum at Gd ($Z = 64$).

*This tetrad effect was first reported by D. F. Peppard and G. W. Mason in the Argonne National Laboratory's Physical Research Monthly Report for September, 1968. A note has been submitted by D. F. Peppard, G. W. Mason, and S. Lewey to *J. Inorg. Nucl. Chem.* to be followed by a comprehensive report. This effect is also reported by Peppard, Mason, and Lewey in the Proceedings of the Fifth International Conference on Solvent Extraction Chemistry (5th ICSEC) held in Jerusalem, Israel, September 16–18, 1968.

Some of the examples of the tetrad effect are shown in Fig. 6.* The half-filled shell effect has been joined by the quarter-filled and three-quarter-filled effects. We make no attempt at explanation.

Interestingly, in several systems in which the K for gadolinium is a minimum in the log K *vs.* Z plot for lanthanides(III), the K for curium is also a minimum in the corresponding plot for actinides(III). We have been unable to make a full test of the "tetrad" hypothesis for actinides(III) because of the non-existence of some members in a trivalent state and because of the unavailability in our laboratory of others. However, we hope to examine in detail the relative K values for elements 94 to 99 and compare the log K *vs.* Z plot with that for lanthanides(III).

The tetrad effect has important implications for the mutual separation of lanthanides(III), and presumably actinides(III), since in even the plots with no maximum the beginning of a tetrad differs markedly from the end of the same tetrad with respect to r, the ratio of K_{Z+1}/K_Z. In curve (A) of Fig. 6, for example, the r for the Ce–La pair is 7.8 while that for the Nd–Pr

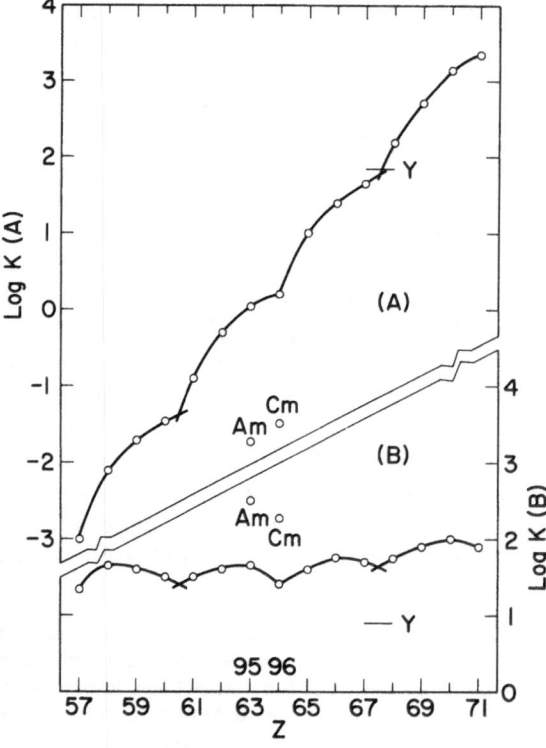

Fig. 6. Variation of log K with Z for lanthanides(III) in the systems: (A) 0.3 F H[DOP] (in benzene) *vs.* 0.05 F HCl; (B) 0.6 F DEH[ClMP] (in benzene) *vs.* 11.4 F LiBr + 0.5 F HBr. H[DOP] is $(n\text{-}C_8H_{17})_2PO(OH)$; DEH[ClMP] is $(ClCH_2)PO(OC_8H_{17})_2$ where C_8H_{17} is 2-ethyl hexyl. Comparative Y, Am, and Cm data.

*See footnote on p. 299

pair is 2.0. Considering the end of the second tetrad and the beginning of the third, the value of 1.5 for Gd–Eu is to be compared with the value of 7.8 for Tb–Gd.

It should be realized that the field of liquid–liquid extraction of metallic cations, basic and applied, is a very large one—even if the systems are restricted to those employing an acidic phosphorus-based extractant and the rare earths are the only metals considered. So the picture of the field presented here should not, in any sense, be considered representative. It is, rather, a reflection of our particular interests, directed by our dual love of coordination chemistry and of the rare earths.

ACKNOWLEDGMENT

For the implanting of these loves we wish to thank the Chemistry Dept. of the University of Illinois and in particular Prof. John C. Bailar, Jr. How many of the phenomena reported in this paper involve true coordination in the University of Illinois sense is questionable. But there is no doubt that the studies have been materially aided and the interpretation of data greatly facilitated by application of the principles of coordination chemistry so clearly delineated by Prof. Bailar in his research and so ably taught in his lectures and in his conversations with his graduate students.

Here is a man to be honored equally for his outstanding contributions to his field of research and for his inspiring direction of the thesis work of so many who have found in this field a lifetime of interest. May he receive the international thanks to which he is so richly entitled through his dedicated and fruitful career.

REFERENCES

1. D. F. Peppard, G. W. Mason, J. L. Maier and W. J. Driscoll, *J. Inorg. Nucl. Chem.* **4**: 334 (1957).
2. Figure 1 is taken from "Liquid–Liquid Extraction of Metal Ions" chapter by D. F. Peppard, in *Advances in Inorganic and Radiochemistry*, Vol. 9, Academic Press, New York (ed. H. J. Emeleus and A. G. Sharpe). The separate curves are adapted from:
    (A) D. F. Peppard, G. W. Mason, and I. Hucher, *J. Inorg. Nucl. Chem.* **18**: 245 (1961).
    (B) E. Hesford, E. E. Jackson, and H. A. C. McKay, *ibid.* **9**: 279 (1959).
    (C) D. F. Peppard, G. W. Mason, and G. Giffin, *ibid.* **27**: 1683 (1965).
    (D) D. F. Peppard, W. J. Driscoll, R. J. Sironen, and S. McCarty, *ibid.* **4**: 326 (1957).
    (E) G. W. Mason, S. McCarty, and D. F. Peppard, *ibid.* **24**: 967 (1962).
3. D. F. Peppard, J. R. Ferraro, and G. W. Mason, *ibid.* **4**: 371 (1957).
4. D. F. Peppard, J. R. Ferraro, and G. W. Mason, *ibid.* **7**: 231 (1958).
5. J. R. Ferraro, G. W. Mason, and D. F. Peppard, *ibid.* **22**: 285 (1961).
6. G. M. Kosolapoff and J. S. Powell, *J. Chem. Soc.* 3535 (1950).

7. a. D. F. Peppard, E. P. Horwitz, and G. W. Mason, "Comparative Liquid–Liquid Extraction Behavior of Europium(II) and Europium(III)," chapter in : *Rare Earth Research*, Gordon and Breach, New York (1962). *Proceedings of the Second Conference on Rare Earth Research*.

   b. D. F. Peppard, E. P. Horwitz, and G. W. Mason, *J. Inorg. Nucl. Chem.* **24** : 429 (1962).

8. D. F. Peppard, G. W. Mason, and S. W. Moline, *ibid.* **5** : 141 (1957).

9. D. F. Peppard, S. W. Moline, and G. W. Mason, *ibid.* **4** : 344 (1957).

10. D. F. Peppard, G. W. Mason, and R. J. Sironen, *ibid.* **10** : 117 (1959).

11. D. F. Peppard, *Advances in Inorganic and Radiochemistry*, Vol. 9, Academic Press Inc., New York, 1966, pp. 1–80.

12. G. W. Mason, S. McCarty, and D. F. Peppard, *J. Inorg. Nucl. Chem.* **24** : 967 (1962).

13. D. F. Peppard, G. W. Mason, W. J. Driscoll, and R. J. Sironen, *ibid.* **7** : 276 (1958).

14. J. R. Ferraro, G. W. Mason, and D. F. Peppard, *ibid.* **22** : 285 (1961).

15. D. F. Peppard, G. W. Mason, and I. Hucher, *ibid.* **24**, 881 (1962).

16. D. F. Peppard, G. W. Mason, and S. McCarty, *ibid.* **13** : 138 (1960).

17. D. F. Peppard, M. N. Namboodiri, and G. W. Mason, *ibid.* **24** : 979 (1962).

18. M. A. Mandil, G. W. Mason and D. F. Peppard, *I and EC Process Design and Development* **2** : 106 (1963).

REACTIONS OF COORDINATION COMPOUNDS IN THE SOLID PHASE

H. E. LeMay, Jr.

University of Nevada
Reno, Nevada

INTRODUCTION

Many coordination compounds have been found to undergo chemical reactions in the solid phase under relatively mild conditions of pressure and temperature. Some of these reactions are well known and provide convenient synthetic procedures. For example, the common method of preparing cis-$[Cr(en)_2Cl_2]Cl$ is by thermal deamination of $[Cr(en)_3]Cl_3$ in the presence of a catalytic amount of NH_4Cl.[1] Few detailed studies of these solid-phase reactions have been made, but at least two general reasons can be cited for giving such reactions further consideration. First, there is the question of the mechanisms involved in ligand exchange, racemization, and isomerization in the solid phase in comparison with the ways in which these processes proceed in solution. In some cases, mechanisms proposed for solid-phase reactions clearly differ from mechanisms proposed for corresponding solution reactions. What causes the mechanisms to differ (if, indeed, they do) is not known, but this matter is clearly of interest to our understanding of the factors governing the reaction mechanisms of coordination compounds. Secondly, some of the reactions are of interest as possible model systems for studying the chemical behavior of solids.

Some examples of solid-state reactions involving coordination compounds are presented below:

1. Anation
$$[Co(NH_3)_5(H_2O)]Cl_3 \longrightarrow [Co(NH_3)_5Cl]Cl_2 + H_2O$$

2. Ligand Exchange
$$trans\text{-}[Co(en)_2Cl_2]\overset{*}{C}l \longrightarrow trans\text{-}[Co(en)_2\overset{*}{C}l_2]Cl$$

3. Racemization
$$l\text{-}K_3[Co(C_2O_4)_3] \cdot nH_2O \longrightarrow dl\text{-}K_3[Co(C_2O_4)_3] \cdot nH_2O$$

4. Geometric Isomerization

$$trans\text{-}[Co(NH_3)_4Cl_2]IO_3 \cdot 2H_2O \longrightarrow cis\text{-}[Co(NH_3)_4Cl_2]IO_3 + 2H_2O$$

5. Linkage Isomerization

$$[Co(NH_3)_5ONO]Cl_2 \longrightarrow [Co(NH_3)_5NO_2]Cl_2$$

Although all of these examples involve cobalt(III) complexes, reactions involving other metal ions could be cited as examples. This paper will deal mainly with the complexes of cobalt(III) and chromium(III) and will be limited to the anation, racemization, and geometric isomerization processes. Linkage isomerization will not be discussed since such reactions are fairly familiar and have been recently reviewed elsewhere.[2,4] Ligand exchange will be discussed only as it is encountered in racemization and geometric isomerization.

ANATION

Many reactions are known in which a volatile ligand is removed from the coordination sphere and replaced by an anion. The volatile ligands involved in these anation reactions have included water, ammonia, pyridine, ethylenediamine, 2,2′-bipyridine and dimethylformamide.[5,6] Any complex cation containing one or more volatile ligands might be expected to undergo a solid-state anation reaction. The recent book by Wendlandt and Smith on the thermal properties of transition-metal complexes contains numerous examples.[6]

Of particular interest and familiarity are reactions of the type:

$$[Co(NH_3)_5(H_2O)]X_3 \longrightarrow [Co(NH_3)_5X]X_2 + H_2O$$

Wendlandt and his co-workers[7] and Mori and co-workers[8,9] have made quantitative studies of some of these reactions. It has been found that the ease of anation depends strongly on the anion involved. Furthermore, activation energies appear to parallel the basicities of the anions toward cobalt(III). Activation energies of 19 ± 2, 25 ± 2, and $31 \pm 3\ kcal\ mole^{-1}$ have been reported for the chloride, bromide and nitrate salts, respectively.[7] The sulfate and perrhenate salts also undergo anation, but the perchlorate salt does not.[9,10] As a result, it has been suggested that such solid-phase anations are essentially associative in nature, taking place by a concerted reaction in which the entering group exerts considerable influence.[10,11] In contrast, corresponding solution processes appear to be dissociative in nature, although they possibly involve some bond-making in the transition state.[12] The question of the mechanism involved in these solid-phase reactions is certainly not settled, and further studies are needed. It is perhaps worth noting that the rates of the solid-phase nitrito-to-nitro isomerizations of compounds of the type $[Co(NH_3)_5ONO]X_2$ depends on the anion X^- even though such processes are intramolecular.[13]

The ease with which $[Cr(en)_3]X_3$ compounds undergo thermal deamination has also been found to depend on anion. These thermal deaminations have long been used as convenient synthetic routes to cis-$[Cr(en)_2Cl_2]Cl$ and trans-$[Cr(en)_2(SCN)_2]SCN$.[1] It was found by Rollinson and Bailar[14] that these reactions are catalyzed by traces of NH_4Cl or NH_4NCS. O'Brien and Bailar[15] studied the effects of heat on other $[Cr(en)_3]X_3$ and $[Cr(pn)_3]X_3$ complexes to which small amounts of NH_4X had been added, but found only $[Cr(en)_2Cl_2]Cl$, $[Cr(pn)_2Cl_2]Cl$, and $[Cr(en)_2(SCN)_2]SCN$ could be formed. Wendlandt and his co-workers[16–19] have studied these reactions by a number of thermoanalytical techniques and found that a large number of anions could be introduced into the coordination sphere by proper choice of $[Cr(en)_3]^{3+}$ salt and ammonium salt if the ammonium salt was present in a large excess. Interestingly, cis isomers were obtained in all cases. For example, the thermal dissociation of $[Cr(en)_3]Br_3$ in an NH_4Cl matrix gave cis-$[Cr(en)_2Cl_2]Cl$. Similarly, cis-$[Cr(en)_2F_2]F$ and cis-$[Cr(en)_2(SCN)_2]SCN$ were obtained using NH_4F and NH_4SCN, respectively.[19]

By studying several $[Cr(en)_3]X_3$ complexes in various NH_4X matrices, a suggested order of ease of anion coordination was found[18] to be

$$F^- > SCN^- > Cl^- > Br^-$$

This order holds, regardless of whether the coordinated anion is initially bound in the complex or in the matrix salt. The order found for the reaction temperatures of the mixtures approximately follows the minimum dissociation temperatures for the ammonium salts: $NH_4SCN < NH_4F < NH_4HCO_2 < NH_4C_2H_3O_2 < NH_4Cl < (NH_4)_2C_2O_4 < NH_4Br < NH_4I$.[19] The order for ease of coordination might be taken as evidence that anation in these reactions is associative in character, depending on bond-making with an anion as well as bond-breaking with the nitrogen of an amine.

Wendlandt and Sveum[19] have postulated that the first step in these reactions is dissociation of the ammonium salt to form HX. This is followed by breaking of the Cr—N bond and protonation of the ethylenediamine nitrogen. Protonation of the nitrogen prohibits reforming of the Cr—N bond, and an anion is free to enter the coordination sphere. The second Cr—N bond is broken similarly. The question of why cis-$[Cr(en)_2(SCN)_2]SCN$ is formed in the presence of a large excess of NH_4SCN while catalytic amounts of NH_4SCN yield the trans isomer remains unanswered.

RACEMIZATION

Several compounds have been reported to racemize in the solid phase, some when heated or subjected to high pressure, others under ordinary temperatures and pressures. The earliest reports of solid-state racemization were made by Johnson and Mead[20-22] who studied the racemization of salts of $[Cr(C_2O_4)_3]^{3-}$ and $[Co(C_2O_4)_3]^{3-}$. Since each compound showed somewhat different behavior, no general mechanism was established. However, it was suggested that lattice water plays an important role in these reactions, possibly by entry into the coordination sphere to give an unstable "expanded" complex which may be either symmetrical or rapidly lose its optical activity.[22]

In an effort to shed further light on the mechanism(s) of these processes, the effect of pressure on the racemization of d-$K_3[Co(C_2O_4)_3]\cdot nH_2O$ was recently investigated.[23,24] Studies of this type have also been conducted on l-$[Fe(phen)_3](ClO_4)_2\cdot nH_2O$ and d-$[Ni(phen)_3](ClO_4)_2\cdot nH_2O$.[25] All exhibit an enhancement of racemization rates when subjected to high pressure (10,000 to 40,000 atm). The negative activation volumes for these processes have been interpreted as indicating a trigonal (Bailar) twist mechanism. An associative process involving lattice water would also appear to be consistent with the negative activation volumes, but in the case of $K_3[Co(C_2O_4)_3]\cdot nH_2O$, water of hydration seemed to decrease the rate of racemization.[24]

Although the results do not appear consistent with a dissociative mechanism, such a mechanism cannot be completely ruled out. The negative activation volumes could arise from lattice changes or distortions rather than volume changes of the complex ion alone. The importance of the nature of the crystal lattice in such reactions is suggested by preliminary studies[26] which indicate that d-$[Ni(phen)_3]I_2\cdot nH_2O$ cannot be made to racemize at a pressure of 3.81×10^4 atm for 70 hr even though the corresponding perchlorate salt does racemize under such conditions. Studies in aqueous solution indicate racemization involving one-ended dissociation of an oxalate ligand in the cases of the $[Cr(C_2O_4)_3]^{3-}$ and $[Co(C_2O_4)_3]^{3-}$ complexes.[27,28] Racemization of $[Fe(phen)_3]^{2+}$ appears to occur mainly via an intramolecular process,[29] while $[Ni(phen)_3]^{2+}$ racemizes entirely by a dissociative mechanism.[30,31]

Preliminary studies[25,26] have shown that d-$[Co(en)_3]I_3$ cannot be racemized by application of high pressure (3900 atm over a period of 70 hr). However, this compound does undergo racemization when heated,[32] perhaps indicating a dissociative mechanism in this case. A more detailed study of the thermal racemization of d-$[Co(en)_3]I_3$ is underway.[33] Studies are also underway on racemization of d-cis-$[Co(en)_2(NH_3)_2]Cl_3$.[33] If this later compound racemizes by a dissociative mechanism, anation would be expected. Racemization without anation could be interpreted as evidence for an intramolecular mechanism, possibly the trigonal (Bailar) twist.

The racemization of l-cis-$[Cr(en)_2Cl_2]Cl\cdot H_2O$ occurs mainly while H_2O is escaping from the crystal lattice.[32] On the other hand, d-cis-$[Co(en)_2$-

Cl$_2$]Cl racemizes when heated even though water is not involved in the reaction.[32] An equation-anation pathway has been proposed for the first reaction, but it has not been ascertained whether the aquation and anation reactions are dissociative or associative in character. The role of water in this racemization is possibly due to the high effective concentration of water molecules in the system as a result of their high mobility relative to charged species in the crystal lattice. This gives rise to a large number of encounters between H$_2$O and l-cis-[Cr(en)$_2$Cl$_2$]$^+$ or a reactive intermediate.

The racemization of d-cis-[Co(en)$_2$Cl$_2$]Cl has not been studied in any detail. However, it has been found that cis-[Co(en)$_2$Cl$_2$]$\overset{*}{\text{C}}$l undergoes chloride exchange when heated,[34,35] and it seems likely that the racemization of d-cis-[Co(en)$_2$Cl$_2$]Cl occurs via the same intermolecular process. In the exchange studies, Cl$^-$ vacancies in the crystal lattice were found to be of primary importance in determining reaction rate. The rate of Cl$^-$ exchange was increased by irradiation with X-rays or with ^{60}Co-γ-rays which increased the number of Cl$^-$ lattice vacancies. Virtually identical rates and activation energies were exhibited by $trans$-[Co(en)$_2$Cl$_2$]$\overset{*}{\text{C}}$l, $trans$-[Co(en)$_2$Br$_2$]$\overset{*}{\text{B}}$r and cis-[Co(en)$_2$Cl$_2$]$\overset{*}{\text{C}}$l indicating that bond-making or breaking is not rate-determining. A detailed study of the racemization of d-cis-[Co(en)$_2$Cl$_2$]Cl is currently planned.

GEOMETRIC ISOMERIZATION

Examples of geometric isomerization are not as numerous as examples of anation or racemization. The compounds $trans$-[Co(NH$_3$)$_4$Cl$_2$]IO$_3$· 2H$_2$O,[36,37] $trans$-[Co(NH$_3$)$_4$Cl$_2$]Cl,[38] and $trans$-[Co(pn)$_2$Cl$_2$](H$_5$O$_2$)Cl$_2$[39] all reportedly undergo $trans$-to-cis isomerization in the solid state. In addition, trans-[Co(pn)$_2$Br$_2$](H$_5$O$_2$)Br$_2$ has been found to isomerize.[40] The complexes cis-K$_3$[Rh(C$_2$O$_4$)$_2$Cl$_2$]·nH$_2$O and cis[K$_3$Ir(C$_2$O$_4$)$_2$Cl$_2$]·nH$_2$O have been reported to undergo cis-to-$trans$ isomerizations when dehydrated.[41] The compound [Co(en)$_2$(NH$_3$)SO$_3$]Cl has been found to undergo a change from yellow-brown to bright orange at 60°.[42] The exact nature of this change is not known, but it might involve either linkage or cis-$trans$ isomerization. On cooling, this compound reverts to its original color.

Some interesting solid-phase isomerizations involving nickel(II) have been reported. The compounds [NiBr$_2$R$_2$], where R is P(C$_2$H$_5$)(C$_6$H$_5$)$_2$ or P(i-C$_3$H$_7$)(C$_6$H$_5$)$_2$, undergo a change from a square-planar configuration to a tetrahedral configuration at room temperature.[43] The compound [(C$_2$H$_5$)$_4$As]$_2$[Ni(NCS)$_4$] undergoes a change from an octahedral form (with thiocyanate bridges) to an approximately tetrahedral geometry at 155°.[44]

It has been proposed that the $trans$-to-cis isomerization of $trans$-[Co(NH$_3$)$_4$Cl$_2$]IO$_3$ · 2H$_2$O occurs via an aquation-anation pathway.[37] It was found that this compound could not be dehydrated without accompanying isomerization. Furthermore, the rate of dehydration and the rate of disappearance of trans-$[Co(NH_3)_4Cl_2]^+$ in static air are equal. Likewise, activation parameters for' these two processes agree within experimental error. Reduction of pressure increases the rates of isomerization[39] and dehydration.[45] In fact, isomerization can be brought about at room temperature by vacuum desiccation.

The isomerization of trans-$[Co(NH_3)_4Cl_2]IO_3\cdot2H_2O$ occurs at lower temperatures than the isomerization of trans-$[Co(NH_3)_4Cl_2]Cl$. The first compound isomerizes readily below 100°,[37] while the second reportedly isomerizes at about 210°.[38] Study of the second isomerization is complicated by loss of ammonia which begins in the same temperature range as the isomerization.[5]

The isomerization of trans-$[Co(pn)_2Cl_2](H_5O_2)Cl_2$ is accompanied by loss of a molecule of HCl and two molecules of H_2O (cis-$[Co(pn)_2Cl_2]Cl$ is formed). Interestingly, the corresponding dehydration and dehydrohalogenation of trans-$[Co(en)_2Cl_2](H_5O_2)Cl_2$ is not accompanied by isomerization. A recent study[46] of the dehydration and dehydrohalogenation reactions of these two compounds failed to reveal any major differences which might explain why the one compound isomerizes while the second does not. It was found that the propylenediamine complex shows a greater resistance to loss of HCl and H_2O from the prominent (100) faces of the crystals, but a more recent study[40] involving trans-$[Co(en)_2Cl_2](D_5O_2)Cl_2$, trans-$[Co(en)_2Br_2]$-$(H_5O_2)Br_2$, trans-$[Rh(en)_2Cl_2](H_5O_2)Cl_2$, trans-$[Co(l-pn)_2Cl_2](H_5O_2)Cl_2$ and trans-$[Co(pn)_2Br_2](H_5O_2)Br_2$ suggests that this difference is largely one of degree. All of these compounds seem to show some resistance to dehydration from the (100) faces. All of the chloride salts exhibit activation energies of 16 ± 1 kcal mole^{-1} for dehydration and dehydrohalogenation in vacuo, regardless of whether isomerization occurs. The activation energies of the bromide salts are in the 24 to 29 kcal mole^{-1} range suggesting that there may be some contribution from HBr and H_2O diffusion to these activation energies.

Radiochloride exchange studies[40] have been more revealing in explaining why trans-$[Co(pn)_2Cl_2](H_5O_2)Cl_2$ undergoes isomerization while trans-$[Co(en)_2Cl_2] (H_5O_2)Cl_2$ does not. These studies have indicated that the dehydration and dehydrohalogenation of trans-$[Co(pn)_2Cl_2](H_5O_2)Cl_2$ are accompanied by complete scrambling of ionic and covalent chlorides. Under the same experimental conditions, trans-$[Co(en)_2Cl_2](H_5O_2)Cl_2$ exhibits no scrambling. Table 1 shows the amount of $\overset{*}{Cl}^-$ exchange occurring when these compounds are heated in static air at 120° for 90 min. Clearly, the isomerization of trans-$[Co(pn)_2Cl_2](H_5O_2)Cl_2$ occurs by an intermolecular ligand exchange. Interestingly, trans-$[Co(pn)_2Cl_2]\overset{*}{Cl}$ undergoes $8.6 \pm 2.5\%$

chloride exchange while heated under these same conditions. Hence, although isomerization occurs most readily while H_2O and HCl are escaping from the crystal lattice, it also occurs at a slow rate even after dehydration and dehydrohalogenation are complete. Only 1 % exchange has been found to accompany heating trans-[Co(en)$_2$Cl$_2$]Cl at 125° for 90 min.[34,35] No isomerization accompanies this process.

It has been found that trans-[Co(pn)$_2$Cl$_2$]$^+$ undergoes more rapid acid hydrolysis than trans-[Co(en)$_2$Cl$_2$]$^+$.[47] Also, chloride exchange in methanol occurs more rapidly with trans-[Co(pn)$_2$Cl$_2$]$^+$ than with trans-[Co(en)$_2$Cl$_2$]$^+$.[48] These reactions are believed to occur by a dissociative mechanism, the rate differences being due to greater steric crowding at the reaction site when the chelate involved is proplenediamine. The different rates at which trans-[Co(pn)$_2$Cl$_2$](H$_5$O$_2$)Cl$_2$ and trans-[Co(en)$_2$Cl$_2$](H$_5$O$_2$)-Cl$_2$ (also trans-[Co(pn)$_2$Cl$_2$]Cl and trans-[Co(en)$_2$Cl$_2$]Cl) undergo chloride exchange may be due to the same effect. The acceleration of chloride exchange and isomerization which occurs when H_2O and HCl are escaping from the crystal lattice of trans-[Co(pn)$_2$Cl$_2$](H$_5$O$_2$)Cl$_2$ is reminiscent of the acceleration by water of the cis-trans isomerization of cis-[Co(pn)$_2$Cl$_2$]$^+$ in alcohol solutions.[49] Presumably, this isomerization occurs by a dissociative process. The isomerization which accompanies the ligand exchange of trans-[Co(pn)$_2$Cl$_2$](H$_5$O$_2$)Cl$_2$ and trans-[Co(pn)$_2$Cl$_2$]Cl in the solid phase could be a result of topochemical factors. It could also arise if cis-[Co(pn)$_2$Cl$_2$]Cl possessed a greater thermal stability than trans-[Co(pn)$_2$Cl$_2$]Cl.

Table 1

Radiochloride Exchange Occurring During the Dehydration and Dehydrohalogenation of
trans-[Co(pn)$_2$Cl$_2$](H$_5$O$_2$)Cl$_2$ and trans-[Co(en)$_2$Cl$_2$](H$_5$O$_2$)Cl$_2$

Compound	% $\overset{*}{Cl}^-$ Retention[a]		% $\overset{*}{Cl}^-$ in covalent positions[a]	
	Expt.	Theory	Expt.	Theory
trans-[Co(pn)$_2$Cl$_2$](H$_5$O$_2$)$\overset{*}{Cl}_2$	77.8 ± 3.5	75.0	51.2 ± 2.6	50.0[b]
trans-[Co(pn)$_2\overset{*}{Cl}_2$](H$_5$O$_2$)Cl$_2$	71.8 ± 2.2	75.0	49.5 ± 1.6	50.0[b]
trans-[Co(en)$_2$Cl$_2$](H$_5$O$_2$)$\overset{*}{Cl}_2$	48.0 ± 2.2	50.0	0.2 ± 0.4	0.0[c]
trans-[Co(en)$_2\overset{*}{Cl}_2$](H$_5$O$_2$)Cl$_2$	97.0 ± 4.7	100.0	97.2 ± 4.7	100.0[c]

[a]Percentage of $\overset{*}{Cl}^-$ in [Co(AA)$_2$Cl$_2$]Cl product based on amount of $\overset{*}{Cl}^-$ in starting trans-[Co(AA)$_2$Cl$_2$](H$_5$O$_2$)Cl$_2$. Samples heated in static air at 120° for 90 min.
[b]Complete scrambling of chlorides. [c]No scrambling of chlorides.

REFERENCES

1. C. L. Rollinson and J. C. Bailar, Jr., *Inorg. Synth.* **2**: 200 (1946).
2. J. L. Burmeister, *Coord. Chem. Rev.* **3**: 225 (1968).

3. R. T. M. Fraser in "Werner Centenial," *Advances in Chemistry Series No. 62*, American Chemical Society, Washington, D.C., 1967, pp. 295–305.

4. F. Basolo and R. G. Pearson, *Mechanisms of Inorganic Reactions*, Second ed. John Wiley and Sons, Inc., New York, N.Y. 1967, pp. 291–300.

5. H. E. LeMay, Jr., Ph.D. Thesis, University of Illinois, 1966.

6. W. W. Wendlandt and J. P. Smith, *The Thermal Properties of Transition-Metal Ammine Complexes*, Elsevier Publishing Co., New York, 1967.

7. W. W. Wendlandt and J. L. Bear, *J. Phys. Chem.* **65**: 1516 (1961).

8. M. Mori, R. Tsuchiya, and Y. Okano, *Bull. Chem. Soc. Japan* **32**: 1029 (1959).

9. M. Mori and R. Tsuchiya, *ibid.* **33**: 841 (1960).

10. E. Lenz and R. K. Murman, *Inorg. Chem.* **7**: 1880 (1968).

11. W. W. Wendlandt and J. P. Smith, *Nature* **201**: 291 (1964).

12. C. H. Langford and H. B. Gray, *Ligand Substitution Processes*, W. A. Benjamin, Inc., New York, 1965, pp. 85–87.

13. B. Adell, *Z. Anorg. Allg. Chem.* **271**: 49 (1952).

14. C. L. Rollinson and J. C. Bailar, Jr., *J. Am. Chem. Soc.* **66**: 641 (1944).

15. T. D. O'Brien and J. C. Bailar, Jr., *ibid.* **67**: 1856 (1945).

16. J. L. Bear and W. W. Wendlandt, *J. Inorg. Nucl. Chem.* **17**: 286 (1961).

17. W. W. Wendlandt and C. H. Strembridge, *ibid.* **27**: 569 (1965).

18. W. W. Wendlandt and C. H. Strembridge, *ibid.* **27**: 575 (1965).

19. W. W. Wendlandt and L. V. Sveum, *ibid.* **28**: 393 (1966).

20. C. H. Johnson and A. Mead, *Trans. Faraday Soc.* **29**: 626 (1933).

21. C. H. Johnson, *ibid.* **31**: 1612 (1935).

22. C. H. Johnson and A. Mead, *ibid.* **31**: 1621 (1935).

23. J. Brady, F. Dachille, and C. D. Schmulbach, *Inorg. Chem.* **2**: 803 (1963).

24. C. D. Schmulbach, J. Brady, and F. Dachille, *ibid.* **7**: 287 (1968).

25. C. D. Schmulbach, F. Dachille, and M. E. Bunch, *ibid.* **3**: 808 (1964).

26. J. Brady, Ph.D. Thesis, Pennsylvania State University, 1963, pp. 38–40.

27. A. J. McCaffery and S. F. Mason, *Proc. Chem. Soc.* 388 (1962).

28. See D. R. Stranks and R. G. Wilkins, *Chem. Rev.* **57**: 743 (1957).

29. F. Basolo, J. Hayes, and H. M. Newmann, *J. Am. Chem. Soc.* **76**: 3807 (1954).

30. F. Basolo, J. Hayes, and H. M. Newmann, *ibid.* **75**: 5102 (1953).

31. R. G. Wilkins and M. J. G. Williams, *J. Chem. Soc.* 1763 (1957).

32. H. E. LeMay, Jr. and J. C. Bailar, Jr., *J. Am. Chem. Soc.* **90**: 1729 (1968).

33. C. Kutal and J. C. Bailar, Jr., University of Illinois, personal communication, 1968.

34. G. B. Schmidt and K. Rossler, *Radiochim. Acta* **5**: 123 (1966).

35. K. Rossler and W. Herr, *Angew. Chem. (Int.)* **6**: 993 (1967).

36. N. I. Labanov, *Russ. J. Inorg. Chem.* **4**: 151 (1959).

37. H. E. LeMay, Jr. and J. C. Bailar, Jr., *J. Am. Chem. Soc.* **89**: 5577 (1967).

38. G. W. Watt and D. A. Butler, *Inorg. Chem.* **5**: 1106 (1966).

39. A. Werner and A. Frohlich, *Chem. Ber.* **40**: 2228 (1907).

40. H. E. LeMay, Jr., unpublished results.

41. J. P. Mathieu and H. Poulet, *J. Chim. Phys.* **59**: 369 (1962).

42. M. E. Baldwin, *J. Chem. Soc.* 3122 (1961).

43. R. G. Hayler and F. C. Humiec, *Inorg. Chem.* **4**: 1701 (1965).

44. D. Forster and D. M. L. Goodgame, *ibid.* **4**: 823 (1965).

45. H. E. LeMay, Jr. and J. Sheen, unpublished results.

46. H. E. LeMay, Jr., *Inorg. Chem.* **7**: 2531 (1968).

47. R. G. Pearson, C. R. Boston and F. Basolo, *J. Am. Chem. Soc.* **75**: 3089 (1953).

48. R. G. Pearson, P. M. Henry and F. Basolo, *ibid.* **79**, 5379 (1957).

49. R. C. Brasted and C. Hirayama, *ibid.* **80**: 788 (1958).

A SIMPLE TRIATOMIC MODEL FOR CALCULATING METAL–LIGAND VIBRATIONS

B. K. W. Baylis

University of Illinois
Urbana, Illinois

INTRODUCTION

The low-frequency spectra of a great many complexes have been reported in the last few years, and there has been much interest in assigning the metal–ligand vibrations.[1] Normal coordinate analyses have been performed on some simple complexes, e.g., ammine complexes,[2–4] hexahalo complexes,[5] oxalato complexes,[6] and carbonato complexes,[7] and the assignments made for these spectra are presumably correct. Most assignments which have been made, however, result from the study of a series of homologous complexes. Neither of these methods are especially practical for complexes of lower symmetry containing complicated ligands: the first method, because the calculation becomes too arduous, and the second, because often too many bands present themselves as possibilities.

The triatomic model presented here was developed because it is simple enough to be applied by a chemist having no access to a computer and in hopes that it would prove sophisticated enough to yield useful values. For complexes of the symmetries that have been studied, agreement is quite good (within 6%) between values determined using the triatomic model and those calculated by a normal coordinate analysis. Some metal-halogen stretching frequencies are assigned using this model.

THEORY

The formulas for the frequencies of the symmetric and asymmetric stretches of a linear triatomic molecule of the type AMB are

$$v_s = \kappa \left\{ \frac{1}{2}\left(\frac{k_1}{\mu_{AM}} + \frac{k_2}{\mu_{MB}} \right) - \sqrt{\left[\frac{1}{2}\left(\frac{k_1}{\mu_{AM}} - \frac{k_2}{\mu_{MB}} \right) \right]^2 + \frac{k_1 k_2}{m_M^2}} \right\}^{1/2} \quad (1)$$

and

$$v_a = \kappa \left\{ \frac{1}{2}\left(\frac{k_1}{\mu_{AM}} + \frac{k_2}{\mu_{MB}}\right) + \sqrt{\left[\frac{1}{2}\left(\frac{k_1}{\mu_{AM}} - \frac{k_2}{\mu_{MB}}\right)\right]^2 + \frac{k_1 k_2}{m_M^2}} \right\}^{1/2} \quad (2)$$

where v_s = frequency in K (cm^{-1}) of the symmetric stretch, v_a = frequency in K of the asymmetric stretch, κ = conversion factor numerically = to 1304, k_1 = force constant in md/Å for the bond between A and M, k_2 = force constant between M and B, μ_{AM} = reduced mass of A and M = $\dfrac{m_A m_M}{m_A + m_M}$ where m_A = atomic weight of A in g/mole and m_M = atomic weight of M, and μ_{MB} = reduced mass of M and B. These formulas can be derived from equations (II, 198) and (II, 199), p. 173 of Herzberg.[8]

If atoms A and B are identical, the above equations reduce to

$$v_s = \kappa \sqrt{\frac{k}{m_A}} \quad (3)$$

and

$$v_a = \kappa \sqrt{k\left(\frac{1}{m_A} + \frac{2}{m_M}\right)} \quad (4)$$

because $m_A = m_B$ and $k_1 = k_2$ (denoted k). In this case the vibrations have the forms

$$\leftarrow A - M - A \rightarrow \qquad \text{and} \qquad \leftarrow A - M \rightarrow \leftarrow A$$
$$v_s \qquad\qquad\qquad\qquad\qquad v_a$$

In the transition from this specific case to the preceding general case, one of the A atoms becomes a B atom. If B is heavier than A, it becomes necessary during a symmetric stretch for M to move with A in order to compensate for the movement of B, since the center of gravity of the molecule must remain constant during a vibration. As the weight of B increases, this vibration tends more and more to be essentially a M—B stretch. As for the asymmetric stretch, the lighter A must move a greater distance than the heavier B in order to hold the center of gravity constant. Therefore, as B increases in weight, the asymmetric vibration tends more and more to be essentially a M—A stretch:

$$\leftarrow A - \leftarrow M - B \rightarrow \qquad \text{and} \qquad \leftarrow A - M \rightarrow \leftarrow B$$
$$v_s \text{ or } v(M-B) \qquad\qquad\qquad v_a \text{ or } v(A-M)$$

if B is always taken to be heavier than A.

However, if the force constant for the M—B bond is much larger than that for the M—A bond, the symmetric stretch becomes essentially a M—A stretch. In this case the asymmetric stretch becomes the $v(MB)$. If, however, the weight of B is very much greater than that of A, the effect of the force constant is neutralized and v_s can be $v(MB)$.

Octahedral Symmetry

For this model to be generally useful for complexes, three-dimensional fundamental modes must be resolved into this one-dimensional model. For example, the three fundamental stretching modes of a complex MA_6 of octahedral symmetry are

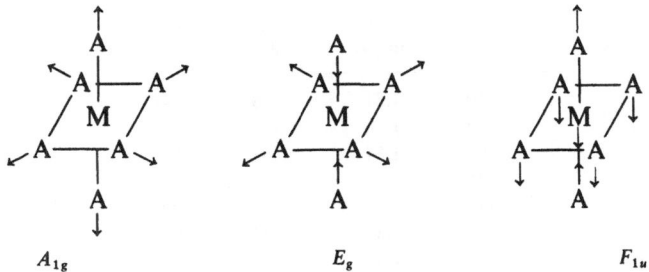

of which only F_{1u} is infrared active.[9] These could be expressed in a one-dimensional model as

$$A_{1g}\ \&\ E_g \qquad\qquad F_{1u}$$

$$\leftarrow A - M - A \rightarrow \qquad \text{and} \qquad \leftarrow A - M \rightarrow \leftarrow A$$

$$v_s = \kappa \sqrt{\frac{k}{m_A}} \qquad\qquad v_a = \kappa \sqrt{k\left(\frac{1}{m_A} + \frac{2}{m_T}\right)}$$

where $m_T = m_M + 4m_A$. One problem with this model becomes immediately obvious in that it cannot distinguish between A_{1g} and E_g, but one cannot expect a one-dimensional model to be sensitive to motions in the two other dimensions. Other weaknesses of this model are that it does not include any repulsion terms nor does it allow for any bending modes to mix in with these stretching modes. In a normal coordinate analysis, consideration of both of these effects is usually necessary in order to get a good fit to the experimental frequencies. Therefore, the most dependable results expected from this model are the ratios of frequencies of identical vibrations of similar molecules. Comparison of ratios rather than actual frequencies should cancel out some of the effects which are not included in the model.

Effective force constants f have been calculated using this model and the experimental frequencies (see Table 1). It must be emphasized that these force constants do not simply indicate the stiffness of the metal-ligand bond, but they include additional repulsions and also vibrations involving atoms of the ligand. For instance, the calculated force constant for $[Co\ en_3]^{+3}$ is 2.51 md/Å larger than that of $[Co(NH_3)_6]^{+3}$. Certainly any stretching of the Co—N bond in $[Co\ en_3]^{+3}$ is also going to cause distortions in the ethylenediamine ring, and these distortions are presumably absorbing the additional 2.51 md/Å.

Table 1
O_h Complexes

Complexes	$v_a/\sqrt{k}$*	k,† md/Å	v_a, K	F_{1u}, K	Ref.	f, md/Å
$[PdCl_6]^=$	250	1.51	308	340	5	1.89
$[PtCl_6]^=$	243	1.86	331	343	5	2.23
$[PtBr_6]^=$	167	1.54	207	244	5	1.95
$[PtI_6]^=$	135			186	10	1.87
$[IrCl_6]^{-3}$	242			296	10	1.50
$[RhCl_6]^{-3}$	249	1.20	273	322	11	1.78
$[Co(NH_3)_6]^{+3}$	356	1.05	362	503	12	2.00
$[Co(ND_3)_6]^{+3}$	329	1.05	334	465	13	2.00
$[Co(NH_3)_6]^{+2}$	356	0.33	202	323	12	0.84
$[Cr(NH_3)_6]^{+3}$	359	0.94	335	470	14	1.72
$[Ni(NH_3)_6]^{+2}$	356	0.34	208	334	11, 13	0.88
$[Ni(ND_3)_6]^{+2}$	329			318	13	0.93
$[Co\ en_3]^{+3}$	275			585	15	4.51
$[Co\ en_3]^{+2}$	275			502	15	3.51
$[Cr\ en_3]^{+3}$	277			567	15	4.21
$[Ni\ en_3]^{+2}$	275			515	15	3.34
$[Cr(NCS)_6]^{-3}$	203			364	11	3.22
$[Cr(H_2O)_6]^{+3}$	350	1.31	400	490	14	1.97
$[Co(NO_2)_6]^{-3}$	226	1.16	243	418	14	3.43
$[Co(CN)_6]^{-3}$	294	2.31	446	565	14	3.71

*The total molecular weight of monodentate ligands is used as m_A. For bidentate ligands, one-half the total molecular weight is used.
†These force constants are from Ref. 5, 11, 12, and 14.

Because of the number of effects which this model does not consider, it was thought necessary to check it using some complexes for which normal coordinate analyses have been performed. Using force constants from the normal coordinate analyses in the literature, ratios of analogous metal–ligand stretches were calculated for many pairs of complexes, and these ratios were compared with ratios of the appropriate infrared bands. Of 32 ratios calculated, 23 agreed with the ratios of experimental F_{1u} frequencies to 6% or better. If a few of the force constants were modified, all but one of the ratios could be brought to within 6% of the experimental ratios.

C_{4v} Symmetry

A monosubstituted six-coordinate complex of the type A_5MB has C_{4v} symmetry. The three stretching vibrations of the O_h point group become five $(3A_1 + B_1 + E)$ in the C_{4v} point group, and of these the three A_1 and the E are active in the infrared.

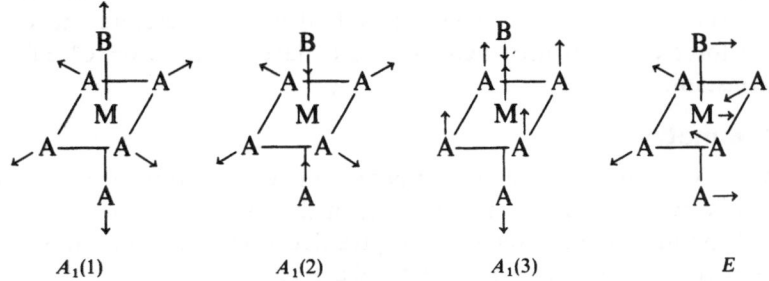

$$A_1(1) \qquad\qquad A_1(2) \qquad\qquad A_1(3) \qquad\qquad E$$

These can be expressed in the one-dimensional model as

$A_1(2)$: $\leftarrow A - M - B \rightarrow$

$$v_s = \kappa \left\{ \frac{1}{2}\left(\frac{k_1}{\mu_{AT}} + \frac{k_2}{\mu_{TB}} \right) - \sqrt{\left[\frac{1}{2}\left(\frac{k_1}{\mu_{AT}} - \frac{k_2}{\mu_{TB}} \right) \right]^2 + \frac{k_1 k_2}{m_T^2}} \right\}^{1/2} \tag{5}$$

$A_1(3)$: $\leftarrow A - M \rightarrow \leftarrow B$

$$v_a = \kappa \left\{ \frac{1}{2}\left(\frac{k_{1.}}{\mu_{AT}} + \frac{k_2}{\mu_{TB}} \right) + \sqrt{\left[\frac{1}{2}\left(\frac{k_1}{\mu_{AT}} - \frac{k_2}{\mu_{TB}} \right) \right]^2 + \frac{k_1 k_2}{m_T^2}} \right\}^{1/2} \tag{6}$$

$$E: \leftarrow A - M \rightarrow \leftarrow A \qquad v_a = \kappa \sqrt{k_1 \left(\frac{1}{m_A} + \frac{2}{m_S} \right)} \qquad m_S = m_M + m_A + m_B$$

where $\mu_{AT} = \dfrac{m_A m_T}{m_A + m_T}, \mu_{TB} = \dfrac{m_T m_B}{m_T + m_B}$, and $m_T = m_M + 4m_A$. $A_1(1)$ and

$A_1(2)$ are symmetric stretches and one of them, $A_1(2)$, is most sensitive to the heavy atom. The other, $A_1(1)$, might reasonably be expected to be basically a $v_s(AM)$ since one might think of these three-dimensional stretches as consisting of three one-dimensional stretches, two of the $v_s(AM)$ type and one of the $v_s(AMB)$ type, combined in different ways. $A_1(3)$ is an asymmetric stretch $v_a(AMB)$, and E is basically an asymmetric stretch $v_a(AM)$. Table 2 lists the force constants used to calculate $v_s(AMB)$, $v_a(AMB)$, $v_s(AM)$, and $v_a(AM)$ and the values obtained along with the experimental frequencies. The assignments in the reference were of the type $v(MN) = 498, 490$, and 466 K and $v(M—Cl) = 284$ K. These frequencies were assigned to the four normal modes by comparison to the model. The Co—N force constant is from Table 1; various Co—Cl force constants were tried until $v_s(AMB)$ was approximately equal to $A_1(2)$, the experimental frequency. Use of this same value, 1.35 md/Å, for the Co—Br force constant results in reasonably good agreement between $v_s(AMB)$ and $A_1(2)$ for $[Co(NH_3)_5Br]^{+2}$. The Co—I force constant reported by Nakagawa, Shimanouchi, and Hiraishi[11] is 0.85 md/Å, whereas their Co—Cl force constant is 1.00 md/Å. Therefore the Co—I effective force constant used here is 85% as large as the Co—Cl effective force constant.

D_{4h} Symmetry

A *trans* disubstituted six-coordinate complex of the type A_4MB_2 has D_{4h} symmetry. Some of the degeneracies in the three stretching vibrations of the O_h point group are removed to give five stretching vibrations in the D_{4h} point group, of which two are infrared active.

$A_{2u}, v_a(MB)$ $E_u, v_a(AM)$

$$v_a = \kappa\sqrt{k_2\left(\frac{1}{m_B} + \frac{2}{m_T}\right)}$$ $$v_a = \kappa\sqrt{k_1\left(\frac{1}{m_A} + \frac{2}{m_S}\right)}$$

As can be seen from the drawings, A_{2u} is chiefly an asymmetric M—B stretch for which a formula of type (4) applies (where $m_T = m_M + 4m_A$), and E_u is chiefly an asymmetric A—M stretch with the same formula but with $m_S = m_M + 2m_B$.

Table 3 lists some of the complexes which have been studied, and includes calculated force constants and experimental frequencies. It should be noted that the effective force constant obtained, using Shimanouchi and Nakagawa's data,[12] for t-$[Co(NH_3)_4Cl_2]^+$ is 1.99 md/Å for the Co—N stretch, showing remarkable agreement with the Co—N effective force constant calculated for the O_h case. The effective force constant for the Co—Cl stretch, however, is 0.32 md/Å higher than that for the C_{4v} case (1.35 md/Å). Watt and Klett's assignment[16] gives a force constant 0.23 md/Å lower than 1.35 md/Å.

Table 2

C_{4v} **Complexes**

Complexes	f_1, md/Å	f_2, md/Å	ν_s(AMB), K	ν_a(AMB), K	ν_s(AM), K	ν_a(AM), K	$A_1(2)$, K	$A_1(3)$, K	$A_1(1)$, K	E, K	Ref.
[Co(NH₃)₅Cl]⁺²	2.00	1.35	281.8	478.8	446	510	284	490	466	498	13
[Co(ND₃)₅Cl]⁺²	2.00	1.35	278.5	444.5	411	478	274	452	425		13
[Co(NH₃)₅Br]⁺²	2.00	1.35	209.9	478.2	446	506	205	480	455	493	13
[Co(NH₃)₅I]⁺²	2.00	1.15	169.3	477.6	446	472	164	480	422	464	13

Table 3
D_{4h} Complexes

Complexes	E_u, K	A_{2u}, K	Ref.	f_1, md/Å	f_2, md/Å
$[PtCl_4(NH_3)_2]$	352	554, 517	17	1.99	2.60
$[PtCl_4(PEt_3)_2]$	340	410	17	2.08	6.82
$[PtCl_4(AsEt_3)_2]$	343	328	17	2.16	5.23
$[PtBr_4(NH_3)_2]$	267	531, 506	17	2.00	2.53
$[PtBr_4(PEt_3)_2]$	247	411	17	2.09	8.06
$[Co(NH_3)_4Cl_2]^+$		290	16		1.12
$[Co(NH_3)_4Cl_2]^+$	501	353	12	1.99	1.67
$[Co(NH_3)_4Br_2]^+$	486	318	2	2.05	2.11
$[Co\,en_2Cl_2]^+$	584, 546, 510	355–366	18	4.12	1.93
$[Co\,en_2Br_2]^+$	571, 552, 510	230	18	4.55	1.31
$[Cr\,en_2Cl_2]^+$	541, 485, 444	351	19	3.49	1.84
$[Cr\,en_2Br_2]^+$	528, 480, 437	289	19	3.84	2.02
$[Cr\,en_2(NCS)_2]^+$	542, 490, 442	351	19	3.81	2.53
$[Cr(NCS)_4(PEt_3)_2]^-$	362	305	20	3.19	3.51
$[Rh\,en_2Cl_2]^+$		343	1		1.85
$[Rh\,en_2Br_2]^+$		223	1		1.36
$[Rh\,en_2I_2]^+$		177	1		1.08
$[Co\,diars_2Cl_2]^{+*}$		380–388	21		2.77
$[Cr\,diars_2Cl_2]^+$		375(?)	21		2.64
$[Rh\,diars_2Cl_2]^+$		349–358	21		2.36
$[Ir\,diars_2Cl_2]^+$		320–335	21		2.06

*The ligand o-phenylene-bisdimethylarsine is denoted diars.

D_{2h} Symmetry

The infrared-active fundamental vibrations in the D_{2h} case are

B_{2u}: $\nu_a(MB) = \kappa\sqrt{k_2\left(\dfrac{1}{m_B} + \dfrac{2}{m_T}\right)}$

where $m_T = m_M + 2m_A$

B_{3u}: $\nu_a(MA) = \kappa\sqrt{k_1\left(\dfrac{1}{m_A} + \dfrac{2}{m_S}\right)}$

where $m_S = m_M + 2m_B$

Table 4
Four-Coordinate Complexes

Complexes	ν_s, K	ν(Pd—Cl), K	ν_a, K	ν(Pd—N), K
t-$[Pd(NH_3)_2Cl_2]$	306	333	452	496
c-$[Pd(NH_3)_2Cl_2]$	279.6	327, 306	449.9	495, 476
$[Pd\,enCl_2]$	268.3	307, 296	367.1	516, 446

C_{2v} Symmetry

For four-coordinate C_{2v} complexes the infrared-active fundamental vibrations are

$A_1(1): \nu_s(AMB)$ $A_1(2): \nu_a(AMB)$ $B_2(1): \nu_a(AMB)$ $B_2(2): \nu_s(AMB)$

where $A_1(1)$ and $B_2(2)$ are represented by equation (5) with $m_T = m_M$, and $A_1(2)$ and $B_2(1)$ are represented by equation (6) with $m_T = m_M$.

Table 4 shows calculated (force constants used were those given by Nakagawa, Shimanouchi, and Hiraishi[11]) and experimental frequencies (from Durig et al.[22]) for a series of planar complexes. From this table it can be seen that one of the ν_a tends to have the same value as the $\nu(Pd-N)$ of the *trans* complex. The calculation of the asymmetric stretches predicts this. The other asymmetric stretch occurs at a lower energy. Comparison of the $A_1(2)$ and $B_2(1)$ vibrations with the B_{3u} vibration in the D_{2h} case indicates that the $B_2(1)$ mode is somewhat more like the B_{3u} mode than is the $A_1(2)$. Hence, the higher energy asymmetric stretch, which should occur at about the same frequency as the corresponding *trans* mode, is considered to be the $B_2(1)$ vibration; and the lower energy asymmetric stretch, the $A_1(2)$ vibration. One of the symmetric vibrations is also found to occur at about the same energy as the $\nu(Pd-Cl)$ of the *trans* complex. This is again the higher-energy transition. The $B_2(2)$ vibration appears to more clearly correspond to the B_{2u} mode, and hence the higher-energy symmetric transition is considered to be the $B_2(2)$ vibration. The value of the lower-energy transition, the $A_1(1)$ vibration, is that predicted by the model. This can be shown using comparisons of ratios.

For six-coordinate C_{2v} complexes there are six infrared-active fundamental vibrations:

$A_1(1); \nu_s(AM)$ $A_1(2); \nu_s(AMB)$ $A_1(3); \nu_a(AMB)$

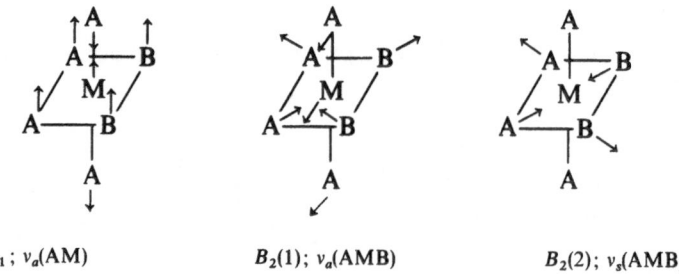

$$B_1;\ \nu_a(\text{AM}) \qquad\qquad B_2(1);\ \nu_a(\text{AMB}) \qquad\qquad B_2(2);\ \nu_s(\text{AMB})$$

It is difficult to reduce these fundamental vibrations to a one-dimensional model. The frequency of the totally symmetric vibration $A_1(1)$ is assumed, by analogy with the C_{4v} case, to be given by $\nu_s = \kappa\sqrt{k_1/m_A}$. The frequency of the $A_1(1)$ vibration in the C_{4v} case was found to vary with change in B, this B-dependency being the same as that of the E vibration. A similar dependency is assumed for the C_{2v} case, using the percentage change in frequency of the B_1 vibration as B changes. The B_1 vibration can clearly be represented as $\nu_a = \kappa\sqrt{k_1(1/m_A + 2/m_S)}$, where $m_S = m_M + 2m_A + 2m_B$. The $A_1(3)$ and $B_2(1)$ vibrations are quite similar to the E_u vibration in D_{4h} symmetry, except that in C_{2v} symmetry the atoms in the plane differ. Hence equation (6) for the general asymmetric stretch applies, where $m_T = m_M + 2m_A$. The $A_1(2)$ and $B_2(2)$ vibrations are the symmetric analogues and are represented by equation (5) with $m_T = m_M + 2m_A$. If the A—M and B—M force constants are approximately equal, the two ν_a frequencies are determined by the light atom and the ν_s frequencies are determined by the heavy atom. The $A_1(2)$, $A_1(3)$, $B_2(1)$, and $B_2(2)$ vibrations are similar to the $A_1(1)$, $A_1(2)$, $B_2(1)$, and $B_2(2)$ vibrations, respectively, of the four-coordinate complex.

Table 5 is a summary of calculated and experimental frequencies for six-coordinate C_{2v} complexes. The force constants used are those which have been determined from other symmetries using the one-dimensional model. The symbols $A_1(1)$, $A_1(2)$ etc. denote the calculated frequencies, which are immediately followed by the corresponding experimental frequencies. With a few exceptions the agreement is quite good. That is, an experimental frequency could be found which corresponds rather well to the calculated frequency. Often no frequency mentioned in the reference corresponded to the calculated $A_1(1)$ frequency, and the frequency entered in the experimental column is that proposed in the reference. It is apparent from the lack of agreement noted here that either the band representing the $A_1(1)$ vibration is of very weak intensity or the assumptions made in applying the one-dimensional model to the $A_1(1)$ vibration are invalid.

Table 5
C_{2v} Complexes

Complexes	f_1, md/Å	f_2, md/Å	$A_1(1)$		$A_1(2)$		$A_1(3)$		B_1		$B_2(1)$		$B_2(2)$		Ref.
			K	K	K	K	K	K	K	K	K	K	K	K	
[PtCl$_4$(NH$_3$)$_2$]	2.01	2.55	3.10	206	328.1	330	<522	518	346	344	522–554	558	352	353	17
[PtCl$_4$(PEt$_3$)$_2$]	2.01	7.8–6.8	298	291	311–314	298	<415	428	332	345	392–415	444	314–340	336	17
[PtCl$_4$(AsEt$_3$)$_2$]	2.01	6.5–5.2	296	304	274–293	285	<363	337	328	321	343–363	346	274–328	294	17
[PtBr$_4$(NH$_3$)$_2$]	2.00	2.55	206	183	226.8	219	<517	485	246	238	517–531	492	226–250	255	17
[PtBr$_4$(PEt$_3$)$_2$]	2.00	7.8–8.0	194	144	220.2	192	<397	425	232	246	391–411	444	220–247	209	17
[Co en$_2$Cl$_2$]$^+$	2.78	1.35	505	518	278.4	281	<570	578	456	550	442	566	281–366	291	18
	4.51	1.35							580		–570				
[Cr en$_2$Cl$_2$]$^+$	2.78	1.84	488	460	321.9	321	<558	544	457	477	444	530	327–351	346	19
	4.21	1.84							563		–558				
[Cr en$_2$Br$_2$]$^+$	2.78	2.02	468	426	252.9	245	<557	548	439	477	437	534	255–289	282	19
	4.21	2.02							540		–557				
[Cr en$_2$(NCS)$_2$]$^+$	2.78	2.45	477	482	306.6	323	<560	543	446	537	442	537	312–351	359	19
	4.21	2.45							550		–560				
[Cr(NCS)$_4$B$_2$]$^-$	3.21	3.50	306		234.5		<377		340	342	362–377	361	234–296		20

B = PEt$_2\varphi$

Two metal–nitrogen force constants were used in calculating the frequencies of the enthylenediamine complexes. The higher force constant was used in predicting the $A_1(1)$ and $A_1(3)$ frequencies because these modes would appear to involve considerable stretching of the ethylenediamine ring. The lower force constant was used for the $A_1(2)$ frequency in which the ethylenediamine ring should at most be bent slightly. The force constants for the B_1, $B_2(1)$, and $B_2(2)$ should lie between these values, because the ethylenediamine ring would appear to be more distorted than in the $A_1(2)$ vibration but less than in the $A_1(1)$ and $A_1(3)$ vibrations. Again, frequencies were reported in the references which corresponded to most of the calculated frequencies. It should be emphasized that the assignments for the ethylene-diamine complexes should not be taken too literally, in view of the number of assumptions that had to be made.

Table 6
M—B Stretching Frequencies for *trans*-Dihalo Complexes

t-MA$_4$B$_2$	Source of f	f, md/Å	v(MB), in K calc.	v(MB), in K found	Ref.
[Rh en$_2$Cl$_2$]ClO$_4$	[RhCl$_6$]$^{-3}$	1.78	336	343	1
[Rh en$_2$Cl$_2$]ClO$_4$		1.85*		343	1
[Rh diars$_2$Cl$_2$]ClO$_4$		2.43*		358	21
[Ir diars$_2$Cl$_2$]ClO$_4$	[IrCl$_6$]$^{-3}$	+0.65 2.15	336	335	21
[Cr diars$_2$Cl$_2$]ClO$_4$	[Cr en$_2$Cl$_2$]$^+$	+0.58 2.42	359	375(?)	21
[Co(NH$_3$)$_4$Cl$_2$]Cl		1.12*		290	16
[Co(NH$_3$)$_4$Br$_2$]Br		1.08*		228	2
[Co en$_2$Cl$_2$]Cl	[Co(NH$_3$)$_5$Cl]$^{+2}$	1.35	301	295	18
[Co en$_2$Br$_2$]Br	[Co(NH$_3$)$_5$Br]$^{+2}$	1.35	234	230	18
[Co diars$_2$Cl$_2$]Cl	[Co(NH$_3$)$_5$Cl]$^{+2}$	+0.65 2.00	326	324	23
[Co diars$_2$Cl$_2$]ClO$_4$	[Co(NH$_3$)$_5$Cl]$^{+2}$	+0.65 2.00	326	325	23
[Co diars$_2$Br$_2$]Br	[Co(NH$_3$)$_5$Br]$^{+2}$	+0.65 2.00	232	243	23
[Co diars$_2$I$_2$]I	[Co(NH$_3$)$_5$I]$^{+2}$	+0.65 1.80	184	198	23
[Co(NH$_3$)$_4$Cl$_2$]Cl		1.67*		353	12
[Co(NH$_3$)$_4$Br$_2$]Br		2.11*		318	2
[Co en$_2$Cl$_2$]Cl		2.00*		366	18
[Co en$_2$Br$_2$]Br	[Co en$_2$Cl$_2$]$^+$	2.00	284	290	18
[Co diars$_2$Cl$_2$]Cl	[Co en$_2$Cl$_2$]$^+$	+0.65 2.65	376	384 ± 4	21
[Co diars$_2$Cl$_2$]ClO$_4$	[Co en$_2$Cl$_2$]$^+$	+0.65 2.65	276	391	23
[Co diars$_2$Br$_2$]Br	[Co en$_2$Br$_2$]$^+$	+0.65 2.65	267	264	23
[Co diars$_2$I$_2$]I	$2.00\left(\dfrac{0.85}{1.00}\right)$	+0.65 2.35	2.10	198	23

*Calculated using experimental frequency.

Cobalt-Halogen Stretches

On Table 3 two assignments are given for the Co—Cl stretches : around 350 and around 290 K. If the force constants determined from the $[Co(NH_3)_5X]^{+2}$ calculations are used, the assignments in the center section of Table 6 seem quite reasonable. The force constants used for the diarsine complexes were determined by comparison of the effective force constants of $[RhCl_6]^{-3}$ and $[Rh \, diars_2Cl_2]^+$ (see the first section of Table 6). The iridium complexes, $[IrCl_6]^{-3}$ and $[Ir \, diars_2Cl_2]^+$, were used as a check. If, however, the higher-energy bands are chosen, a Co—Cl force constant of 2.00 md/Å can be calculated for $[Co \, en_2Cl_2]^+$, from which a series of cobalt-halogen stretches can be calculated which also corresponds to bands in the spectra (see the third section of Table 6). Some of the bands in both series are fairly weak; for instance, the spectra of $[Co \, diars_2Cl_2]Cl$ and $[Co \, diars_2Br_2]Br$ are almost identical in the region 350 to 400 K, where Rodley[21] proposed that the Co—Cl stretch must be found.

Similarly, using both sets of force constants, the Co—Cl stretches for cis-$[Co \, en_2Cl_2]^+$ and cis-$[Co \, diars_2Cl_2]^+$ have been calculated (see Table 7). The agreement with the data of Hughes and McWhinnie[18] is distinctly better for the lower force constants. Hence, this model would indicate that the Co—Cl stretches are to be found in the 280 to 330 K region of the spectrum, rather than from 350 to 390 K, as has been proposed by other authors.

Table 7
M—B Stretching Frequencies for cis-Dichloro Complexes

c-MA$_4$B$_2$	f_1, md/Å	f_2, md/Å	ν(Co—Cl) calc. K	ν(Co—Cl) found K	ν(Co—Cl) calc. K	ν(Co—Cl) found K
$[Co \, en_2Cl_2]Cl$	2.78	1.35	278.4	281		
	4.51	1.35			294–281	291
$[Co \, diars_2Cl_2]ClO_4$	0.87	1.93			326–320	328
	1.8	1.93	<322	320		
$[Co \, en_2Cl_2]Cl$	2.78	2.00	333.3	360		
	4.51	2.00			366–339	367
$[Co \, diars_2Cl_2]ClO_4$	0.87	2.58			391–371	379
	1.8	2.58	<372	354		

ACKNOWLEDGMENTS

I wish to express my sincere gratitude to Prof. John C. Bailar, Jr., for his helpful suggestions and especially for his constant encouragement and support. Further, I should like to express my appreciation to Prof. E. O. Fischer of the Technische Hochschule, Munich, Germany, for the use of his laboratory facilities in preparing several of these complexes and in obtaining their spectra. I also wish to thank my husband, Dr. William E. Baylis, for his helpful discussions.

REFERENCES

1. D. M. Adams, *Metal-Ligand and Related Vibrations: a Critical Survey of the Infrared and Raman Spectra of Metallic and Organometallic Compounds*, Edward Arnold Ltd., London, 1967.
2. I. Nakagawa and T. Shimanouchi, *Spectrochim. Acta* **22**: 759 (1966).
3. K. Nakamoto, P. J. McCarthy, J. Fujita, R. A. Condrate, and G. T. Behnke, *Inorg. Chem.* **4**: 36 (1965).
4. M. G. Miles, J. H. Patterson, C. W. Hobbs, M. J. Hopper, J. Overend, and R. S. Tobias, *ibid.* **7**: 1721 (1968).
5. J. Hiraishi, I. Nakagawa, and T. Shimanouchi, *Spectrochim. Acta* **20**: 819 (1964).
6. J. Fujita, A. E. Martell, and K. Nakamoto, *J. Chem. Phys.* **36**: 324, 331 (1962).
7. J. Fujita, A. E. Martell, and K. Nakamoto, *ibid.* **36**: 339 (1962).
8. G. Herzberg, *Molecular Spectra and Molecular Structure, Vol. II. Infrared and Raman Spectra of Polyatomic Molecules*, Van Nostrand Co., Inc., Princeton, N.J., 1945.
9. K. Nakamoto, *Infrared Spectra of Inorganic and Coordination Compounds*, John Wiley & Sons, Inc., New York, 1963.
10. D. M. Adams and H. A. Gebbie, *Spectrochim. Acta* **19**: 925 (1963).
11. I. Nakagawa, T. Shimanouchi, and J. Hiraishi, "Infrared Spectra and Metal-Ligand Force Constants in Coordination Compounds" in: *Proceedings of the 8th International Conference on Coordination Chemistry, Vienna, Sept. 7–11, 1964*, (ed. V. Gutmann) Springer-Verlag, Wien, 1964, p. 21.
12. T. Shimanouchi and I. Nakagawa, *Inorg. Chem.* **3**: 1805 (1964).
13. L. Sacconi, A. Sabantini, and P. Gans, *ibid.* **3**: 1772 (1964).
14. I. Nakagawa and T. Shimanouchi, *Spectrochim. Acta* **20**: 429 (1964).
15. D. B. Powell and N. Sheppard, *J. Chem. Soc.* **1961**: 1112 (1961).
16. G. W. Watt and D. S. Klett, *Inorg. Chem.* **3**: 782 (1964).
17. D. M. Adams and P. J. Chandler, *J. Chem. Soc.* (*A*) **1967**: 1009 (1967).
18. M. N. Hughes and W. R. McWhinnie, *J. Inorg. Nuclear Chem.* **28**: 1659 (1966).
19. M. N. Hughes and W. R. McWhinnie, *J. Chem. Soc.* (*A*) **1967**: 592 (1967).
20. M. A. Bennett, R. J. H. Clark, and A. D. J. Goodwin, *Inorg. Chem.* **6**: 1625 (1967).
21. G. A. Rodley, Ph.D. Thesis, University College, London, 1963.
22. J. R. Durig, R. Layton, D. W. Sink, and B. R. Mitchell, *Spectrochim. Acta* **21**: 1367 (1965).
23. B. K. W. Baylis, unpublished data.

INDEX OF CONTRIBUTORS

SUBJECT INDEX